ELEMENTS OF ADVANCED QUANTUM THEORY

In memory of my Mother

ELEMENTS OF ADVANCED QUANTUM THEORY

by J. M. ZIMAN, F.R.S.

Professor of Theoretical Physics in the University of Bristol

CAMBRIDGE
AT THE UNIVERSITY PRESS
1969

Published by the Syndics of the Cambridge University Press
Bentley House, 200 Euston Road, London N.W.1
American Branch: 32 East 57th Street, New York, N.Y.10022

Library of Congress Catalogue Card Number: 69–16290
Standard Book Number: 521 07458 4

Printed in Great Britain
at the University Printing House, Cambridge
(Brooke Crutchley, University Printer)

PREFACE

'Sir, I have found you an argument: I am not obliged to find you an understanding.'

SAMUEL JOHNSON

Perhaps there never really was a time when the more mathematical physicists did not mystify their contemporaries with their abstract language. But in the past half century, quantum mechanics has soared to such rarefied heights that most research workers in physics can no longer comprehend the published theory of their own discipline. Surely it is unhealthy to put all the observing into the hands of 'experimenters' and leave all the thinking to those arrogant and plausible experts, the 'theoreticians'?

The task of providing advanced instruction, both for graduate students and mature research workers, is common to all fields of learning. But quantum theory presents special difficulties because of its hierarchical structure. The more abstract formalisms and techniques are quite meaningless until one has mastered the earlier stages. New generalizations do not, in practice, supersede old empiricisms and approximations. There is no short cut to the upper levels of the pyramid.

Even this image is misleading if it calls to mind a uniformly sloping edifice up which one may laboriously clamber hand over foot. To my mind, quantum theory is much more like a *ziggurat*, with sudden high cliffs to be surmounted before one can move freely on the next plane of abstraction. The mental leap upward at each of these barriers is as demanding as, say, the mastery of the differential calculus, or of the Euclidean method in geometry.

The first great step, from classical to quantum physics, is to accept the experimental evidence for the wave nature of matter, and the probabilistic character of microscopic events. With the aid of the Schrödinger equation, the student can then account for atomic and nuclear energy levels, electron and neutron diffraction, tunnelling, etc.

The second phase is to learn the language of Hilbert space—states and operators, observables, matrix elements, perturbation theory—into which may be translated a vast range of physical phenomena involving elementary particles, nuclei, atoms or molecules. This language is, indeed, so well founded in principle, and so rich in applications, that most physicists end their formal education in these gently sloping fields, where so much understanding may be acquired with so little extra exertion.

But when the graduate student begins active research, he is often confronted, in the theoretical literature, by yet another barrier of mysterious symbols and concepts—*field operators*, *graphs*, *propagators*, *Green functions*, *spinors*, *the S-matrix*, *irreducible representations*, *continuous groups*, and so on. Although these ideas are not all closely connected mathematically, their concurrent applications in many different branches of physics make this step seem even higher and steeper than the ones already surmounted. It is not surprising that those physicists who are not specialists in mathematical theory abandon the ascent, despite the obvious value of what they can glimpse in the clouds above their heads.

This book is an attempt to explain, in the simplest possible terms, the inwardness of these varied concepts and techniques. After he has read it, the student who encounters in the literature a Kubo formula, say, or an appeal to the Bethe–Salpeter equation, or a table of group characters, should be able to say to himself 'I see what he would be after'; and then have the confidence to study the argument further. It is not a grand staircase, but I hope it will prove a scaling ladder for these intellectual heights.

The starting point that I have assumed is the standard of a good physics graduate of a British university in his first or second year of research—and still with some thirst for knowledge. The reader need only have an elementary acquaintance with functional analysis and abstract algebra, and of the basic physical principles of electromagnetism, statistical mechanics and special relativity; but he should by now be quite at home with quantum physics in the Dirac–von Neumann formalism—matrix representations, orthogonal functions, operators, eigenvalues, etc. I also assume, naturally, that he is familiar with the basic facts about elementary particles, nuclei, atoms, and crystalline solids, at the level attained after a few dozen lectures on each of these topics.

The main subjects with which I deal are, of course, expounded in many other books. But the best account, in each case, is often to be found only in a treatise where the new method is applied in some particular branch of physics. It is not at all easy to disentangle the key principle from a mass of minor detail, nor is there then the occasion to emphasize its wide applicability. For example, the graphical method of Feynman is usually taught either in the specialized context of the many-body problem of solid state physics, or else with all the additional complexities of relativistic invariance that arise in the theory of elementary particles. The important idea here, surely, is that a graph exactly represents a certain type of term in a perturbation series, and that whole classes of such terms can be summed by topological arguments; the rest is detail that can easily be filled in when we come to deal with a specific problem.

To avoid irrelevancies, I have therefore given only very simple examples of the use of each technique, trying to refer only to principles and phenomena that would be fairly familiar to any physics graduate. This is not the place for an exposition of the experimental and theoretical basis of, say, superconductivity, or the structure of nuclei, or the classification of hyperons. My purpose is to make it possible for the student to read proper books on such topics for himself, in due course.

On the other hand, excessive mathematical abstraction would quite defeat my intention. Nothing is more repellent to normal human beings than the clinical succession of definitions, axioms, and theorems generated by the labours of pure mathematicians. The logical rigour achieved by such investigations is of the highest value, but can seldom come before we have grasped the idea in itself. Geometry existed before Euclid, and analysis before Cauchy; the significance of an irreducible representation of a group can be understood without a proof of Schur's Lemma in all its generality. I have tried to give a connected mathematical derivation of the important results, and to avoid making statements that are patently false; but for all incidental questions of exactitude, existence and exceptions the reader is referred to genuine treatises. If you do not understand or believe what I write on a particular point, do not blame yourself but look elsewhere for the truth.

Within these physical and mathematical limitations, it is scarcely possible to give workable 'exercises' to strengthen the reader's comprehension; each new problem would have to expound more than it would teach. But the book should be used in conjunction with other, more specialized works, whose main arguments and set piece 'examples' should prove sufficiently athletic for the most serious student.

I have also decided not to attempt a bibliography of relevant texts. A mere catalogue of all the books and review articles on all aspects of quantum physics would not be of much value. My own experience is limited to those I happen to have at hand, and I do not feel competent to judge fairly between these and other less familiar works.

Indeed, it must be confessed that I am no expert on most of the subject-matter of this book. For nearly twenty years I have contrived to give the appearance of doing research on the theory of solids, with little more serious analytical equipment than could be learnt from the 3rd edition of Dirac's *Quantum Mechanics*. But this imposture could not be sustained indefinitely. When I moved to Bristol in 1964, I offered to lecture on field-theoretical methods in solid state physics, as much for my own benefit as for the instruction of our experimental and theoretical research students. When I came to repeat these lectures the following year, I decided to write them out more precisely, perhaps for publication. From these first three chapters I learned much that I had previously only half understood; I was tempted to enlarge the scope of the enterprise and teach myself many other things about which I really knew nothing. This agreeable task of self-education has kept me busy for the winter months of each of the last two years, writing, and lecturing from, the remaining chapters. Even the labour of copying the mathematics into the admirable typescript prepared by Miss Jane Farmer and Mrs Lilian Murphy has not been too tedious; many a somnolent afternoon of academic committee has been made tolerable by this undemanding activity.

In defence of the impertinence of thus offering for sale such an inept work, let me say that, precisely because I have myself had to master the successive difficulties as I wrote each chapter, I am more aware of them than if I had had long experience in these techniques. Looking back now, I can see how trivial some of these obstacles really were—but I hope that I have been able to give, in the text, the reasoning that carried me up each step of the ladder. At this stage of the journey, sympathetic personal guidance may be more sustaining than the harsh scholarly exactitude of a *Baedeker*.

J.M.Z.

Bristol
April 1968

CONTENTS

7 The algebra of symmetry

CHAPTER 1

BOSONS

The more we are together the merrier we'll be.

1.1 The simple harmonic oscillator

The algebra of advanced quantum theory is dominated by what we often call the *occupation-number representation*. A quantum state is characterized by a set of integers which tell us how many particles (or more tentatively, 'quasi-particles' or 'excitations') there are in each of a basic set of wave functions. At the heart of the mathematics of such representations we find one of the most elementary and familiar topics in physics—the theory of the simple harmonic oscillator.

The classical equation of motion for a mass m moving in one dimension, x, under a force $-gx$, say, is

$$m\ddot{x} = -gx. \tag{1.1}$$

This has solution

$$x = x_0 e^{i\omega t}, \tag{1.2}$$

where the frequency is

$$\omega = \sqrt{(g/m)}. \tag{1.3}$$

These equations would be derivable from a classical Hamiltonian function

$$\mathscr{H} = \frac{1}{2m}p^2 + \tfrac{1}{2}gx^2, \tag{1.4}$$

which contains the momentum $p = m\dot{x}$.

In quantum theory the energy of the oscillator is quantized; the energy levels are given by

$$\mathscr{E}_n = (n+\tfrac{1}{2})\hbar\omega, \tag{1.5}$$

where n is an integer. How is this result derived? The elementary method is to express the momentum as a differential operator, and to solve the Schrödinger equation by analytical techniques. But there is a more elegant procedure, starting from the commutation relation of the conjugate operators x and p:

$$[x, p] \equiv xp - px = i\hbar. \tag{1.6}$$

What we do, in effect, is to try to factorize the sum of squares (1.4) into the product of two special operators a and a^* defined as follows:

$$\left.\begin{aligned} a &= \frac{1}{\sqrt{(2\hbar\omega)}}\left(\frac{1}{\sqrt{m}}p - \mathrm{i}\sqrt{g}.x\right), \\ a^* &= \frac{1}{\sqrt{(2\hbar\omega)}}\left(\frac{1}{\sqrt{m}}p + \mathrm{i}\sqrt{g}.x\right). \end{aligned}\right\} \tag{1.7}$$

The coefficients are chosen so that these operators may satisfy the following commutation relation, derived directly from (1.6), using (1.3):

$$\begin{aligned} [a, a^*] &= \frac{1}{2\hbar\omega}\left\{\frac{1}{m}[p,p] + \mathrm{i}\sqrt{\frac{g}{m}}[p,x] - \mathrm{i}\sqrt{\frac{g}{m}}[x,p] + g[x,x]\right\} \\ &= -\frac{1}{2\hbar}2\mathrm{i}[x,p] \\ &= 1. \end{aligned} \tag{1.8}$$

This relation is almost the simplest conceivable between operators that do not commute with one another: that is why it is so important.

We express the Hamiltonian (1.4) in terms of the new operators. This can be done either by solving (1.7) for p and x in terms of a and a^*, or even by inspection. Thus,

$$\mathscr{H} = \tfrac{1}{2}\hbar\omega(aa^* + a^*a). \tag{1.9}$$

This way of writing $\mathscr{H}$ shows that the attempt simply to factorize it into the product of a with a^* is frustrated by the fact that the operators x and p do not commute.

The next step is really to find a matrix representation of a and a^* that satisfies the commutation relation (1.8) and in which $\mathscr{H}$ is diagonal. It is rather tedious to derive this representation from first principles, but the result is easily written down. Let us define a set of basis functions or ket-vectors by the rules

$$\left.\begin{aligned} a\,|n\rangle &= n^{\frac{1}{2}}\,|n-1\rangle, \\ a^*\,|n\rangle &= (n+1)^{\frac{1}{2}}\,|n+1\rangle. \end{aligned}\right\} \tag{1.10}$$

Each function is labelled by an integer n. The operator a has the effect of transforming the function with label n into the function with label $(n-1)$; the operator a^* has the opposite effect of raising the index by unity.

It is easy to verify that this definition is consistent with the commutation relation (1.8). For example, let us apply the commutator to one of the basis functions:

$$\begin{aligned}[a, a^*]\,|n\rangle &\equiv aa^*\,|n\rangle - a^*a\,|n\rangle \\ &= a(n+1)^{\frac{1}{2}}\,|n+1\rangle - a^*n^{\frac{1}{2}}\,|n-1\rangle \\ &= (n+1)^{\frac{1}{2}}\,a\,|n+1\rangle - n^{\frac{1}{2}}\,a^*\,|n-1\rangle \\ &= (n+1)^{\frac{1}{2}}\,(n+1)^{\frac{1}{2}}\,|n\rangle - n^{\frac{1}{2}}n^{\frac{1}{2}}\,|n\rangle \\ &= (n+1)\,|n\rangle - n\,|n\rangle \\ &= |n\rangle. \end{aligned} \tag{1.11}$$

In other words, the commutator changes every basis function into itself; if the basis functions form a complete set, then $[a, a^*]$ is exactly equivalent to the unit operator, which is what we require.

Looking now at the Hamiltonian (1.9) we readily discover that each function $|n\rangle$ is an eigenvector of $\mathscr{H}$. The algebraic proof is exactly along the lines of (1.11), except for a change of sign of the second term. Thus,

$$\begin{aligned}\mathscr{H}\,|n\rangle &= \tfrac{1}{2}\hbar\omega(2n+1)\,|n\rangle \\ &= \mathscr{E}_n\,|n\rangle; \end{aligned} \tag{1.12}$$

the function $|n\rangle$ is the eigenfunction corresponding to the nth energy level $\mathscr{E}_n$, just as in the elementary formula (1.5). This completes the derivation of the 'quantization' of the energy of a simple harmonic oscillator.

1.2 Annihilation and creation operators

The operators defined by (1.7) have a number of properties and a variety of uses. Since a has the effect of reducing the number of quanta in the system by one, it is called an *annihilation operator*: a^* is a *creation operator* for a quantum of excitation energy. The product a^*a is an operator that measures the *occupation number* of a state of the system and has the functions $|n\rangle$ as its eigenfunctions.

The objective of this attack is to derive all the quantum-mechanics whilst keeping the properties of the state vectors as simple as possible. We do not need to know anything about them as 'wave functions'—i.e. as explicit analytical functions of the variable x; in addition to the rules (1.10) we only need to specify that, being eigenfunctions of $\mathscr{H}$, they may be assumed to be normal and orthogonal to one another, i.e.

$$\langle n|n'\rangle = \delta_{nn'}. \tag{1.13}$$

In fact we can generate the whole set by operating with a^* repeatedly on the *ground state*, or *vacuum* state $|0\rangle$: by (1.10),

$$|n\rangle = (n!)^{-\frac{1}{2}}(a^*)^n|0\rangle. \qquad (1.14)$$

On the other hand, the annihilation operator acting on the ground state gives zero; again, from (1.10) we get

$$a|0\rangle = 0. \qquad (1.15)$$

This means that the occupation number can never be negative—which is surely reasonable in the circumstances.

We can use these rules now to calculate any quantum-mechanical property of the system. Suppose, for example, that we wanted to know the average value of x^4 in the ground state of the oscillator. We can now use (1.7) to define the operator x in terms of a and a^*:

$$x = \mathrm{i}\sqrt{\frac{\hbar\omega}{2g}}\cdot(a-a^*). \qquad (1.16)$$

Then

$$\begin{aligned}\overline{x^4} &\equiv \langle 0|x^4|0\rangle \\ &= \left(\frac{\hbar\omega}{2g}\right)^2\langle 0|(a-a^*)^4|0\rangle \\ &= \frac{\hbar^2}{4mg}\{\langle 0|a^4|0\rangle - \langle 0|a^3a^*|0\rangle - \langle 0|a^2a^*a|0\rangle\ldots \\ &\qquad \ldots+\langle 0|(a^*)^4|0\rangle\}. \end{aligned} \qquad (1.17)$$

To evaluate these matrix elements we may use some elementary rules. Thus, all terms in which the number of annihilators is not equal to the number of creators are automatically zero. This is because by acting to the right on $|0\rangle$ we should generate a state with $n \neq 0$, which would then, by (1.13), be orthogonal to $\langle 0|$.

We must also exclude all terms in which an annihilation operator appears just before $|0\rangle$, or where $\langle 0|$ is followed immediately by a^*. The former rule is trivial, being a consequence of (1.15). But we may also write

$$\langle 0|a^* = 0, \qquad (1.18)$$

because a^* is the Hermitian conjugate of a, as one may easily see from the definitions (1.7). Thus (1.18) is derived from (1.15) by Hermitian conjugation. This is a way of saying that there is no state from which the vacuum may be created. It is important to remember that the operators a and a^* are not themselves Hermitian, and that a^* is not simply the complex conjugate of a, in an elementary algebraic sense.

This is because the momentum operator p is itself 'imaginary'. For consistency we shall always use the 'star' suffix to mean Hermitian conjugation.

We are left now with only two terms in (1.17). These can be evaluated by repeated application of the rules (1.10). But it is more fun to use the commutation relation (1.8). Thus, the term aa^*aa^* can be written

$$\begin{aligned}(aa^*)(aa^*) &= (a^*a+1)(a^*a+1)\\ &= (n+1)^2,\end{aligned} \tag{1.19}$$

because the number operator a^*a is diagonal in this representation. Again,

$$\begin{aligned}aaa^*a^* &= a(a^*a+1)a^*\\ &= aa^*aa^*+aa^*\\ &= (n+1)^2+(n+1),\end{aligned} \tag{1.20}$$

by the same argument. In the ground state these two operators have a total expectation value 3, so that

$$\overline{x^4} = \frac{3}{4}\frac{\hbar^2}{mg}. \tag{1.21}$$

Of course we could have obtained this result from the analytical form of the ground-state wave function, i.e. from

$$|0\rangle = \alpha^{\frac{1}{2}}\pi^{-\frac{1}{4}}\exp\left(-\tfrac{1}{2}\alpha^2x^2\right), \tag{1.22}$$

where

$$\alpha^4 = mg/\hbar^2. \tag{1.23}$$

Our operator algebra is the equivalent of evaluating $\overline{x^4}$ by integration by parts, from the expectation value

$$\langle 0|\,x^4\,|0\rangle = \alpha\pi^{-\frac{1}{2}}\int x^4\,\mathrm{e}^{-\alpha^2x^2}\,\mathrm{d}x. \tag{1.24}$$

Again the formula (1.14) for generating the set of basis states is just the standard theory for the generation of the Hermite polynomials by repeated differentiations of the ground-state function (1.22).

1.3 Coupled oscillators: the linear chain

Suppose now that we have, not just one simple harmonic oscillator but a whole assembly of vibrating masses, interacting with each other. Classical mechanics assures us that if the vibrations are of small amplitude they may be analysed into *normal modes*. We anticipate that the energy in each mode should then be quantized in units of its characteristic frequency.

A formal proof of this is easily devised. We know that the Hamiltonian is transformed in the 'normal mode representation' to a sum of squares, so that the system behaves like a collection of independent oscillators, each with a Hamiltonian like (1.4). The introduction of appropriate annihilation and creation operators for each degree of freedom, as in (1.9), is then trivial.

But there is one special and simple case of such a system that is the prototype of a vast range of models. This is where a large number of identical masses are arranged in a regular array, and coupled by short-range forces. We recognize the prescription for the dynamics of a crystal lattice, but the mathematics of such a system has a much wider scope.

For simplicity, let us suppose that the array is linear—a chain of 'atoms' spaced a distance a apart—and that there is a 'spring', of force constant g, between nearest neighbours. The Hamiltonian may be written

$$\mathscr{H} = \frac{1}{2m}\sum_l p_l^2 + \tfrac{1}{2}g\sum_l (u_l - u_{l+a})^2. \tag{1.25}$$

In this formula, u_l denotes the displacement, and p_l the momentum of the mass at the point l in the chain. We suppose that the chain is closed on itself, to eliminate end effects.

The classical theory of such a system is well known. Let us work it out by the operator technique, starting from the commutation relations

$$[u_l, p_{l'}] = \mathrm{i}\hbar\,\delta_{ll'}. \tag{1.26}$$

This goes a little beyond (1.6); it is convenient to assert explicitly that the displacement operator for the particle at l does not interfere with the momentum of a different particle, at l', so these operators must commute.

To diagonalize the Hamiltonian we make a Fourier transformation to a set of new operators:

$$U_k = \frac{1}{\sqrt{N}}\sum_l \mathrm{e}^{\mathrm{i}kl}\,u_l; \quad P_k = \frac{1}{\sqrt{N}}\sum_l \mathrm{e}^{-\mathrm{i}kl}\,p_l. \tag{1.27}$$

The introduction of the wave-number k is associated mathematically with the translational invariance of the lattice (see § 7.7). If the chain has N links—i.e. if it is of length $L = Na$—then k must belong to the set of 'allowed values'

$$k_n = \frac{2\pi n}{Na} = \frac{2\pi n}{L}, \tag{1.28}$$

where n is an integer. In practice we take N to be so large that this is a continuous set. By elementary arguments of Fourier analysis we can then invert the definitions (1.27), and write

$$u_l = \frac{1}{\sqrt{N}} \sum_k e^{-ikl} U_k; \quad p_l = \frac{1}{\sqrt{N}} \sum_k e^{ikl} P_k, \tag{1.29}$$

where the sum is over all essentially distinct values of k in the allowed set. Thus, the sum might be over all values of the integer n in (1.28), from $-\frac{1}{2}N$ to $\frac{1}{2}N$, with a convention to exclude one of these end points if N happens to be even.

The main point about these definitions is that they preserve the commutation relations (1.26). We can easily demonstrate this:

$$\begin{aligned}[U_k, P_{k'}] &= \frac{1}{N} \sum_{l,l'} e^{i(kl-k'l')} [u_l, p_{l'}] \\ &= \frac{1}{N} \sum_{l,l'} e^{i(kl-k'l')} i\hbar\, \delta_{ll'} \\ &= i\hbar . \frac{1}{N} \sum_l e^{i(k-k')l} \\ &= i\hbar\, \delta_{kk'},\end{aligned} \tag{1.30}$$

by the elementary properties of Fourier series. Thus, our new 'displacements' and 'momenta' are canonically conjugate, and non-commuting, if they are of the same wave number; otherwise they are dynamically independent operators.

We wish now to substitute from (1.29) into the Hamiltonian. But a difficulty arises, which is sometimes a source of confusion: the new operators, U_k and P_k, are not Hermitian. They satisfy more complicated conjugation relations,

$$\left.\begin{aligned} U_k^* &= \frac{1}{\sqrt{N}} \sum_l e^{-ikl} u_l^* = \frac{1}{\sqrt{N}} \sum_l e^{i(-k)l} u_l = U_{-k}, \\ P_k^* &= P_{-k}. \end{aligned}\right\} \tag{1.31}$$

We could have avoided this by deliberately symmetrizing the definition of these operators. But it is easier to start from a form of the Hamiltonian in which each product is made explicitly Hermitian, e.g.

$$p_l^2 = p_l^* p_l, \quad \text{etc.} \tag{1.32}$$

With this convention we can easily get the following result:

$$\mathscr{H} = \tfrac{1}{2} \sum_k \left\{ \frac{1}{m} P_k^* P_k + G(k)\, U_k^* U_k \right\}, \tag{1.33}$$

where

$$G(k) = 2g(1 - \cos ka). \tag{1.34}$$

We recognize this as the reduction to a sum of squares of momenta and displacements of the normal co-ordinates. The formula for the frequency of the mode of wave-number k is then

$$\omega_k = \sqrt{\frac{G(k)}{m}}$$
$$= 2\sqrt{(g/m)}\sin\tfrac{1}{2}|ka|. \tag{1.35}$$

Again, by convention, we make all frequencies positive, so that ω_{-k} is identical with ω_k.

The final step is the introduction of annihilation and creation operators, just as in (1.7). This is slightly complicated by the demands of Hermitian conjugation: we write

$$\left.\begin{aligned} a_k &= (2\hbar\omega_k m)^{-\frac{1}{2}}(P_k - im\omega_k U_k^*),\\ a_k^* &= (2\hbar\omega_k m)^{-\frac{1}{2}}(P_k^* + im\omega_k U_k). \end{aligned}\right\} \tag{1.36}$$

It can readily be verified that these operators satisfy the commutation relations

$$[a_k, a_{k'}^*] = \delta_{kk'}, \tag{1.37}$$

which are the analogues of (1.8). One can also show, using (1.31), that all other combinations of operators commute, i.e.

$$[a_k, a_{k'}] = 0; \quad [a_k^*, a_{k'}^*] = 0. \tag{1.38}$$

The equations (1.36) can be solved for the operators P_k, etc. by use of the conjugation relation (1.31); thus, for example,

$$P_k = (2\hbar\omega_k m)^{\frac{1}{2}}\,.\,\tfrac{1}{2}(a_k^* + a_{-k}). \tag{1.39}$$

We can then substitute into the Hamiltonian (1.33), to get

$$\mathscr{H} = \tfrac{1}{4}\sum_k \hbar\omega_k(a_k^* a_k + a_k a_k^* + a_{-k}^* a_{-k} + a_{-k} a_{-k}^*)$$
$$= \tfrac{1}{2}\sum_k \hbar\omega_k(a_k^* a_k + a_k a_k^*) \tag{1.40}$$

(since a summation over 'allowed values of $-k$' merely duplicates the sum over k).

The discussion now follows the same lines as in §§1.1, 1.2. The operators a_k and a_k^* have the effect of annihilating or creating quanta in the mode of wave-number k. Any dynamical properties of the lattice may be expressed in terms of these operators, from the chain of definitions (1.27) and (1.36). We might show, for example, that the excitations correspond to waves travelling to right or left round the ring of atoms, with frequency ω_k and wave-number k. The dispersion formula (1.35) would then give their phase velocity—and so on. Such excitations are, of course, the *phonons* of solid state physics. The eigenstates of $\mathscr{H}$ would be written in the form

$$|n_1, n_2, \ldots, n_k \ldots\rangle$$

corresponding to there being n_k quanta in the kth mode.

1.4 Three-dimensional lattices and vector fields

The linear chain, although it demonstrates many of the major principles of quantum field theory, is a very artificial model, even in solid state physics. Fortunately, the generalization to two or three dimensions is very simple. All that we need to do is to specify the position of each lattice site by a vector $\boldsymbol{l}$, instead of by a distance l, and to introduce *wave-vectors*, $\mathbf{k}$, to play the role of the wave-number k. The general theorems of Fourier analysis that were used in the previous section are then still valid, with the product kl interpreted as the scalar product, $\mathbf{k}\cdot\boldsymbol{l}$.

The only complication is in the range of 'allowed values' of $\mathbf{k}$. If the three-dimensional lattice is simple cubic, this is easy; each of the components of $\mathbf{k}$ must satisfy a condition like (1.28), e.g.

$$k_x = \frac{2\pi n_x}{N_x a} = \frac{2\pi n_x}{L_x}, \tag{1.41}$$

where n_x is an integer in the range $-\frac{1}{2}N_x$ to $\frac{1}{2}N_x$, there being N_x lattice spacings in the length L_x of the side of a unit cube of the 'crystal'. We say that the allowed distinct values of $\mathbf{k}$ lie in the *Brillouin zone* of the lattice—a cube in $\mathbf{k}$-*space*.

This is a rather special choice of lattice; in real crystals we may find more complicated arrangements of sites, such as the body-centred, face-centred, hexagonal, etc. structures. The geometrical properties of these different arrangements are, of course, extremely important in the theory of real solids. But here, for simplicity, we shall avoid all such distinctions. If a lattice comes into the calculation at all, we shall assume that it is simple cubic and that it has the Brillouin zone defined by (1.41). This is not so arbitrary as it sounds. It can easily be shown that the density of allowed wave-vectors in $\mathbf{k}$-space is independent of the crystal structure, but is just equal to $(V/8\pi^3)$, where V is the total volume of the crystal. It is also true, in general, that the total number of distinct wave-vectors in the zone is just N. From these rules we can write down relations like

$$\sum_{\mathbf{k}} e^{i\mathbf{k}\cdot(\boldsymbol{l}-\boldsymbol{l}')} \to \frac{V}{8\pi^3}\int_{\text{zone}} e^{i\mathbf{k}\cdot(\boldsymbol{l}-\boldsymbol{l}')}\, d^3\mathbf{k}$$

$$= N\,\delta_{\boldsymbol{l}\boldsymbol{l}'} \tag{1.42}$$

(assuming that N is so large that the sum tends to an integral), without having to discuss the detailed arrangement of allowed $\mathbf{k}$-values, or the shape of the volume of integration. We shall usually take $V = 1$ to avoid an unnecessary symbol.

A more tiresome complication of the transition to three dimensions is that we can no longer suppose that the 'displacement' or 'momentum' of one of the masses is a simple scalar variable, u_l or p_l. If the dynamical lattice model is still to seem realistic, the displacement at the site $\boldsymbol{l}$ must be at least a vector, $\mathbf{u}_l$, with conjugate momentum vector $\mathbf{p}_l$. This means that the commutation relations between such operators must be more general; we have to write

$$[\mathbf{u}_l, \mathbf{p}_{l'}] = i\hbar\mathbf{I}\,\delta_{ll'}, \tag{1.43}$$

with a unit cartesian tensor $\mathbf{I}$ to indicate that components of $\mathbf{u}_l$ and $\mathbf{p}_l$ along different axes also commute.

This leads us to a rather more complicated form of Hamiltonian. We might as well allow for coupling between atoms that are further apart in the lattice than nearest neighbours, and write

$$\mathscr{H} = \frac{1}{2m}\sum_{l}\mathbf{p}_l^*\cdot\mathbf{p}_l + \tfrac{1}{2}\sum_{l,l'}\mathbf{u}_l^*\cdot\mathbf{G}_{l-l'}\cdot\mathbf{u}_{l'}. \tag{1.44}$$

The cartesian tensor $\mathbf{G}_{l-l'}$ plays the role of the 'force constant' for the effect at the site $\boldsymbol{l}'$ of a displacement at the site $\boldsymbol{l}$; we use Hermitian products throughout, just as in (1.32).

The first steps in the reduction of this Hamiltonian to a phonon representation follow exactly the lines of § 1.3. Thus, we introduce new (vector) operators,

$$\mathbf{U}_{\mathbf{k}} = \frac{1}{\sqrt{N}}\sum_{l} e^{i\mathbf{k}\cdot\boldsymbol{l}}\,\mathbf{u}_l, \quad \mathbf{P}_{\mathbf{k}} = \frac{1}{\sqrt{N}}\sum_{l} e^{-i\mathbf{k}\cdot\boldsymbol{l}}\,\mathbf{p}_l, \tag{1.45}$$

just as in (1.27). We then find that the Hamiltonian retains, superficially, the form deduced in (1.33):

$$\mathscr{H} = \tfrac{1}{2}\sum_{\mathbf{k}}\left\{\frac{1}{m}\mathbf{P}_{\mathbf{k}}^*\cdot\mathbf{P}_{\mathbf{k}} + \mathbf{U}_{\mathbf{k}}^*\cdot\mathbf{G}(\mathbf{k})\cdot\mathbf{U}_{\mathbf{k}}\right\}, \tag{1.46}$$

where we have generalized (1.34) to read

$$\mathbf{G}(\mathbf{k}) = \sum_{\mathbf{h}} \mathbf{G}_{\mathbf{h}} e^{-i\mathbf{k}\cdot\mathbf{h}}. \tag{1.47}$$

Unfortunately we cannot proceed directly from here to the annihilation and creation operators. The trouble is that $\mathbf{G}(\mathbf{k})$ is a tensor, and is not necessarily diagonal. In the theory of lattice dynamics, this is where the real work must be done; one must solve the eigenvalue problem for this tensor before one can proceed.

Nevertheless, for any given value of $\mathbf{k}$, this is a well-defined and essentially elementary task. Rather than plunging into a forest of new symbols let us give a verbal description of the results. First we find the principal axes of the tensor $\mathbf{G}(\mathbf{k})$ and refer our operators $\mathbf{U}_{\mathbf{k}}$ and $\mathbf{P}_{\mathbf{k}}$ to those axes. The diagonal elements of the tensor may now be written $m(\omega_{\mathbf{k}}^{(1)})^2$, $m(\omega_{\mathbf{k}}^{(2)})^2$, $m(\omega_{\mathbf{k}}^{(3)})^2$, by analogy with (1.35). We now choose each principal axis in turn, and introduce annihilation and creation operators according to the prescription (1.36). For example, we might write

$$a_{\mathbf{k}}^{(1)} = (2\hbar\omega_{\mathbf{k}}^{(1)} m)^{-\frac{1}{2}} (P_{\mathbf{k}}^{(1)} - im\omega_{\mathbf{k}}^{(1)} U_{\mathbf{k}}^{(1)*}) \tag{1.48}$$

for components along the first axis—and so on. Eventually we shall arrive at a more general form of (1.40):

$$\mathscr{H} = \tfrac{1}{2} \sum_{\mathbf{k},p} \hbar\omega_{\mathbf{k}}^{(p)} (a_{\mathbf{k}}^{(p)*} a_{\mathbf{k}}^{(p)} + a_{\mathbf{k}}^{(p)} a_{\mathbf{k}}^{(p)*}), \tag{1.49}$$

with a commutation relation

$$[a_{\mathbf{k}}^{(p)}, a_{\mathbf{k}'}^{(p')*}] = \delta_{\mathbf{k}\mathbf{k}'}\delta_{pp'}. \tag{1.50}$$

The index (p), taking the values 1, 2, 3, of course refers to the *polarization* of the mode; it tells us the direction of the ionic displacements associated with an excitation of that type.

The main point to be emphasized is that this *phonon representation* of the dynamics of the lattice is quite precise and specific. Just as in the linear chain, just as for the simple harmonic oscillator, any state of the system can be constructed out of phonon states—i.e. the eigenstates of the occupation number operators

$$n_{\mathbf{k}}^{(p)} = a_{\mathbf{k}}^{(p)*} a_{\mathbf{k}}^{(p)}. \tag{1.51}$$

Any matrix element of any dynamical variable may also be calculated in the representation, simply by working through the succession of transformations such as (1.45) and (1.48). This is a very powerful principle and technique—so powerful, indeed, that many people forget that the original form of the Hamiltonian, in terms of local displacement operators, is sometimes a better starting-point for a calculation.

The general argument that we have given here is by no means confined to simple lattices with one atom per unit cell: there is no difficulty, in principle, in extending it to a solid with a vibrating molecule, or several different atoms, in each cell. One can also set up an equivalent formalism in which the local operators (our $\mathbf{u}_l$ and $\mathbf{p}_l$) are not displacement and momentum operators but are components of the angular momentum of the object (usually a transition-metal ion) at the lth site. If these spins interact with one another by exchange forces, then we can get '*spin waves*' propagated through the crystal. These excitations can be represented approximately by a Hamiltonian of our standard form (1.49), with operators for the annihilation and creation of *magnons*. We again refer to texts on solid state physics for further details.

1.5 The continuum limit

Up to this point, we have been dealing with precisely defined systems, and the transformations have been exact. But for many purposes the underlying sub-microscopic structure of the lattice is of little interest; we look upon it simply as a medium in which waves are propagated.

It is often convenient, therefore, to smear out all the structure and to treat the solid as a continuum. We replace the lattice vector $\boldsymbol{l}$, which goes through discrete values, by a continuous position-vector $\mathbf{r}$. Sums over all values of $\boldsymbol{l}$ are then replaced by integrals over $\mathbf{r}$, and so on. Thus, we might define a function $\mathbf{u}(\mathbf{r})$, which is a vector representing the displacement of the lattice at the point $\mathbf{r}$, and a corresponding 'momentum density'

$$\mathbf{p}(\mathbf{r}) = \rho_0 \mathbf{v}(\mathbf{r}), \tag{1.52}$$

where ρ_0 is the mean density and $\mathbf{v}(\mathbf{r})$ is the local velocity of the medium in some mode of vibration.

In such a case, we might have a Hamiltonian of the same form as (1.44),

$$\mathscr{H} = \frac{1}{2\rho_0}\int \mathbf{p}^*(\mathbf{r})\cdot\mathbf{p}(\mathbf{r})\,d^3\mathbf{r} + \frac{1}{2}\iint \mathbf{u}^*(\mathbf{r})\cdot\mathbf{G}(\mathbf{r}-\mathbf{r}')\cdot\mathbf{u}(\mathbf{r}')\,d^3\mathbf{r}\,d^3\mathbf{r}'. \tag{1.53}$$

In this expression of course the tensor $\mathbf{G}(\mathbf{r}-\mathbf{r}')$ represents the interaction between the displacements of a unit volume about the point $\mathbf{r}$

and the displacement at $\mathbf{r}'$. The integration is over the whole volume V of the solid.

Now we make a Fourier integral transform, equivalent to (1.45), i.e.

$$\mathbf{U}_\mathbf{k} = \frac{1}{\sqrt{V}}\int e^{i\mathbf{k}\cdot\mathbf{r}}\,\mathbf{u}(\mathbf{r})\,d^3\mathbf{r}; \quad \mathbf{P}_\mathbf{k} = \frac{1}{\sqrt{V}}\int e^{-i\mathbf{k}\cdot\mathbf{r}}\,\mathbf{p}(\mathbf{r})\,d^3\mathbf{r}. \tag{1.54}$$

The wave-vector $\mathbf{k}$ that appears in these formulae is defined as in (1.41); it must belong to the allowed set of points in reciprocal space. But because our real space is no longer dissected into discrete lattice points, the range of values of $\mathbf{k}$ is no longer limited to a finite Brillouin zone. The lattice constant a has become infinitesimally small; reciprocally, N_x, the number of sites along the side L_x of our cube of material, has become infinitely large. This means that n_x in (1.41) is unrestricted in its range, although it must still be an integer if the usual boundary conditions on $\mathbf{u}(\mathbf{r})$ are to be satisfied.

The inversion of (1.54) is an elementary case of the standard Fourier theorem. From the rule

$$\frac{1}{V}\int e^{i(\mathbf{k}-\mathbf{k}')\cdot\mathbf{r}}\,d^3\mathbf{r} = \delta_{\mathbf{k}\mathbf{k}'}, \tag{1.55}$$

we get

$$\mathbf{u}(\mathbf{r}) = \frac{1}{\sqrt{V}}\sum_\mathbf{k} e^{-i\mathbf{k}\cdot\mathbf{r}}\,\mathbf{U}_\mathbf{k}. \tag{1.56}$$

These can be substituted into the Hamiltonian (1.53); the result is obviously exactly similar to (1.46), i.e.

$$\mathscr{H} = \tfrac{1}{2}\sum_\mathbf{k}\left\{\frac{1}{\rho_0}\mathbf{P}^*_\mathbf{k}\cdot\mathbf{P}_\mathbf{k} + \mathbf{U}^*_\mathbf{k}\cdot\mathbf{G}(\mathbf{k})\cdot\mathbf{U}_\mathbf{k}\right\}, \tag{1.57}$$

with a modified form of (1.47), i.e.

$$\mathbf{G}(\mathbf{k}) = \int \mathbf{G}(\mathbf{R})\,e^{-i\mathbf{k}\cdot\mathbf{R}}\,d^3\mathbf{R}. \tag{1.58}$$

From here on, we want to use the theory of the previous sections—diagonalization of the tensor $\mathbf{G}(\mathbf{k})$, introduction of annihilation and creation operators, etc. But this depends upon the symbols $\mathbf{P}_\mathbf{k}$ and $\mathbf{U}_\mathbf{k}$ in (1.57) being operators with the proper dynamical properties, that is, obeying commutation relations like (1.30), i.e.

$$[\mathbf{U}_\mathbf{k}, \mathbf{P}_{\mathbf{k}'}] = i\hbar\mathbf{I}\,\delta_{\mathbf{k}\mathbf{k}'}. \tag{1.59}$$

This forces upon the symbols $\mathbf{u}(\mathbf{r})$, $\mathbf{p}(\mathbf{r})$ the following condition:

$$\begin{aligned}[\mathbf{u}(\mathbf{r}), \mathbf{p}(\mathbf{r}')] &= i\hbar\mathbf{I}\,\frac{1}{V}\sum_\mathbf{k} e^{i\mathbf{k}\cdot(\mathbf{r}-\mathbf{r}')} \\ &= i\hbar\mathbf{I}\,\delta(\mathbf{r}-\mathbf{r}').\end{aligned} \tag{1.60}$$

This step causes us to introduce a Dirac delta-function in place of the Kronecker delta that appears in (1.43). No difficulties are occasioned by the infinity at $\mathbf{r} = \mathbf{r}'$, provided that we treat the function consistently as the limit of the infinite series represented by the sum over 'all values of $\mathbf{k}$'.

What we are saying, in effect, is that the *field operators* $\mathbf{u}(\mathbf{r})$, $\mathbf{p}(\mathbf{r})$ commute with one another unless they refer to 'exactly the same point in the medium'. Since this is an infinitesimal region we must make up for it with an infinite commutator. Alternatively, we could recall that $\mathbf{p}(\mathbf{r})$ is a *density* of momentum, so that we should integrate over some small volume in order to define an actual momentum. Our relation (1.60) amounts then to saying that this momentum will not commute with any displacement operator measured inside that volume.

We now have a *quantized field*. Any state of the system can be generated by applying the creation operator to the 'vacuum' state $|0\rangle$, and the energy can be expressed in terms of the occupation numbers of the various modes. We could calculate the local displacement in any such state by means of the transformations (1.56), etc.—and so on. In fact, the results would be quite consistent with the ordinary classical theory of waves in an elastic continuous medium, except that it is less complicated to prescribe the deformation of the material by a displacement vector $\mathbf{u}(\mathbf{r})$ than to introduce strain tensors.

But we have made an untenable postulate—that we may indeed go to the limit of a continuous field. In the case of a crystal lattice, such a postulate is justified only as an approximation; we know that there really is a lattice structure, and in the last resort we may check our continuum calculations by carrying out an exact analysis as in § 1.4.

The basic assumption of *quantized field theory* is that we may make just this sort of continuum postulate about fundamental particles. We assume that they may be described in terms of local field operators, which are quantized via a delta-function commutator as in (1.60). The interesting thing is that such an operator—the analogue of our 'displacement vector' $\mathbf{u}(\mathbf{r})$—looks very like what we have learned to call the *wave function* of the particle that it generates. This is the elegant and subtle theory that we shall now discuss.

1.6 Classical field theory

What we need is a proper mathematical scheme for defining field operators that will generate particles (more correctly 'particle states') having prescribed properties. The properties of phonons are derived from the dynamics of a lattice, or of an elastic continuum; the properties of magnons are derived from the basic quantum equations of an arrangement of spins. But (at present anyway) elementary particles are, so to speak, just themselves; we cannot derive their equations from more basic principles, but must devise a formalism that shows the observed phenomena and that is not internally contradictory.

The starting point for most such theories is the same as for the other parts of quantum theory—classical Hamiltonian dynamics. Just as there is a canonical theory of point particles, centred on Lagrange's equation, so there is a canonical theory of a classical field—for example, just such a field as the elastic displacement vector $\mathbf{u}(\mathbf{r})$ discussed above.

For generality, let us use the symbol $\phi(\mathbf{r})$ (not necessarily a scalar quantity) to represent the amplitude of a field at the point $\mathbf{r}$. A field has an infinity of degrees of freedom. To specify it we must know its value at each of a whole set of points, $\mathbf{r}_1$, $\mathbf{r}_2$, $\mathbf{r}_3$, etc. in the region being considered. For a continuous space, the series is endless. The values of $\phi(\mathbf{r}_1)$, $\phi(\mathbf{r}_2)$, $\phi(\mathbf{r}_3)$, etc. may be thought of as the generalized co-ordinates of the system (just like the components of the position vectors of the particles in elementary dynamics) but this set is infinite.

It follows, as in § 1.5, that we must replace the symbol for 'a sum over all the co-ordinates of all the particles' by the symbol for 'an integral over the whole region'. Once this is understood, the mathematics of classical field theory becomes quite intelligible, and presents no special difficulties.

Thus, for example, in order to define the total Lagrangian L of the system, we must introduce a *Lagrangian density*, $\mathscr{L}$ such that

$$L = \int \mathscr{L}(\mathbf{r})\, \mathrm{d}^3\mathbf{r}. \tag{1.61}$$

$\mathscr{L}(\mathbf{r})$ must obviously depend on the amplitude of the field at or near $\mathbf{r}$. For example, it might be a function of $\phi(\mathbf{r})$ itself. It must also contain the time derivative of ϕ, just as the Lagrangian of a particle contains the kinetic energy, which is a function of the velocity. To make a sensible theory, $\mathscr{L}$ must also depend on spatial derivatives of $\phi(\mathbf{r})$, otherwise there would be no connection between the field amplitudes at neighbouring points in space. In the passage to a continuum limit, differences like $(u_l - u_{l+a})$ in (1.25) become derivatives with respect to distance.

In general, therefore, we have

$$\mathscr{L}(\mathbf{r}) = \mathscr{L}\left(\phi(\mathbf{r}), \frac{\partial\phi(\mathbf{r})}{\partial x}, \frac{\partial\phi(\mathbf{r})}{\partial y}, \frac{\partial\phi(\mathbf{r})}{\partial z}, \frac{\partial\phi(\mathbf{r})}{\partial t}\right), \tag{1.62}$$

which we shall write

$$\mathscr{L} = \mathscr{L}(\phi, \phi_{,i}), \tag{1.63}$$

using the symbol

$$\phi_{,i} = \frac{\partial\phi}{\partial X_i} = \left(\frac{\partial\phi}{\partial y}, \frac{\partial\phi}{\partial y}, \frac{\partial\phi}{\partial z}, \frac{\partial\phi}{\partial t}\right) \tag{1.64}$$

to stand for a derivative with respect to one or other co-ordinate of space or time. Of course, our field itself may be more complicated; the symbol ϕ may stand for a vector or a tensor (as in the proper theory of elastic waves). The expression for $\mathscr{L}(\mathbf{r})$ may then look very complicated with a variety of different derivatives of the different components. But $\mathscr{L}$ must itself be a scalar, which is an important restriction on the form of field that may be generated in this way.

To construct an equation of motion, we use Hamilton's principle:

$$\delta\int_{t_0}^{t_1} L\,\mathrm{d}t = 0. \tag{1.65}$$

This tells us that the action integral between two fixed times is an extremum for the actual path of the motion. In our case, this means

$$\delta\iint \mathscr{L}(\phi, \phi_{,i})\,\mathbf{d}^3\mathbf{r}\,\mathrm{d}t = 0,$$

i.e.

$$\delta\int \mathscr{L}(\phi, \phi_{,i})\,\mathrm{d}^4X = 0, \tag{1.66}$$

where the integration is over the four-dimensional continuum of space and time.

The meaning of a variational expression such as (1.66) is explained in detail in textbooks on the calculus of variations. But we may get the result we need here by treating the operator δ as some sort of 'differentiator' and then manipulating the integral. Thus

$$\begin{aligned}\delta\int \mathscr{L}(\phi, \phi_{,i})\,\mathrm{d}^4X &= \int\left\{\frac{\partial\mathscr{L}}{\partial\phi}\delta\phi + \sum_{i=1}^{4}\frac{\partial\mathscr{L}}{\partial\phi_{,i}}\delta\phi_{,i}\right\}\mathrm{d}^4X \\ &= \int\left\{\frac{\partial\mathscr{L}}{\partial\phi} - \sum_{i=1}^{4}\frac{\partial}{\partial X_i}\left(\frac{\partial\mathscr{L}}{\partial\phi_{,i}}\right)\right\}\delta\phi\,\mathrm{d}^4X \\ &\qquad + \text{boundary term.}\end{aligned} \tag{1.67}$$

The first step is an ordinary differentiation of a function of several variables. The second step is a generalized integration by parts, or a four-dimensional application of Green's theorem. It expresses everything in terms of the varied function ϕ, and eliminates the partial derivatives $\phi_{,i}$, but introduces a boundary term that must also vanish. The main point, however, is that, for (1.66) to hold for arbitrary variations of ϕ, the function appearing with $\delta\phi$ in the integrand must be zero everywhere. We thus arrive at the *Euler equation*

$$\frac{\partial \mathscr{L}}{\partial \phi} - \sum_i \frac{\partial}{\partial X_i}\left(\frac{\partial \mathscr{L}}{\partial \phi_{,i}}\right) = 0. \tag{1.68}$$

This is a necessary condition for an extremum of the action integral, and is the equation of motion of the field.

This piece of formalism is given to show how one can link a field equation with a local energy function like a Lagrangian density. The ordinary equations of particle dynamics are derivable as an elementary case of (1.68). Suppose that none of the space derivatives of ϕ appear in $\mathscr{L}$. Then we have

$$\frac{\partial \mathscr{L}}{\partial \phi} - \frac{\partial}{\partial t}\left(\frac{\partial \mathscr{L}}{\partial \dot{\phi}}\right) = 0 \tag{1.69}$$

which is immediately recognizable as Lagrange's equation for a system with one degree of freedom in the generalized co-ordinate ϕ.

An elementary example of a field is described by the Lagrangian density

$$\mathscr{L} = \tfrac{1}{2}\rho_0\left(\frac{\partial \phi}{\partial t}\right)^2 - \tfrac{1}{2}G\left\{\left(\frac{\partial \phi}{\partial x}\right)^2 + \left(\frac{\partial \phi}{\partial y}\right)^2 + \left(\frac{\partial \phi}{\partial z}\right)^2\right\}. \tag{1.70}$$

We might think of ϕ as the measure of the 'displacement' of an elastic medium, so that the first term, containing the density ρ_0 and the 'velocity' $(\partial\phi/\partial t)$ would be a kinetic energy density. The second term, quadratic in derivatives of the displacement might then be minus the potential energy of the elastic 'strain'.

The meaning of (1.68), in this case is readily established. Since ϕ itself does not appear, $\partial \mathscr{L}/\partial \phi = 0$. For the other derivatives we have, say,

$$\frac{\partial \mathscr{L}}{\partial \phi_{,i}} \equiv \frac{\partial \mathscr{L}}{\partial(\partial\phi/\partial x)} = -G\left(\frac{\partial \phi}{\partial x}\right), \quad \text{etc.,} \tag{1.71}$$

so that the equation of motion reads

$$G\left(\frac{\partial^2 \phi}{\partial x^2} + \frac{\partial^2 \phi}{\partial y^2} + \frac{\partial^2 \phi}{\partial z^2}\right) - \rho_0 \frac{\partial^2 \phi}{\partial t^2} = 0, \tag{1.72}$$

which is just the equation for the propagation of waves with velocity $\sqrt{(G/\rho_0)}$.

The next step is to define a 'momentum' field, conjugate to ϕ. The analogy with particle dynamics directs us at once to define this as

$$\pi(\mathbf{r}) = \frac{\partial \mathscr{L}}{\partial \dot{\phi}(\mathbf{r})}. \tag{1.73}$$

Then we can define the *Hamiltonian density*

$$\mathscr{H}(\mathbf{r}) = \pi(\mathbf{r})\,\dot{\phi}(\mathbf{r}) - \mathscr{L}(\mathbf{r}). \tag{1.74}$$

The total Hamiltonian of our system must then be

$$H = \int \mathscr{H}(\mathbf{r})\,\mathbf{d}^3\mathbf{r}. \tag{1.75}$$

We should, of course, use (1.73) to eliminate the function $\dot{\phi}(\mathbf{r})$ from $\mathscr{H}(\mathbf{r})$, which should now depend only on the space functions $\phi(\mathbf{r})$ and $\pi(\mathbf{r})$. To see how this works, let us work out the case of our simple example—the 'scalar elasticity model' of (1.70). From (1.73) we get

$$\pi(\mathbf{r}) = \rho_0\dot{\phi}(\mathbf{r}), \tag{1.76}$$

and from (1.74)

$$\mathscr{H}(\mathbf{r}) = \frac{1}{2\rho_0}\pi^2(\mathbf{r}) + \tfrac{1}{2}G(\nabla\phi)^2. \tag{1.77}$$

We might have written this down at once, as the total energy of the field but would not then have a rule defining conjugate variables.†

1.7 Second quantization

In particle dynamics, the transition from classical to quantum mechanics occurs when the conjugate variables are changed into operators with a standard commutation relation such as (1.6). The corresponding transition in field theory is quite similar; we treat $\phi(\mathbf{r})$ and $\pi(\mathbf{r})$ as operators with the commutation relation

$$[\phi(\mathbf{r}), \pi(\mathbf{r}')] = i\hbar\,\delta(\mathbf{r}-\mathbf{r}'). \tag{1.78}$$

† It is interesting to compare (1.77) with (1.44) and (1.53). In going from a lattice model to a continuum limit it was natural to retain a *non-local interaction* such as might be defined by the function $G(\mathbf{r}-\mathbf{r}')$. But if we had used this form of potential energy in our model Lagrangian (1.70) the equation of motion would be much more complicated than the wave equation (1.72). In a field theory it is much simpler to use strictly local functions. The reader may amuse himself by showing that the equations of standard elasticity theory are equivalent to replacing $G(\mathbf{r}-\mathbf{r}')$ in (1.53) by a differential operator of the form

$$G\delta(\mathbf{r}-\mathbf{r}')\frac{\partial}{\partial\mathbf{r}}\frac{\partial}{\partial\mathbf{r}'},$$

with appropriate tensor indices. Our simple Hamiltonian (1.77) is thus a 'scalarized' and 'localized' version of the lattice dynamical theory.

The necessity and significance of a relation of this form has already been discussed in the special case of (1.59), which was derived from a lattice model.

The function (1.75) that was called the Hamiltonian in the classical theory has now itself become an operator. The familiar rules of quantum theory suggest that the 'state' of the system should now be defined by some state vector, or state function, $|\,\rangle$, say, satisfying the equation of motion

$$H|\,\rangle = \frac{\hbar}{\mathrm{i}}\frac{\partial}{\partial t}|\,\rangle. \tag{1.79}$$

These two relations, (1.78) and (1.79), are supposed to govern all the physics of the system.

In previous sections we have been learning how to deal with such equations by a transformation to $\mathbf{k}$-space and the introduction of annihilation and creation operators. It is easy enough to show, for example, that the simple scalar elasticity model (1.70) has solutions corresponding to the occupation of phonon modes of energy

$$\mathscr{E}_{\mathbf{k}} = (n_{\mathbf{k}} + \tfrac{1}{2})\hbar\omega_{\mathbf{k}}, \tag{1.80}$$

where

$$\omega_{\mathbf{k}} = \sqrt{(G/\rho_0)}\,k. \tag{1.81}$$

These states are, of course, equivalent to solutions of the classical wave equation (1.72): the function

$$\phi(\mathbf{r}, t) = \phi_0 \exp \mathrm{i}(\mathbf{k}\cdot\mathbf{r} - \omega_{\mathbf{k}} t) \tag{1.82}$$

satisfies the equation. The effect of turning the classical field variable into an operator is not much more than to 'quantize' the energy in units of $\hbar\omega_{\mathbf{k}}$—that is, to prescribe a set of values for the amplitude ϕ_0. In this case, our physical interpretation of ϕ as the displacement of a material medium makes sense of the commutation relations (1.78), and guarantees the validity of the results as being no more than what we should expect for a lattice of negligibly fine grain.

But suppose that we knew nothing of the existence of such a lattice. Suppose that the function $\phi(\mathbf{r}, t)$, was originally conceived as a 'wave function' for some particle or excitation. If it satisfied some partial differential equation of the canonical form (1.68), then we could carry out this same quantization scheme, call $\phi(\mathbf{r})$ an operator, define a state vector satisfying (1.79), etc.

For example, consider the time-dependent Schrödinger equation for an electron:

$$-\frac{\hbar^2}{2m}\nabla^2\psi + V(r)\psi = \frac{\hbar}{\mathrm{i}}\frac{\partial\psi}{\partial t}. \tag{1.83}$$

This is a second-order partial differential equation in the field variable ψ. It is easy enough to show that it is the Euler equation for the variation of a Lagrangian density

$$\mathscr{L} = i\hbar\dot{\psi}^*\dot{\psi} - \frac{\hbar^2}{2m}(\nabla\psi^* \cdot \nabla\psi) - V(\mathbf{r})\psi^*\psi. \tag{1.84}$$

From this, by mechanical algebraic procedures, we can derive a Hamiltonian density, canonical operators, etc. We could, so to speak, 'quantize' the system at a higher level; the wave function that was introduced as the operand for the observables of particle dynamics becomes, in its turn, an operator acting on a more general state function.

Actually the *second quantization* of the Schrödinger equation does not proceed precisely as in (1.78) because electrons are fermions, whereas the algebra of previous sections has always produced eigenstates with any number of excitations in each mode, as if these 'particles' were bosons. This difficulty will be dealt with in the next chapter.

But we can perhaps see that the second quantization procedure is a way of dealing with systems of many particles. It is a way of generalizing the equation for the wave function of a single excitation so as to describe what happens when there are a number of such excitations present simultaneously in the field region. In elementary quantum theory this is difficult because one has to add a complete set of three dimensions to the abstract space of the wave function for every new particle that we bring in. It is much easier to introduce the abstract field operator $\phi(\mathbf{r})$, whose physical significance is not so elementary as the position and momentum observables of particle dynamics, but which can be thought of as measuring the probability of finding a particle or particles at $\mathbf{r}$.

One can demonstrate this mathematically. One can show that the state functions satisfying (1.79) correspond to symmetrized combinations of single particle functions satisfying this wave function (1.68), just as required for an assembly of non-interacting bosons. The general lines of such a proof may be deduced from the arguments to be given in the next chapter, when we deal with the case of fermions.

From a physical point of view, however, one ought perhaps to treat the second quantization scheme, (1.78) and (1.79), as primary—as constituting axiomatic relations or hypotheses for a theory of quantum-mechanical fields. Again it is not difficult to derive the equations for a single excitation of such a field, and to show that it may be described by a wave function satisfying an equation such as (1.68). The advantage of this point of view is that it allows us to quantize other fields, such as the electromagnetic and gravitational fields, whose single-particle states are not so manifest. It is also a natural starting point for the definition of interactions between different particles of the same field, or between different fields.

1.8 Klein–Gordon equation

The 'scalar elasticity' field defined by (1.70) is a very useful simple model of a boson system. We may think of its excitations as being 'phonons' or even 'photons', stripped of their gorgeous trappings of polarization vectors, tensor components, gauge transformations, etc. But this is a field whose 'particles' have zero rest-mass; all their energy is dynamical.

It is therefore useful to have available another model whose wave equation is similar to (1.72), except for an additional term proportional to the field itself. Let ϕ satisfy the *Klein–Gordon equation*,

$$\nabla^2\phi - \frac{\partial^2\phi}{\partial t^2} - m^2\phi = 0, \tag{1.85}$$

where m has the dimensions of mass in the scheme of units where $\hbar = c = 1$.

It is easy to construct wave-like solutions of this equation, with frequency

$$\omega_{\mathbf{k}} = (k^2 + m^2)^{\frac{1}{2}}. \tag{1.86}$$

Thus, the energy of a single quantum of the field is

$$\hbar\omega_{\mathbf{k}} = \pm mc^2 \left(1 + \frac{\hbar^2 k^2}{m^2c^2}\right)^{\frac{1}{2}}. \tag{1.87}$$

We put in $\hbar$ and c explicitly to show that this is just the relativistic energy of a particle of rest mass m moving with velocity $\hbar k/m$. In the present context we shall absolutely exclude the solutions of negative energy, which are discussed in § 6.4.

This equation derives from a Lagrangian density like (1.70) with an extra term $-\frac{1}{2}m^2\phi^2$. The corresponding Hamiltonian density then takes the form

$$\mathscr{H} = \tfrac{1}{2}\{\pi^2 + (\nabla\phi)^2 + m^2\phi^2\}. \tag{1.88}$$

We may quantize this system by taking $\phi(\mathbf{r})$ and $\pi(\mathbf{r})$ to be conjugate field operators, as in (1.78). It is easy to make a Fourier transformation of these fields, as in (1.45), and then to introduce annihilation and creation operators, $a_{\mathbf{k}}$ and $a_{\mathbf{k}}^*$, just as in (1.48). The eigenstates of the Hamiltonian then belong to the set $|n_{\mathbf{k}}\rangle$ where $n_{\mathbf{k}}$ is the number of excitations, each of the energy $\hbar\omega_{\mathbf{k}}$ given by (1.87), in the mode of wave-vector **k**.

These equations therefore describe a field of free particles, each of rest mass m. The wave equation (1.85) is relativistically invariant, and the energy therefore transforms correctly in a Lorentz transformation (see § 6.4). These particles are bosons, because there is no restriction on the number that may occupy a given one-particle state. They cannot be electrons or nucleons, which are fermions. The wave-function ϕ is a real, scalar quantity, which means that they have neither charge nor spin (see § 1.12). We may think of them as neutral *mesons* of spin zero.

The point here is not to identify Klein–Gordon particles with some observable physical objects, but to define a simple model of a field of particles of finite mass. This equation might have arisen in some different context—for example in the study of the excitations of a dense plasma, or of vortex motions in liquid helium.

1.9 Sources of a field, and interactions between fields

It is not enough to describe the properties of a system of free, non-interacting particles; we must know how to do something to them. The most primitive thing one can do to anything is to create or destroy it. The question to be asked about our particles is how they got there in the first place.

A phonon field is easy enough to create; one takes hold of one or other of the atoms, and one jiggles it. In practice this is done with an electromagnetic field, or by contact with another solid. What we need is some other field or object coupled to the phonon field. To express this idea, we must put a new term into the Hamiltonian, typically of the form

$$H_I = \int X(\mathbf{r})\,\phi(\mathbf{r})\,\mathbf{d}^3\mathbf{r}. \tag{1.89}$$

If $\phi(\mathbf{r})$ measures 'displacement' then $X(\mathbf{r})$ is just the 'force' applied to the medium at the point **r**.

In a consistent theory, $X(\mathbf{r})$ must itself be an operator on its own field. But for the moment let us treat this external force as classical, and consider only the consequences for the quantized field ϕ. It is evident that the interaction term can be expressed as a sum of annihilation and creation operators in the phonon field. Thus, we follow the lines of (1.54) and write

$$\phi(\mathbf{r}) = \frac{1}{\sqrt{V}} \sum_{\mathbf{k}} \mathrm{e}^{-\mathrm{i}\mathbf{k}\cdot\mathbf{r}}\, \Phi_{\mathbf{k}}; \quad \pi(\mathbf{r}) = \frac{1}{\sqrt{V}} \sum_{\mathbf{k}} \mathrm{e}^{\mathrm{i}\mathbf{k}\cdot\mathbf{r}}\, \Pi_{\mathbf{k}}, \tag{1.90}$$

followed by

$$\left.\begin{aligned} \Phi_{\mathbf{k}} &= -\mathrm{i}(\tfrac{1}{2}\hbar/\rho_0 \omega_{\mathbf{k}})^{\frac{1}{2}}\, (a^*_{\mathbf{k}} - a_{-\mathbf{k}}), \\ \Pi_{\mathbf{k}} &= (\tfrac{1}{2}\hbar\rho_0 \omega_{\mathbf{k}})^{\frac{1}{2}}\, (a^*_{-\mathbf{k}} + a_{\mathbf{k}}), \end{aligned}\right\} \tag{1.91}$$

which reduce the Hamiltonian (1.77) to a standard form like (1.40). Substituting into (1.89) we get

$$H_I = -\mathrm{i} \sum_{\mathbf{k}} (\tfrac{1}{2}\hbar/V\rho_0 \omega_{\mathbf{k}})^{\frac{1}{2}}\, X(\mathbf{k})\, (a^*_{\mathbf{k}} - a_{-\mathbf{k}}), \tag{1.92}$$

where

$$X(\mathbf{k}) = \int X(\mathbf{r})\, \mathrm{e}^{-\mathrm{i}\mathbf{k}\cdot\mathbf{r}}\, \mathbf{d}^3\mathbf{r}. \tag{1.93}$$

If we treat H_I as a perturbation acting on the vacuum state of the field, then it is easy to see that it will create excitations. The production of particles in the **k**th mode will depend upon $X(\mathbf{k})$, i.e. upon the corresponding Fourier component in the force field. In fact the rate of production will depend upon $|X(\mathbf{k})|^2$, which is a measure of the energy in the power spectrum at this wavelength.

The external field can also take energy from the field (if this is not empty of particles) through the annihilation operators in (1.92). Thus, this term is a genuine representation of processes by which energy is transferred to and from the phonon (or photon) field in quantized units.

We can use just the same argument for our general field of Klein–Gordon particles; a term such as (1.89) would be capable of changing the number of 'mesons' in various states. Since there are, indeed, observable processes in which elementary particles are created and destroyed, it is reasonable to postulate the existence of such terms in the total field Hamiltonian.

What then, would the 'force' $X(\mathbf{r})$ represent? If the meson arose through the direct decay of some other particle then $X(\mathbf{r})$ could be proportional to the wave function of that other field. Thus, we might write

$$\mathscr{H}_I = g\psi(\mathbf{r})\,\phi(\mathbf{r}) \tag{1.94}$$

with g a parameter measuring the coupling between, say, 'psions' and 'phions'. It is evident that such an interaction would describe, quite nicely, the annihilation of an excitation of the ψ field simultaneously with the creation of a particle of the ϕ field. It is easy to show, also, that the momentum, **k**, would be conserved in every process.

Of course, such direct conversion phenomena are almost always forbidden, because of energy conservation. In real processes we have three or more particles involved. But the argument can easily be generalized. Thus, an interaction Hamiltonian density of the form

$$\mathscr{H}_I = g\{\psi(\mathbf{r})\}^p \{\phi(\mathbf{r})\}^q \tag{1.95}$$

describes processes in which there are p particles of the ψ field and q particles of the ϕ field, counting all that appear on both sides of the transformation equation. It is easy enough to show, also, that the total $\mathbf{k}$-vector of the two fields will be conserved in all such transformations; in (1.92), the creation operator carries the wave-vector $\mathbf{k}$, whilst the annihilation operator carries $-\mathbf{k}$. Interactions between the fields of relativistic particles are discussed in § 6.9.

Examples of such interaction terms can also be found in solid state theory. Thus, a term that is cubic in the lattice displacement field—an *anharmonic term*—gives rise to interactions between the phonons, typically of the sort in which two phonons combine to make a third. Again, a term of the type $\psi\psi^*\phi$ generates *electron–phonon interactions*, in which an electron in the ψ field (which must be a charged fermion field, of course) is scattered with the production of a phonon. The main difference in the case of excitations of a solid is that the $\mathbf{k}$-vector is not absolutely conserved, but may change by a reciprocal lattice vector.

1.10 Example: Rayleigh scattering of phonons

To show how we might use all this theory, consider the following elementary example. Suppose that we introduce a point mass ΔM into a continuous medium of density ρ_0. What would be its effect on the phonon field?

For simplicity, take the mass to be located at $\mathbf{r} = 0$. Then we have a density change

$$\Delta\rho = \Delta M\, \delta(\mathbf{r}), \tag{1.96}$$

which would introduce an extra term in the Hamiltonian density

$$\begin{aligned} \mathscr{H}_I &= \Delta\rho \left(\frac{\partial \phi}{\partial t}\right)^2 \\ &= \frac{\Delta M}{\rho_0^2} \{\pi(\mathbf{r})\}^2\, \delta(\mathbf{r}). \end{aligned} \tag{1.97}$$

Substituting now from (1.90) and (1.91), we get

$$H_I = \frac{\Delta M}{\rho_0^2 V} \sum_{\mathbf{k}} (\tfrac{1}{2}\hbar\rho_0\omega_{\mathbf{k}})^{\frac{1}{2}} (a^*_{-\mathbf{k}} + a_{\mathbf{k}}) \sum_{\mathbf{k}'} (\tfrac{1}{2}\hbar\rho_0\omega_{\mathbf{k}'})^{\frac{1}{2}} (a^*_{-\mathbf{k}'} + a_{\mathbf{k}'})$$

$$= \frac{\Delta M\hbar}{2\rho_0 V} \sum_{\mathbf{k},\mathbf{k}'} (\omega_{\mathbf{k}}\omega_{\mathbf{k}'})^{\frac{1}{2}} \{a_{\mathbf{k}}a_{\mathbf{k}'} + (a^*_{-\mathbf{k}}a_{\mathbf{k}'} + a_{\mathbf{k}}a^*_{-\mathbf{k}'}) + a^*_{-\mathbf{k}}a^*_{-\mathbf{k}'}\}. \quad (1.98)$$

We are interested in the following situation: initially there is a phonon in $\mathbf{k}$ and none in $\mathbf{k}'$; afterwards there is one in $\mathbf{k}'$, but not in $\mathbf{k}$. Thus we consider transitions with the following matrix element

$$T_{\mathbf{k}}^{\mathbf{k}'} = \langle 0_{\mathbf{k}}, 1_{\mathbf{k}'} | H_I | 1_{\mathbf{k}}, 0_{\mathbf{k}'} \rangle. \quad (1.99)$$

Obviously, the terms with two annihilators or two creators do not contribute; but the others do, so we get

$$T_{\mathbf{k}}^{\mathbf{k}'} = \frac{\Delta M\hbar}{2\rho_0 V} (\omega_{\mathbf{k}}\omega_{\mathbf{k}'})^{\frac{1}{2}} \langle 0_{\mathbf{k}}, 1_{\mathbf{k}'} | a^*_{\mathbf{k}'}a_{\mathbf{k}} + a_{\mathbf{k}}a^*_{\mathbf{k}'} | 1_{\mathbf{k}}, 0_{\mathbf{k}'} \rangle$$

$$= \frac{\Delta M\hbar}{\rho_0 V} (\omega_{\mathbf{k}}\omega_{\mathbf{k}'})^{\frac{1}{2}}. \quad (1.100)$$

In this algebra it is important to notice that both the a^*a terms in (1.98) contribute, because $\mathbf{k}$ and $\mathbf{k}'$ go over all positive and negative values.

To calculate the scattering cross-section we need to square this matrix element and then multiply by the density of states at the scattering energy. It is easy to show that the latter will be proportional to $\omega^2 V$. We thus get a transition rate of the form

$$P_{\mathbf{k}}^{\mathbf{k}'} \propto \left(\frac{\Delta M}{\rho_0}\right)^2 \omega^4, \quad (1.101)$$

which is the well-known formula for Rayleigh scattering.

This is trivial. But it is an opportunity for pointing out an important consequence of the fact that we are dealing with bosons. Let us make the same calculation for the transition rate in the more general case where the modes $\mathbf{k}$ and $\mathbf{k}'$ initially contain n and n' particles respectively. Instead of (1.99) we must then calculate the matrix element

$$\langle n-1, n'+1 | H_I | n, n' \rangle = (n'+1)^{\frac{1}{2}} n^{\frac{1}{2}} T_{\mathbf{k}}^{\mathbf{k}'}, \quad (1.102)$$

by the rules (1.10). Thus, when we calculate the transition rate, we shall get

$$Q_{\mathbf{k}}^{\mathbf{k}'} = (1+n')nP_{\mathbf{k}}^{\mathbf{k}'}. \quad (1.103)$$

The extra factor n is easy enough to understand; the rate of scattering from the mode $\mathbf{k}$ must surely be proportional to the number of excitations already in that mode. But the factor $(1+n')$ shows that the transition rate also depends on the state of occupation of the mode $\mathbf{k}'$ into which the phonon is going. This is none other than the phenomenon of *stimulated emission*, which is known to be typical of particles obeying Bose–Einstein statistics. This argument from field theory (see § 2.4) justifies the familiar derivation given originally by Einstein, from the phenomenon of spontaneous emission of radiation.

1.11 Example: Yukawa force

Suppose we have a number of simple point sources of bosons in a field; what is their effect upon its energy? For example, consider a number of heavy 'psions' (i.e. nucleons) capable of generating light 'phions' (i.e. mesons) through an interaction of the form

$$\mathscr{H}_I = g\psi(\mathbf{r})\,\psi^*(\mathbf{r})\,\phi(\mathbf{r}). \tag{1.104}$$

In this interaction the 'psions' are scattered, but not changed in number; it is reasonable to ignore recoil effects and to treat the operator $\psi(\mathbf{r})$ as a static wave function. Then $\psi\psi^*$ measures the local number density of these particles, which is concentrated like a delta function at each of their centres, $\mathbf{R}_j$. Thus

$$\mathscr{H}_I = g\sum_j \delta(\mathbf{r}-\mathbf{R}_j)\,\phi(\mathbf{r}). \tag{1.105}$$

Using (1.90) and (1.91) we can transform this to the particle representation in the phion field. Taking $\rho_0 = \hbar = 1$, we get

$$H_I = -\mathrm{i}g\sum_{\mathbf{k}}(2V\omega_{\mathbf{k}})^{-\frac{1}{2}}\sum_j\left(a^*_{\mathbf{k}}\,\mathrm{e}^{-\mathrm{i}\mathbf{k}\cdot\mathbf{R}_j} - a_{\mathbf{k}}\,\mathrm{e}^{\mathrm{i}\mathbf{k}\cdot\mathbf{R}_j}\right). \tag{1.106}$$

It is easy enough to calculate the effect of such a term as a perturbation. We see immediately that in any eigenstate of the free field the expectation value of H_I is zero. But there are evidently second-order terms, because H_I can first create an excitation in any mode $\mathbf{k}$, and then, on second application, destroy it.

Most of the interesting results are obtained if we consider just two sources, at $\mathbf{R}_1$ and $\mathbf{R}_2$, acting on the vacuum of the boson field. Then the change of energy is

$$\Delta\mathscr{E} = \sum_{\mathbf{k}} \frac{|\langle 0|\,H_I\,|1_{\mathbf{k}}\rangle|^2}{\mathscr{E}(0)-\mathscr{E}(\mathbf{k})}; \tag{1.107}$$

substituting from (1.106), we get

$$\begin{aligned}\Delta\mathscr{E} &= \sum_{\mathbf{k}} \frac{1}{-\omega_{\mathbf{k}}} \frac{g^2}{2V\omega_{\mathbf{k}}} \left|\sum_j \mathrm{e}^{\mathrm{i}\mathbf{k}\cdot\mathbf{R}_j}\right|^2 \\ &= -\frac{g^2}{V}\sum_{\mathbf{k}} \frac{1}{\omega_{\mathbf{k}}^2}\{1+\cos\mathbf{k}\cdot(\mathbf{R}_1-\mathbf{R}_2)\}. \end{aligned} \tag{1.108}$$

The first term in this sum is independent of the positions of the sources, and would occur just N times for N sources. It is obviously a measure of the self-energy of each source due to the emission and reabsorption of virtual bosons. Suppose we try to evaluate it. By (1.42), we get

$$\Delta\mathscr{E}_{\text{self}} = -\frac{g^2}{V}\left(\frac{V}{8\pi^3}\right)\int_0^{k_m} \frac{4\pi k^2}{\omega_{\mathbf{k}}^2}\,\mathrm{d}k. \tag{1.109}$$

It is satisfactory that this expression is independent of the volume of the box. Unfortunately, it is infinite. We may use (1.86) for $\omega_{\mathbf{k}}^2$, whether or not the bosons have any rest mass. Thus

$$\begin{aligned}\Delta\mathscr{E}_{\text{self}} &= -\frac{g^2}{2\pi^2}\int_0^{k_m} \frac{k^2}{k^2+m^2}\,\mathrm{d}k \\ &= -\frac{g^2}{2\pi^2}\left(k_m - m\tan^{-1}\frac{k_m}{m}\right), \end{aligned} \tag{1.110}$$

which evidently tends to infinity with k_m, whatever the value of m. This singularity could be avoided, no doubt, by cutting off the range of allowed values of k, but this is difficult to do consistently in relativistic systems. Here we see a typical case of the infinities that arise in field theoretical calculations, and that can only be removed by complete and drastic schemes (see § 3.9).

Now turn to the part of (1.108) depending on the relative coordinates of the sources, $\mathbf{R} = \mathbf{R}_1 - \mathbf{R}_2$; we get

$$\begin{aligned}\Delta\mathscr{E}_{\text{interaction}} &= -\frac{g^2}{8\pi^3}\int \frac{\cos(\mathbf{k}.\mathbf{R})}{k^2+m^2}\,\mathrm{d}^3\mathbf{k} \\ &= -\frac{g^2}{4\pi R}\exp(-mR), \end{aligned} \tag{1.111}$$

by an elementary Fourier transform. Thus, the energy of two sources depends upon their distance apart, as if there were a force between them with this expression as its potential energy. This is the famous *Yukawa force*. It is coulombic at short distances, but falls off exponentially with a range of the order of $\hbar/mc$. It is well known that range of nuclear forces is about 10^{-13} cm, which can be matched if we make m about 200 times the mass of an electron—which is appropriate for one of the 'mesons'.

Notice also that if m had been zero we should have a pure coulomb interaction between the sources. Schematically, this is the origin of ordinary electrostatic forces, which may be described as the exchange of virtual photons between charges (evidently e then takes the place of g). But the full theory of electromagnetism (see § 6.3) is much more complicated, because photons belong to a vector field, and cannot be described by a simple scalar wave-function ϕ. A similar case is the exchange of phonons between electrons in a metal—the interaction leading to superconductivity. But here, as shown in §§ 3.5 and 5.13, the electron recoil is very important.

A consequence of this type of calculation is that any interaction between particles can be attributed to another field of particles bearing the force between them. It would almost seem as if any general field theory would thus generate a hierarchy of such fields and particles. But of course the shorter the range of the forces—the more it approximates to a contact interaction such as postulated in (1.104)—the larger the mass of the exchanging particles and the less we need consider the possibility of their being really excited. The general theory implicit in these remarks is that of the analytic S-matrix, discussed in § 6.11.

1.12 Charged bosons

In elementary quantum theory, the electron wave-function is complex. What happens if we try to set up a complex *boson* field? Let us write

$$\phi = \frac{1}{\sqrt{2}}(\phi_1 + i\phi_2), \tag{1.112}$$

where ϕ_1 and ϕ_2 are supposed to be two independent real fields.

By the ordinary arguments of mathematics, there must be a field ϕ^* which is in some sense the complex conjugate of ϕ. Let us define this relation as follows: if ϕ undergoes a *gauge transformation*, $\phi \to \phi\, e^{i\alpha}$, then automatically ϕ^* undergoes the complex conjugate transformation, $\phi^* \to \phi^*\, e^{-i\alpha}$. We can then readily construct physical quantities as products of ϕ with ϕ^* which will be invariant to this transformation. The significance of the gauge transformation itself will be discussed in § 6.3, where we consider relativistic electrodynamics.

As an obvious generalization of the theory of § 1.8 for the real scalar field, we postulate a Lagrangian density of the form

$$\mathscr{L} = -\tfrac{1}{2}(\nabla\phi\cdot\nabla\phi^* - \dot{\phi}\dot{\phi}^* + m^2\phi\phi^*). \tag{1.113}$$

Because ϕ and ϕ^* are effectively independent (or at least depend on the two independent field functions ϕ_1 and ϕ_2), they can be varied separately so that each satisfies its own Klein–Gordon equation

$$(\square^2 - m^2)\,\phi = 0; \quad (\square^2 - m^2)\,\phi^* = 0. \tag{1.114}$$

But when we come to define the 'momentum field variables' we get

$$\pi = \frac{\partial \mathscr{L}}{\partial \dot{\phi}} = \dot{\phi}^* \quad \text{and} \quad \pi^* = \dot{\phi}, \tag{1.115}$$

whence

$$\mathscr{H} = \pi\pi^* + \nabla\phi\cdot\nabla\phi^* + m^2\phi\phi^*. \tag{1.116}$$

The effort of making this Hamiltonian density real (i.e. gauge invariant) imposes a dynamical linkage between the two fields.

We now quantize these fields by introducing annihilation and creation operators for ϕ_1 and ϕ_2. Suppose we work by analogy with (1.90) and (1.91), and write

$$\left.\begin{aligned} \phi_1(\mathbf{r}) &= -\mathrm{i}\,V^{-\frac{1}{2}} \sum_{\mathbf{k}} (2\omega_{\mathbf{k}})^{-\frac{1}{2}}\,(a_{\mathbf{k}}^{(1)*} - a_{-\mathbf{k}}^{(1)})\,\mathrm{e}^{\mathrm{i}\mathbf{k}\cdot\mathbf{r}}, \\ \phi_2(\mathbf{r}) &= -\mathrm{i}\,V^{-\frac{1}{2}} \sum_{\mathbf{k}} (2\omega_{\mathbf{k}})^{-\frac{1}{2}}\,(a_{\mathbf{k}}^{(2)*} - a_{-\mathbf{k}}^{(2)})\,\mathrm{e}^{\mathrm{i}\mathbf{k}\cdot\mathbf{r}}. \end{aligned}\right\} \tag{1.117}$$

These fields are independent, so that all commutators will vanish except the usual ones

$$[a_{\mathbf{k}}^{(1)}, a_{\mathbf{k}'}^{(1)*}] = \delta_{\mathbf{k}\mathbf{k}'}; \quad [a_{\mathbf{k}}^{(2)}, a_{\mathbf{k}'}^{(2)*}] = \delta_{\mathbf{k}\mathbf{k}'}. \tag{1.118}$$

From (1.112) and (1.117) we should find $\phi(\mathbf{r})$ as a linear combination of all four types of operator. It is instructive to group these as follows: Let us write

$$\left.\begin{aligned} a_{\mathbf{k}} &= \frac{1}{\sqrt{2}}\{a_{\mathbf{k}}^{(1)} - \mathrm{i}a_{\mathbf{k}}^{(2)}\}; & b_{\mathbf{k}} &= \frac{1}{\sqrt{2}}\{a_{\mathbf{k}}^{(1)} + \mathrm{i}a_{\mathbf{k}}^{(2)}\}, \\ a_{\mathbf{k}}^* &= \frac{1}{\sqrt{2}}\{a_{\mathbf{k}}^{(1)*} + \mathrm{i}a_{\mathbf{k}}^{(2)*}\}; & b_{\mathbf{k}}^* &= \frac{1}{\sqrt{2}}\{a_{\mathbf{k}}^{(1)*} - \mathrm{i}a_{\mathbf{k}}^{(2)*}\}. \end{aligned}\right\} \tag{1.119}$$

These relations define a canonical unitary transformation in the Hilbert space of the operators. Thus, all commutators again vanish except those involving pairs of Hermitian conjugates, $[a_{\mathbf{k}}, a_{\mathbf{k}}^*]$ and $[b_{\mathbf{k}}, b_{\mathbf{k}}^*]$.

Substituting now from (1.119) in (1.112) and (1.117) we get

$$\phi(\mathbf{r}) = -\mathrm{i}V^{-\frac{1}{2}} \sum_{\mathbf{k}} (2\omega_{\mathbf{k}})^{-\frac{1}{2}} \{a^*_{\mathbf{k}} - b_{-\mathbf{k}}\} \mathrm{e}^{\mathrm{i}\mathbf{k}\cdot\mathbf{r}}. \tag{1.120}$$

This is, so far, only a conjecture. But if we write

$$\pi(\mathbf{r}) = V^{-\frac{1}{2}} \sum_{\mathbf{k}} (\tfrac{1}{2}\omega_{\mathbf{k}})^{\frac{1}{2}} \{a_{\mathbf{k}} + b^*_{-\mathbf{k}}\} \mathrm{e}^{-\mathrm{i}\mathbf{k}\cdot\mathbf{r}} \tag{1.121}$$

we shall get the canonical commutation relation

$$[\phi(\mathbf{r}), \pi(\mathbf{r}')] = \mathrm{i}\delta(\mathbf{r}-\mathbf{r}') \tag{1.122}$$

required by the general principle of second quantization of § 1.7.

Suppose now that we apply the 'star' operation to each of (1.120) and (1.121). The rule for this is that we apply an asterisk to every unstarred operator, and remove the star from every operator that has one (thus changing it into its Hermitian conjugate) and we also take the complex conjugate of every complex number in the formula. Thus

$$\left.\begin{aligned} \phi^*(\mathbf{r}) &= \mathrm{i}V^{-\frac{1}{2}} \sum_{\mathbf{k}} (2\omega_{\mathbf{k}})^{-\frac{1}{2}} \{a_{\mathbf{k}} - b^*_{-\mathbf{k}}\} \mathrm{e}^{-\mathrm{i}\mathbf{k}\cdot\mathbf{r}}, \\ \pi^*(\mathbf{r}) &= V^{-\frac{1}{2}} \sum_{\mathbf{k}} (2\omega_{\mathbf{k}})^{\frac{1}{2}} \{a^*_{\mathbf{k}} + b_{\mathbf{k}}\} \mathrm{e}^{\mathrm{i}\mathbf{k}\cdot\mathbf{r}}, \end{aligned}\right\} \tag{1.123}$$

which will be consistent with the corresponding commutation relation

$$[\phi^*(\mathbf{r}), \pi^*(\mathbf{r}')] = \mathrm{i}\delta(\mathbf{r}-\mathbf{r}'). \tag{1.124}$$

(Notice that this does not at first look like the 'star' of (1.122)—until we recall that the Hermitian conjugate of a product reverses the order: $(AB)^* = B^*A^*$.)

The point to notice is that this quantization scheme is more general than the formula for a real field, such as each of (1.117). If our field ϕ had been self-conjugate—the Hermitian equivalent of real—then to make $\phi = \phi^*$ we should need to have $a_{\mathbf{k}}$ and $b_{\mathbf{k}}$ the same. A complex field thus has room for two different types of 'particle' or 'excitation' —those generated by the a and b types of annihilation and creation operator respectively.

We can see this by expressing the Hamiltonian in the form

$$\begin{aligned} H &= \sum_{\mathbf{k}} \omega_{\mathbf{k}}(a^*_{\mathbf{k}} a_{\mathbf{k}} + b^*_{\mathbf{k}} b_{\mathbf{k}} + 1) \\ &= \sum_{\mathbf{k}} \omega_{\mathbf{k}}(n^+_{\mathbf{k}} + n^-_{\mathbf{k}} + 1), \end{aligned} \tag{1.125}$$

where each wave-vector mode may now be occupied by particles of the two different types. Moreover, these are independent excitations in this representation.

But what do they mean? Think of the well-known formula for the current density in electron theory:

$$\mathbf{j} = -\mathrm{i}e(\phi\nabla\phi^* - \phi^*\nabla\phi). \tag{1.126}$$

It is evident that here is another quantity that is gauge invariant. Consider also

$$\rho = -\mathrm{i}e(\pi\phi - \pi^*\phi^*) \tag{1.127}$$

which satisfies the following equation, derivable from the equations of motion (1.114) and (1.115):

$$\nabla\cdot\mathbf{j} - \dot{\rho} = 0. \tag{1.128}$$

We naturally interpret ρ as a charge density, with a continuity equation (1.128).

When these field equations are quantized, we can calculate the total charge in an occupation-number representation:

$$\begin{aligned} Q = \int \rho\, \mathbf{d}^3\mathbf{r} &= e \sum_{\mathbf{k}} (a^*_{\mathbf{k}} a_{\mathbf{k}} - b^*_{\mathbf{k}} b_{\mathbf{k}}) \\ &= e \sum_{\mathbf{k}} (n^+_{\mathbf{k}} - n^-_{\mathbf{k}}). \end{aligned} \tag{1.129}$$

In other words, $a^*_{\mathbf{k}}$ generates particles of charge e and $b^*_{\mathbf{k}}$ generates particles of charge $-e$. The two types differ just by their electric charge; we have learnt to think of them as particle and *antiparticle*, although this concept involves deeper questions to be taken up in due course.

It is worth remarking that our field operator ϕ does not refer solely to one or the other type. According to (1.120), it has the effect of *either* increasing the number of particles with charge $+e$, *or* of decreasing the number with charge $-e$. If we are interested only in electromagnetic interactions of our bosons, then these two processes are effectively the same.

It must be emphasized, however, that the above procedure is not a perfectly general method by which electric charge is brought into quantum theory. More subtle formalisms, applicable to nucleons and other baryons for example, are discussed in § 7.12.

CHAPTER 2

FERMIONS

You gotter accentuate *the positive,* eliminate *the negative...*

2.1 Occupation-number representation

The theory of the previous chapter is quite inappropriate to the description of an assembly of particles, such as electrons or nucleons, that obey Fermi–Dirac statistics. We need a scheme in which there can never be more than one excitation in each mode. This is not just a question of limiting the number of eigenstates of the number operators; it has a more fundamental significance.

The basis of the Pauli principle is the rule that the wave function of a system of indistinguishable fermions must be *antisymmetric* for the exchange of any pair of particle co-ordinates:

$$\Psi(\mathbf{r}_1, \mathbf{r}_2, \ldots \mathbf{r}_n \ldots \mathbf{r}_m \ldots \mathbf{r}_N) = -\Psi(\mathbf{r}_1, \mathbf{r}_2, \ldots \mathbf{r}_m \ldots \mathbf{r}_n \ldots \mathbf{r}_N). \quad (2.1)$$

The corresponding rule for bosons is that the wave function must be *symmetric* for such exchanges—but this is much less powerful.

The Pauli principle is derived from (2.1) by consideration of the sort of wave function we should write down for an assembly of *independent* fermions. Thus, let $\psi_1, \psi_2, \ldots, \psi_N$ be a set of single-particle functions, each a solution of the appropriate one-particle Schrödinger equation. The product of such functions, i.e.

$$\psi_1(\mathbf{r}_1).\psi(\mathbf{r}_2) \ldots \psi_N(\mathbf{r}_N),$$

does not satisfy the antisymmetric principle. But we can easily construct a totally antisymmetric function out of these, by application of the permutation operator P, which exchanges pairs of co-ordinates. Thus we might have

$$\Psi = (N!)^{-\frac{1}{2}} \sum_P (-1)^P \psi_1(P\mathbf{r}_1).\psi_2(P\mathbf{r}_2) \ldots \psi_N(P\mathbf{r}_N), \quad (2.2)$$

where the summation is over all the $N!$ permutations of the co-ordinates, with sign ± 1 according as the permutation is odd or even.

It is well known that (2.2) defines, in fact, a *determinantal function*—a determinant in which the ith element of the jth column is $\psi_i(\mathbf{r}_j)$. It is easy to prove that (2.1) is satisfied, and also that this wave function vanishes identically if any two of the functions ψ_i happen to be the same. This is the familiar interpretation of the Pauli principle: no two particles may be in the same state. One can also show that (2.2) is normalized if the original ψ_i functions are normalized and orthogonal.

But of course we want to construct more complicated many-body wave functions than a simple determinant. This we can do by taking linear combinations of determinants—and the result will then, automatically be antisymmetric. Now suppose our functions $\psi_1, \psi_2, \ldots, \psi_N$ are part of an infinite, complete normal orthogonal sequence of functions. There are an infinity of determinantal functions we might construct by choosing any states from this sequence. Since the $\psi_i(\mathbf{r})$ functions span the space of a single variable $\mathbf{r}$, it follows (or can be proved if one feels like it) that these determinants provide a basis for constructing any antisymmetric function in the N variables $\mathbf{r}_1 \ldots \mathbf{r}_N$.

We need a compact notation for such functions. The obvious convention is to define each determinant by the functions ψ_i, etc. which occur in it, or, as we might say, by the single particle states or *modes* which are 'occupied' in it. Thus, we write

$$|n_1, n_2, \ldots\rangle$$

for the determinantal function with n_1 particles in mode ψ_1, n_2 particles in ψ_2, etc. This is very like what we have done for boson states in § 1.3, except that here we limit n_1 to be just 0 or 1. It must be emphasized that there is nothing mysterious about these symbols. They are just a shorthand for functions out of the set like (2.2). Thus

$$|1_1, 0_2, 0_3 \ldots\rangle \equiv \psi_1(\mathbf{r}_1), \tag{2.3}$$

$$|1_1, 1_2, 0_3 \ldots\rangle \equiv 2^{-\frac{1}{2}}\{\psi_1(\mathbf{r}_1)\,\psi_2(\mathbf{r}_2) - \psi_1(\mathbf{r}_2)\,\psi_2(\mathbf{r}_1)\}, \quad \text{etc.} \tag{2.4}$$

We want to use these to construct *any* more general Ψ that is antisymmetric in the variables. Of course these functions contain the variables $\mathbf{r}_1, \mathbf{r}_2, \ldots$, etc., but these will eventually be absorbed in an integration over all variables when we come to compute expectation values and matrix elements. The particles must be indistinguishable in the final physical results.

2.2 Annihilation and creation operators: anticommutation

All that we have said about fermion states might have applied just as well to boson states, except that we should have been trying to construct only symmetric states. In the boson theory we defined operators a^* and a that had the property of changing the occupation numbers of the various modes. Let us do the same thing for the fermion states. Let us define an operator $b^*_{\mathbf{k}}$ that creates a particle in the $\mathbf{k}$th mode—i.e.

$$b^*_{\mathbf{k}}\,|n_1, n_2, \ldots, n_{\mathbf{k}}, \ldots\rangle = |n_1, n_2, \ldots, n_{\mathbf{k}}+1, \ldots\rangle. \tag{2.5}$$

Similarly, we define a conjugate operator that destroys an excitation, i.e.

$$b_{\mathbf{k}}\,|n_1, n_2, \ldots, n_{\mathbf{k}}, \ldots\rangle = |n_1, n_2, \ldots, n_{\mathbf{k}}-1, \ldots\rangle. \tag{2.6}$$

For the fermion states that actually exist—i.e. those for which $n_{\mathbf{k}}$ is 0 or 1—these symbols have the same effects as the corresponding boson operators. But we have to make some additional formal rules to make sure that we do not exceed the allowed occupation numbers. These rules are obvious: we must have

$$(b^*_{\mathbf{k}})^2 = 0 \quad \text{and} \quad (b_{\mathbf{k}})^2 = 0 \tag{2.7}$$

for any state on which they might try to act.

One of the useful features of the boson theory was our ability to build up all states from the vacuum by the application of a succession of creation operators. We want to do the same thing for fermions. Thus we might write

$$b^*_{\mathbf{k}}\,|0\rangle = |1_{\mathbf{k}}\rangle = \psi_{\mathbf{k}}(\mathbf{r}), \tag{2.8}$$

and then

$$\begin{aligned} b^*_{\mathbf{k}'} b^*_{\mathbf{k}}\,|0\rangle &= |1_{\mathbf{k}}, 1_{\mathbf{k}'}\rangle \\ &= 2^{-\frac{1}{2}}\{\psi_{\mathbf{k}}(\mathbf{r}_1)\,\psi_{\mathbf{k}'}(\mathbf{r}_2) - \psi_{\mathbf{k}}(\mathbf{r}_2)\,\psi_{\mathbf{k}'}(\mathbf{r}_1)\}, \end{aligned} \tag{2.9}$$

and so on.

But we must be careful. Suppose that we had built up our state with particles in $\mathbf{k}$ and $\mathbf{k}'$ in the opposite order—by applying $b^*_{\mathbf{k}'}$ first, in the vacuum, and then $b^*_{\mathbf{k}}$. If (2.9) is to be algebraically consistent we must be able to interchange these (arbitrary) labels, and get

$$\begin{aligned} b^*_{\mathbf{k}} b^*_{\mathbf{k}'}\,|0\rangle &= 2^{-\frac{1}{2}}\{\psi_{\mathbf{k}'}(\mathbf{r}_1)\,\psi_{\mathbf{k}}(\mathbf{r}_2) - \psi_{\mathbf{k}'}(\mathbf{r}_2)\,\psi_{\mathbf{k}}(\mathbf{r}_1)\} \\ &= -\,b^*_{\mathbf{k}'} b^*_{\mathbf{k}}\,|0\rangle. \end{aligned} \tag{2.10}$$

Thus, the order in which these operators are written down is most important. Our result is equivalent to

$$(b^*_{\mathbf{k}'} b^*_{\mathbf{k}} + b^*_{\mathbf{k}} b^*_{\mathbf{k}'})\,|0\rangle = 0; \tag{2.11}$$

applying it more generally, we find the same dependence of the sign on the ordering of this pair of operators when they occur before any state function at all. Moreover, (2.7) covers the case when **k** and **k**′ happen to be identical. Thus, quite generally,

$$b_{\mathbf{k}}^* b_{\mathbf{k}'}^* + b_{\mathbf{k}'}^* b_{\mathbf{k}}^* = 0. \tag{2.12}$$

We say that these two operators *anticommute* and write

$$\{b_{\mathbf{k}}^*, b_{\mathbf{k}'}^*\} = 0, \tag{2.13}$$

by direct analogy with the symbolism for a commutation relation such as (1.38).

There is no doubt that this rule seems mysterious. How can two quite independent modes, $\psi_{\mathbf{k}}$ and $\psi_{\mathbf{k}'}$, interfere with one another in this way? The mystery lies in the antisymmetry principle itself, which we must accept as a fact of life. Our relation (2.10) is really a case of (2.1). The difficulty in the demonstration that we have given is that the variables $\mathbf{r}_1, \mathbf{r}_2, \ldots$ seem to play an explicit part—yet they refer to indistinguishable particles. The point is, simply, that we must establish some basic ordering of these variables, and a convention that the first operator acting on $|0\rangle$ applies to the first of them, and so on. But this is only a convention. The sign of the determinantal function (2.2) is also arbitrary, until we have prescribed a canonical order for the variables corresponding to the identical permutation.

It is obvious that all annihilation operators must similarly anticommute, i.e.

$$\{b_{\mathbf{k}}, b_{\mathbf{k}'}\} = 0. \tag{2.14}$$

Again, if we consider the case where **k** is not equal to **k**′, the conventions imply that

$$\{b_{\mathbf{k}}, b_{\mathbf{k}'}^*\} = 0 \quad \text{if} \quad \mathbf{k} \neq \mathbf{k}'. \tag{2.15}$$

So we are left with the case of $b_{\mathbf{k}}$ and $b_{\mathbf{k}}^*$, whose anticommutator can be constructed by inspection. For consider the following equations, all of which follow from the definitions and the Pauli principle:

$$\left.\begin{aligned} b_{\mathbf{k}}^* b_{\mathbf{k}} |0_{\mathbf{k}}\rangle &= 0; \qquad b_{\mathbf{k}} b_{\mathbf{k}}^* |0_{\mathbf{k}}\rangle = |0_{\mathbf{k}}\rangle, \\ b_{\mathbf{k}}^* b_{\mathbf{k}} |1_{\mathbf{k}}\rangle &= |1_{\mathbf{k}}\rangle; \quad b_{\mathbf{k}} b_{\mathbf{k}}^* |1_{\mathbf{k}}\rangle = 0. \end{aligned}\right\} \tag{2.16}$$

Now add these up, with arbitrary coefficients α and β:

$$(b_{\mathbf{k}}^* b_{\mathbf{k}} + b_{\mathbf{k}} b_{\mathbf{k}}^*)\,(\alpha\,|0_{\mathbf{k}}\rangle) + \beta\,|1_{\mathbf{k}}\rangle = \alpha\,|0_{\mathbf{k}}\rangle + \beta\,|1_{\mathbf{k}}\rangle. \tag{2.17}$$

Thus, for operations on any ket vector referring to the **k**th mode, we have the result

$$b_{\mathbf{k}}^* b_{\mathbf{k}} + b_{\mathbf{k}} b_{\mathbf{k}}^* = 1. \tag{2.18}$$

This result is general, for operations on any state vector. Thus, we may extend (2.15) to read

$$\{b_{\mathbf{k}}, b^*_{\mathbf{k}'}\} = \delta_{\mathbf{k}\mathbf{k}'}. \tag{2.19}$$

From (2.16) it is evident that

$$n_{\mathbf{k}} = b^*_{\mathbf{k}} b_{\mathbf{k}} \tag{2.20}$$

plays the role of the *occupation number operator* for the $\mathbf{k}$th mode. We can also write the relations (2.5) and (2.6) in the form

$$\left.\begin{aligned} b_{\mathbf{k}}\,|n_{\mathbf{k}}\rangle &= \sqrt{n_{\mathbf{k}}}\,|n_{\mathbf{k}}-1\rangle, \\ b^*_{\mathbf{k}}\,|n_{\mathbf{k}}\rangle &= \sqrt{(1-n_{\mathbf{k}})}\,|n_{\mathbf{k}}+1\rangle \end{aligned}\right\} \tag{2.21}$$

which automatically cut off the ladder of quantum numbers above 1 and below zero.

The beauty of the anticommutation relations (2.13), (2.14) and (2.19) is that they are exactly the same as the corresponding commutation relations (1.32) and (1.38) for the boson operators. This is no accident. The boson theory can be developed in just the same way as here, with 'symmetry' playing the role of 'antisymmetry'. The only essential difference in the algebra is that the (-1) in (2.2), giving a change of sign with each permutation of variables, now becomes $(+1)$, and correspondingly the anticommutators become commutators. The change of sign is reflected in the matrix elements of the operators $b^*_{\mathbf{k}}$ and $b_{\mathbf{k}}$, as defined in (2.21), which is otherwise the analogue of (1.10).

We chose to build up the boson theory in another way, deriving our annihilation and creation operators from the simple co-ordinate and momentum operators for a single particle. Unfortunately there is no such elementary representation of the fermion annihilation and creation operators, which have to be defined in this rather abstract way, and which therefore seem a little less familiar.

2.3 Second quantization

This theory of fermion states has been built up from a representation of the wave functions with a set of definite modes $\psi_{\mathbf{k}}$ as a basis. Often enough, these functions are just plane waves labelled by the wave-vector $\mathbf{k}$. In the case of electrons in a perfect crystal we might use Bloch functions as our basis. But the general argument does not depend on whether we are using a *momentum representation*; it would apply just as well to any complete set of functions such as the one-particle eigenstates of an electron in the potential of a complicated organic molecule.

To escape from the particular representation it is useful to build up a theory of a general fermion field, with second-quantized field operators. This we can do, easily enough, by working backwards through a set of equations like (1.90) and (1.91). We define the following operators

$$\psi(\mathbf{r}) = \sum_{\mathbf{k}} \psi_{\mathbf{k}}(\mathbf{r})\, b_{\mathbf{k}}; \quad \psi^*(\mathbf{r}) = \sum_{\mathbf{k}} \psi^*_{\mathbf{k}}(\mathbf{r})\, b^*_{\mathbf{k}}. \tag{2.22}$$

It is easy enough to derive from (2.19) and from the orthogonality of the basis functions, some anticommutation relations for these operators. Thus

$$\{\psi(\mathbf{r}), \psi(\mathbf{r}')\} = 0; \quad \{\psi^*(\mathbf{r}), \psi^*(\mathbf{r}')\} = 0; \tag{2.23}$$

but

$$\{\psi(\mathbf{r}), \psi^*(\mathbf{r}')\} = \delta(\mathbf{r}-\mathbf{r}'). \tag{2.24}$$

These relations are obviously analogues of the boson commutation relations (1.78), except that here we can use the Hermitian conjugate of ψ instead of having to introduce a special field, π, that is canonically conjugate to the field operator. This is a special property of fermions, related to the symmetry between 'particle' descriptions and 'hole' descriptions (see § 2.8).

But we now need some equations of motion. These are derivable from the time-dependent Schrödinger equation,

$$H\,|\rangle = \frac{\hbar}{\mathrm{i}}\frac{\partial}{\partial t}|\rangle, \tag{2.25}$$

once we have a Hamiltonian. In the boson theory it was shown that the canonical Hamiltonian operator for the quantized field was just the same as the Hamiltonian function for the corresponding classical field. It was also shown, in some special cases, that the single-particle modes of the quantized field were solutions of equations such as the wave equation (1.72), or the Klein–Gordon equation (1.85). These equations can thus be interpreted as equations of motion of one-particle states, derivable by the ordinary 'first quantization' procedure of putting

$$\mathbf{p} = \frac{\hbar}{\mathrm{i}}\nabla \tag{2.26}$$

for the one-particle momentum in a one-particle version of the Hamiltonian.

Well, suppose we know the equations of motion for the fermion modes $\psi_{\mathbf{k}}(\mathbf{r},t)$. Suppose these are of the form

$$\mathscr{H}\psi_{\mathbf{k}}(\mathbf{r},t) = \frac{\hbar}{\mathrm{i}}\frac{\partial}{\partial t}\psi_{\mathbf{k}}(\mathbf{r},t), \tag{2.27}$$

where $\mathscr{H}$ is an operator of the usual one-particle type. Then it can be shown that $\mathscr{H}$ is just what we need for a Hamiltonian density, from which to construct the general field Hamiltonian H, to go into (2.25).

This is rather abstract: let us give an example. Consider the elementary Schrödinger equation (1.83). The Hamiltonian operator for this is

$$\mathscr{H}(\mathbf{r}) = -\frac{\hbar^2}{2m}\nabla^2 + \mathscr{V}(\mathbf{r}). \tag{2.28}$$

Then a field theory for an assembly of fermions in the potential $\mathscr{V}(\mathbf{r})$ would have Hamiltonian

$$H = -\frac{\hbar^2}{2m}\int \psi^*(\mathbf{r})\,\nabla^2\psi(\mathbf{r})\,\mathbf{d}^3\mathbf{r} + \int \psi^*(\mathbf{r})\,\mathscr{V}(\mathbf{r})\,\psi(\mathbf{r})\,\mathbf{d}^3\mathbf{r}, \tag{2.29}$$

where we treat ψ and ψ^* as field operators obeying (2.23) and (2.24). In other words, the expectation value of the Hamiltonian for a single particle with wave-function $\psi(\mathbf{r})$ has been 'quantized' again and is now to be treated as an operator. It is now a many-particle operator, acting on a general state of the system, in the abstract *Fock space* of the state vectors $|\;\rangle$.

To show this, let us substitute from (2.22) into (2.29). We get the following:

$$\begin{aligned} H &= \sum_{\mathbf{k},\mathbf{k}'}\left[\int \psi^*_{\mathbf{k}'}(\mathbf{r})\left\{-\frac{\hbar^2}{2m}\nabla^2 + \mathscr{V}(\mathbf{r})\right\}\psi_{\mathbf{k}}(\mathbf{r})\,\mathbf{d}^3\mathbf{r}\right] b^*_{\mathbf{k}'}b_{\mathbf{k}} \\ &= \sum_{\mathbf{k},\mathbf{k}'}\mathscr{E}(\mathbf{k})\int \psi^*_{\mathbf{k}'}(\mathbf{r})\,\psi_{\mathbf{k}}(\mathbf{r})\,\mathbf{d}^3\mathbf{r}\, b^*_{\mathbf{k}'}b_{\mathbf{k}} \\ &= \sum_{\mathbf{k}}\mathscr{E}(\mathbf{k})\, b^*_{\mathbf{k}}b_{\mathbf{k}}, \end{aligned} \tag{2.30}$$

if we assume that each function $\psi_{\mathbf{k}}(\mathbf{r})$ satisfies the Schrödinger equation with energy $\mathscr{E}(\mathbf{k})$.

But by (2.20) this says that the standard states $|n_{\mathbf{k}}\rangle$ of the occupation number representation are eigenstates of the Hamiltonian H, with energy

$$\mathscr{E} = \sum_{\mathbf{k}}\mathscr{E}(\mathbf{k})\, n_{\mathbf{k}}. \tag{2.31}$$

Thus, our choice of (2.29) as Hamiltonian of the field is consistent with the assertion that this field describes an assembly of independent fermions in the one-particle modes, each mode having a Schrödinger equation with the one-particle Hamiltonian (2.28).

Again, it should be emphasized that the line of argument used in this chapter can also apply to bosons; we could set up a second quantized field, with a Hamiltonian operator, by working backwards from the occupation-number representation, without trying to derive the theory directly from a Lagrangian density, etc. Conversely, our fermion theory can be derived by postulating that some canonical Hamiltonian function can be quantized by making all 'wave functions' into anticommuting field operators. These two approaches to field theory are equivalent alternatives; one should keep both in mind when thinking about the physics, and the axioms and postulates that underlie it.

2.4 Scattering: connection with statistical mechanics

In particle physics the excitations are usually supposed to be free, unless they are explicitly bound in chains. Thus, the typical Hamiltonian is the free-field kinetic energy operator

$$\mathscr{H}_0 = \psi^*(\mathbf{r})\left(-\frac{\hbar^2}{2m}\nabla^2\right)\psi(\mathbf{r}). \tag{2.32}$$

In solid state physics there are also devices by which the electron Hamiltonian can be reduced, approximately, to this form, perhaps with an 'effective mass' m^* in place of the free-electron mass. The basic mode functions of $\mathscr{H}_0$ are plane waves.

To this term we may add various types of interaction. The argument is very much as in § 1.9. Thus, a term linear in ψ^* has the effect of creating new excitations of the fermion field, whilst terms in ψ tend to destroy existing excitations. A term containing $\psi^*(\mathbf{r})\,\psi(\mathbf{r})$ has the effect of scattering fermions from one mode into another, but conserves their number—and so on.

For example, consider the effect of an extra potential energy $\mathscr{V}(\mathbf{r})$ in the field. This produces the interaction Hamiltonian

$$H_I = \int \psi^*(\mathbf{r})\,\mathscr{V}(\mathbf{r})\,\psi(\mathbf{r})\,\mathbf{d}^3\mathbf{r}, \tag{2.33}$$

which can be transformed to a plane-wave representation, using (2.22), of the form

$$\begin{aligned} H_I &= \sum_{\mathbf{k},\mathbf{k}'}\int e^{-i(\mathbf{k}'-\mathbf{k})\cdot\mathbf{r}}\,\mathscr{V}(\mathbf{r})\,\mathbf{d}^3\mathbf{r}\,b^*_{\mathbf{k}'}b_{\mathbf{k}} \\ &= \sum_{\mathbf{k},\mathbf{k}'}\mathscr{V}(\mathbf{k}-\mathbf{k}')\,b^*_{\mathbf{k}'}b_{\mathbf{k}}, \end{aligned} \tag{2.34}$$

where $\mathscr{V}(\mathbf{k}-\mathbf{k}')$ is a Fourier transform of the function $\mathscr{V}(\mathbf{r})$. It is immediately obvious that this interaction has the effect of creating a particle in $\mathbf{k}'$ after destroying one in $\mathbf{k}$—or, as we might say, scattering the particle from one state to another. The matrix element for the process is obviously $\mathscr{V}(\mathbf{k}-\mathbf{k}')$; if we ignore multiple scattering, we may put the square of this matrix element into the usual formula for the transition probability, and get the Born approximation.

If we had calculated the effect of one of the terms in H_I upon a general state of the fermion system we should have found matrix elements like

$$\langle n_{\mathbf{k}'}+1, n_{\mathbf{k}}-1 | H_I | n_{\mathbf{k}'}, n_{\mathbf{k}} \rangle = (1-n_{\mathbf{k}'})^{\frac{1}{2}} n_{\mathbf{k}}^{\frac{1}{2}} \mathscr{V}(\mathbf{k}-\mathbf{k}'). \tag{2.35}$$

Upon squaring this, we get a transition rate proportional to

$$n_{\mathbf{k}}(1-n_{\mathbf{k}'}) \, |\mathscr{V}(\mathbf{k}-\mathbf{k}')|^2. \tag{2.36}$$

The probability of a transition is proportional to the occupation number of the initial mode, and also, as required by the Pauli principle, to the 'unoccupation' number of the final mode.

This result is reminiscent of (1.103). The difference between bosons and fermions is only a change of sign; instead of 'stimulated' emission, with an enhancement factor $(1+n_{\mathbf{k}'})$, we now have 'inhibited' emission with a reduction factor $(1-n_{\mathbf{k}'})$.

These formulae can be used to derive the Bose–Einstein and Fermi–Dirac distributions, by the following simple argument. Suppose we have a field of particles (bosons or fermions) in equilibrium with a classical reservoir of energy. Consider two modes, $\mathbf{k}$ and $\mathbf{k}'$, containing n and n' excitations, on the average. (This means that n and n' need not be integral.) Suppose these modes differ in energy by the amount

$$\mathscr{E} - \mathscr{E}' = \Delta. \tag{2.37}$$

We can think of the rate of transition between these two modes, by processes in which the energy Δ is taken from, or given to, the reservoir. These rates must be of the following general form:

$$\left.\begin{aligned} Q(\mathbf{k}\to\mathbf{k}') &= n(1\pm n').f_1.P_{\mathbf{k}\mathbf{k}'}, \\ Q(\mathbf{k}'\to\mathbf{k}) &= n'(1\pm n).f_2.P_{\mathbf{k}'\mathbf{k}}. \end{aligned}\right\} \tag{2.38}$$

Here we designate bosons by the upper sign and fermions by the lower sign in the $\pm$ symbol. The probability that the reservoir is in its lower state, ready to take energy Δ from the field, is designated by f_1, whilst f_2 is the probability that it is ready to give up this energy again in the inverse process.

We know the Boltzmann factor for the ratio of these probabilities, i.e.

$$f_1/f_2 = \exp(\Delta/kT). \tag{2.39}$$

We also know, by the principle of microscopic reversibility, that the transition rates $P_{\mathbf{kk}'}$ and $P_{\mathbf{k}'\mathbf{k}}$, corresponding to simple scattering processes without statistical factors for the occupation numbers of initial and final states, are equal. The principle of detailed balance tells us that the two rates in (2.38) must be equal. It follows that

$$\frac{n(1 \pm n')}{n'(1 \pm n)} = \exp(-\Delta/kT). \tag{2.40}$$

Now finally assume that the occupation number of a mode of the particle field is a function only of the energy of that mode. It is easy then to show that the functional relation (2.40) is satisfied, subject to the conservation of energy (2.37), if

$$n(\mathscr{E}) = \frac{1}{\exp(\mathscr{E}/kT) \mp 1}. \tag{2.41}$$

We recognize at once the Bose–Einstein and Fermi–Dirac distribution functions.

2.5 Interactions between particles: momentum conservation

An important situation, especially for a gas of electrons or nucleons, is where there is a mutual interaction between the particles of the field. If these are conserved in number, we need an interaction Hamiltonian such as

$$H_I = \frac{1}{2}\iint \psi^*(\mathbf{r}')\,\psi^*(\mathbf{r})\,\mathscr{V}(\mathbf{r}'-\mathbf{r})\,\psi(\mathbf{r})\,\psi(\mathbf{r}')\,\mathbf{d}^3\mathbf{r}\,\mathbf{d}^3\mathbf{r}'. \tag{2.42}$$

It must by now be obvious that this describes processes in which two fermions see each other by a potential $\mathscr{V}(\mathbf{r}'-\mathbf{r})$ and are scattered from one another.

To show this explicitly, we use a plane-wave representation in (2.22), and transform the interaction Hamiltonian into

$$H_I = \tfrac{1}{2}\sum_{\mathbf{kk}'\mathbf{k}''\mathbf{k}'''} \mathscr{V}(\mathbf{k}''-\mathbf{k}')\,b^*_{\mathbf{k}'''}b^*_{\mathbf{k}''}b_{\mathbf{k}'}b_{\mathbf{k}}\,\delta(\mathbf{k}+\mathbf{k}'-\mathbf{k}''-\mathbf{k}'''). \tag{2.43}$$

The typical term in this summation removes particles from the modes $\mathbf{k}$ and $\mathbf{k}'$ and transfers them into $\mathbf{k}''$ and $\mathbf{k}'''$. The matrix element of the transition is just the Fourier transform of the interaction potential $\mathscr{V}(\mathbf{r})$, as if for the change $\mathbf{k}''-\mathbf{k}'$ in the momentum of the second particle. But the first particle also changes momentum, from $\mathbf{k}$ to $\mathbf{k}'''$, by an equal and opposite amount.

This conservation of momentum is derived in the algebra from an integral of the form

$$\int e^{i(\mathbf{k}+\mathbf{k}'-\mathbf{k}''-\mathbf{k}''')\cdot\mathbf{r}}\, d^3\mathbf{r}' = \delta(\mathbf{k}+\mathbf{k}'-\mathbf{k}''-\mathbf{k}''') \tag{2.44}$$

which ensures that there is no net change of the total wave-vector of the particles after every scattering process. This is the typical way in which such conditions appear in field theory; it is just the condition for constructive interference between the waves corresponding to the different particle modes. It depends, fundamentally, on the fact that the interaction $\mathscr{V}(\mathbf{r}'-\mathbf{r})$ is a function only of the relative positions of the particles in space, so that H_I is invariant to spatial translations.

It is interesting to note that this condition would not have been quite the same if we had been dealing with electron states in a crystalline solid. We should then have represented our field operators in the form (2.22) with Bloch functions as basis states. As is well known, these are not necessarily plane waves, but each must satisfy a condition

$$\psi_{\mathbf{k}}(\mathbf{r}+\boldsymbol{l}) = e^{i\mathbf{k}\cdot\boldsymbol{l}}\psi_{\mathbf{k}}(\mathbf{r}), \tag{2.45}$$

where $\boldsymbol{l}$ is any translational vector of the lattice. The equivalent of (2.43) is somewhat more complicated; and instead of the integral (2.44) we can only factor out a sum of the form

$$\frac{1}{N}\sum_{\boldsymbol{l}} e^{i(\mathbf{k}+\mathbf{k}'-\mathbf{k}''-\mathbf{k}''')\cdot\boldsymbol{l}} \tag{2.46}$$

which does not vanish when $(\mathbf{k}+\mathbf{k}'-\mathbf{k}''-\mathbf{k}''')$ equals any vector of the reciprocal lattice of the crystal. Thus, instead of a strict principle of the conservation of momentum, we have a weaker principle stating that *crystal momentum* is conserved in interactions between electrons, but only up to the arbitrary addition of a reciprocal lattice vector. The possibility of *Umklapp processes* must always be allowed for.

In (2.42) the order of the operators is significant. If we had not kept all the annihilation operators to the right, we should have found ourselves with an expression in (2.43) which would have made a non-vanishing contribution to the expectation value of H_I, in any state of the assembly, from terms for which $\mathbf{k}$, $\mathbf{k}'$, $\mathbf{k}''$ and $\mathbf{k}'''$ happened to be equal. This would be a self energy of the bare particles, which we usually suppose to be zero. The more complicated consequences of interactions of this sort will be discussed in chapter 5.

It is worth remarking that the basic assumption here of an interaction $\mathscr{V}(\mathbf{r}-\mathbf{r}')$ between the fermions does not preclude the possibility that this force is itself carried by some other field. As we have shown in § 1.11, the exchange of, say, mesons between nucleons, can give rise to an energy contribution which looks very like a potential energy for a force acting at a distance. The only point to worry about here is whether the energy of excitation of the main field particles (in this example, say, the nucleons) might not be large enough to create a real excitation of the interaction field.

2.6 Fermion–boson interaction

Typical phenomena of solid state and particle physics arise from the coupling of a boson field with a fermion field by an interaction Hamiltonian of the form

$$H_I = \sum_{\mathbf{k},\,\mathbf{q}} F(q)\,(a_{\mathbf{q}} - a^*_{-\mathbf{q}})\,b^*_{\mathbf{k}+\mathbf{q}}\,b_{\mathbf{k}}. \qquad (2.47)$$

In the occupation-number representation, this must correspond to a process in which the fermion—an electron or nucleon, say—is knocked from state $\mathbf{k}$ to state $\mathbf{k}+\mathbf{q}$ by the emission or absorption of a boson—a photon, a phonon, or a meson.

The form factors $F(q)$ will depend on the physical system. In the case of a point interaction, such as was studied in § 1.11, the Hamiltonian density from which (2.47) would arise must be of the form

$$\mathscr{H}_I = g\psi^*(\mathbf{r})\,\phi(\mathbf{r})\,\psi(\mathbf{r}), \qquad (2.48)$$

where ϕ is the boson field operator, ψ is the fermion field, and g is a coupling constant. We then get (cf. (1.106))

$$F(q) = -ig/\sqrt{(2\omega_{\mathbf{q}})}, \qquad (2.49)$$

which would depend upon the form of the boson spectrum. This is the type of interactions involving nucleons and mesons—although really there would be further complications arising from the fact that the fields might have several components, and the coupling would be governed by symmetry principles. Moreover, we should properly use the relativistic theory of chapter 6 for particles of such energy.

A standard case in solid state physics is the electron–phonon interaction in a semiconductor. It is plausibly arguable that the energy of an electron in the conduction band will depend on the local state of strain of the crystal. If the lattice is dilated in the neighbourhood of the point $\mathbf{r}$ by an amount $\Delta(\mathbf{r})$, then one would expect an energy change of amount $C\Delta(\mathbf{r})$, where C is called the *deformation potential* parameter.

The Hamiltonian operator for the electron field should contain an extra term:

$$H_I = \int \psi^*(\mathbf{r})\, C\Delta(\mathbf{r})\, \psi(\mathbf{r})\, \mathbf{d}^3\mathbf{r}. \tag{2.50}$$

But this dilatation may be attributable to a lattice wave, represented perhaps by a displacement vector field $\mathbf{u}(\mathbf{r})$ as in § 1.5. In terms of phonon operators, we then have

$$\begin{aligned}\Delta(\mathbf{r}) &= \nabla\cdot\mathbf{u}(\mathbf{r}) \\ &= \sum_{\mathbf{q}} (\tfrac{1}{2}\rho_0\omega_{\mathbf{q}})^{-\frac{1}{2}}\, q\, \mathrm{e}^{-\mathrm{i}\mathbf{q}\cdot\mathbf{r}}\, (a_{\mathbf{q}} - a^*_{-\mathbf{q}})\end{aligned} \tag{2.51}$$

for longitudinally polarized acoustic modes in the continuum limit. In this case

$$F(q) = \frac{Cq}{(\tfrac{1}{2}\rho_0\omega_{\mathbf{q}})^{\frac{1}{2}}} = \frac{C}{(\tfrac{1}{2}\rho_0 s)^{\frac{1}{2}}} q^{\frac{1}{2}}, \tag{2.52}$$

where ρ_0 is the density of the medium and s is the velocity of sound. In a metal the electron–phonon interaction is somewhat similar to this, except that C also varies slowly with q.

As it happens, the form factor for the interaction of electrons with *photons* (in free space), has the same dependence on q as in (2.52). A full derivation of this formula requires the elaborate theory of the quantization of the electromagnetic field, which we postpone to chapter 6. But let us assume that the analogue of the local 'displacement' field $\mathbf{u}(\mathbf{r})$ is the vector potential $\mathbf{A}(\mathbf{r})$, and that this interacts with the electric current $\mathbf{j}(\mathbf{r})$ associated with the electrons through the usual classical term (cf. § 6.9)

$$\begin{aligned}\mathscr{H}_I &= \mathbf{j}\cdot\mathbf{A} \\ &= -e\,\mathrm{i}(\psi^*\nabla\psi - \psi\nabla\psi^*)\cdot\mathbf{A}.\end{aligned} \tag{2.53}$$

In second quantization, the field $\mathbf{A}(\mathbf{r})$ is represented in terms of annihilation and creation operators, exactly as $\mathbf{u}(\mathbf{r})$ is quantized in § 1.5, except that there is no 'density' factor ρ_0. We arrive at an expression just like (2.47), but with

$$F(q) = \frac{e}{\sqrt{(\hbar c)}} q^{\frac{1}{2}}, \tag{2.54}$$

where c is the velocity of light. The coupling constant here is recognizable as the square root of the fine structure constant $e^2/\hbar c$, which is well known to be nearly 1/137. But the above argument is non-relativistic, and ignores the serious problem of eliminating the contribution of the electrostatic potential.

Yet another type of form factor occurs in the case of an electron interacting with the optical modes in a polar crystal. The electric polarization field $\mathbf{P}(\mathbf{r})$ is proportional to the local 'displacement' $\mathbf{u}(\mathbf{r})$, which can be quantized in the usual way. But the electrostatic energy of the electron is given by $e\phi(\mathbf{r})$, where

$$4\pi\mathbf{P}(\mathbf{r}) = \nabla\phi(\mathbf{r}). \tag{2.55}$$

The Fourier transform of this relation introduces a factor $1/q$ for the effect of the mode of wave-vector q. But the frequency of a polar mode, ω_l, is almost independent of q, so that we get

$$F(q) = -\frac{eF}{q}. \tag{2.56}$$

It is instructive to evaluate the proportionality constant F by working out the effective force between two fixed electrons, due to the exchange of optical phonons, along the lines of § 1.11. One should get a coulomb potential $\mathbf{e}^2/\epsilon R$, with the apparent dielectric constant

$$\epsilon = 2\pi\omega_l/F^2, \tag{2.57}$$

which can be determined experimentally.

Now let us consider some phenomena caused by this interaction. For example, let us calculate the density of the *boson cloud* accompanying a fermion. Suppose we start with the state $|\mathbf{k}, 0_\mathbf{q}\rangle$, meaning a fermion of momentum $\mathbf{k}$ in the vacuum of the boson field. Treating H_I as a constant perturbation, we find that this state is modified into

$$\begin{aligned}|\mathbf{k}\rangle' &= |\mathbf{k}, 0_\mathbf{q}\rangle + \sum_\mathbf{q} |\mathbf{k}-\mathbf{q}, 1_\mathbf{q}\rangle \frac{\langle \mathbf{k}-\mathbf{q}, 1_\mathbf{q}| H_I |\mathbf{k}, 0_\mathbf{q}\rangle}{\mathscr{E}_0(\mathbf{k}) - \mathscr{E}_0(\mathbf{k}-\mathbf{q}) - \omega_\mathbf{q}} \\ &= |\mathbf{k}, 0_\mathbf{q}\rangle - \sum_\mathbf{q} \frac{F(\mathbf{q})}{\mathscr{E}_0(\mathbf{k}) - \mathscr{E}_0(\mathbf{k}-\mathbf{q}) - \omega_\mathbf{q}} |\mathbf{k}-\mathbf{q}, 1_\mathbf{q}\rangle \end{aligned} \tag{2.58}$$

because the perturbation has the power to mix in states in which a phonon has been created in the mode $\mathbf{q}$.

The mean number of bosons in the $\mathbf{q}$th mode is the square of this admixture coefficient. Summed over all modes, this gives

$$\langle N\rangle = \frac{m^{*2}}{8\pi^3}\int \frac{|F(q)|^2\, \mathbf{d}^3\mathbf{q}}{\{\frac{1}{2}(2\mathbf{k}\cdot\mathbf{q} - \mathbf{q}^2) - m^*\omega_\mathbf{q}\}^2}, \tag{2.59}$$

assuming that in the unperturbed state, of energy $\mathscr{E}_0(\mathbf{k})$, the fermions are free particles of mass m^*.

This integral may be evaluated for various forms of $F(q)$ and of ω_q

according to the physical situation. In each of the cases considered above, there is no difficulty near $q = 0$. But consider the electron–phonon interaction with form factor (2.52). For the moment suppose that the electron is of low energy, so that the term in $\mathbf{k}\cdot\mathbf{q}$ in the denominator may be neglected. Our integral then takes the following form:

$$\begin{aligned}\langle N\rangle &= \frac{1}{8\pi^3}\frac{m^{*2}C^2}{\frac{1}{2}\rho_0 s}4\pi\int\frac{4q\,.\,q^2\,\mathrm{d}q}{(q^2+2m^*qs)^2}\\ &= \frac{4m^{*2}C^2}{\pi^2\rho_0 s}\int_0^{q_D}\frac{q\,\mathrm{d}q}{(q+2m^*s)^2}\\ &\sim \frac{4}{\pi^2}\frac{m^{*2}C^2}{\rho_0 s}\ln\left(\frac{q_D}{2m^*s}\right).\end{aligned} \tag{2.60}$$

This expression is finite because the Debye wave-number q_D provides automatically an upper cut-off limit. But suppose we make a similar calculation for the electromagnetic interaction (2.54). The integral will be exactly the same; but because there is no natural upper limit to the frequency of a photon, the formula (2.60) will diverge logarithmically at large q. We encounter here one of the standard divergences of quantum electrodynamics.

Now consider what happens if $\mathbf{k}$ is large enough for the denominator of (2.59) to vanish for some values of $\mathbf{q}$. The condition for this is best derived generally from the energy denominator in (2.58); there will be a singularity in the integral if

$$\mathscr{E}_0(\mathbf{k})-\mathscr{E}_0(\mathbf{k}-\mathbf{q}) > q\omega_{\mathbf{q}}, \tag{2.61}$$

i.e. for small values of q, if

$$\mathbf{q}\cdot\frac{\partial\mathscr{E}_0(\mathbf{k})}{\partial\mathbf{q}} > qs. \tag{2.62}$$

This says that the group velocity of the fermion in state $\mathbf{k}$ must exceed the boson velocity s.

What happens then? The integral (2.59) may still converge as a principal value, but at the singularity itself we have the possibility of real scattering processes, so that the state (2.58) would not be stationary in time. In the case of electrons in a semiconductor, these real processes scatter the carriers and produce an electrical resistance. This phenomenon is only allowed in the electromagnetic case if the fermions can travel faster than light, which requires a medium of refractive index greater than unity. This is, of course, the condition for *Cerenkov radiation*.

Another quantity that might be calculated is the self-energy of the fermion in the boson field. Using (2.58) as our perturbed state, we get

$$\mathscr{E}(\mathbf{k}) = \mathscr{E}_0(\mathbf{k}) + \sum_{\mathbf{q}} \frac{|\langle \mathbf{k}-\mathbf{q}, 1_{\mathbf{q}}| H_I |\mathbf{k}, 0_{\mathbf{q}}\rangle|^2}{\mathscr{E}_0(\mathbf{k}) - \mathscr{E}_0(\mathbf{k}-\mathbf{q}) - \omega_{\mathbf{q}}}$$

$$= \mathscr{E}_0(\mathbf{k}) - \frac{m^*}{8\pi^3}\int \frac{|F(q)|^2\, \mathrm{d}^3\mathbf{q}}{-\frac{1}{2}(2\mathbf{k}\cdot\mathbf{q} - \mathbf{q}^2) + m^*\omega_{\mathbf{q}}}. \qquad (2.63)$$

For the nucleon–pion system this is very like the calculation of § 1.11, where we found that the self-energy correction would be infinite. The integral in (2.63) does not diverge quite so badly as (1.110), but this is not really significant because we cannot use a non-relativistic theory to discuss what might happen at very high pion energy.

The self energy is also infinite for the electron–photon interaction (2.54), and depends strongly on q_D for the electron–phonon case, as in (2.60). But suppose we use in (2.63) the interaction (2.56) of an electron with a polar mode of a crystal, of constant frequency ω_l. The integral then converges, for an infinite upper limit. We get, by a straightforward expansion in powers of k,

$$\mathscr{E}(k) - \mathscr{E}_0(k) = -\frac{me^2F^2}{2\pi^2}\int_0^\pi \sin\theta\, \mathrm{d}\theta \int_0^\infty \frac{\mathrm{d}q}{q^2 - 2kq\cos\theta + 2m\omega_l}$$

$$\approx -\alpha\left(\omega_l + \frac{1}{12m}k^2 + \dots\right). \qquad (2.64)$$

The coupling parameter α may be expressed in terms of the effective dielectric constant (2.57) in the form

$$\alpha = (e^2/\epsilon)\sqrt{(\tfrac{1}{2}m\omega_l)}. \qquad (2.65)$$

The correction to the ground state energy of an electron at rest is finite: it represents the energy gained by allowing the lattice to be polarized in the neighbourhood of the particle. But there is also a correction to the relationship between the energy and momentum of the electron; we may say that the particle behaves as if it had the mass

$$m^* \approx \frac{m}{1 - \frac{1}{6}\alpha}. \qquad (2.66)$$

The fact that the electron seems to get heavier because of the optical modes that it must carry with it as it moves through the crystal is physically understandable.

But notice that the theory of the *polaron* would break down if α (which is dimensionless in the proper units) were greater than 6. The above theory is only plausible in the limit of weak coupling, and quite different calculations would need to be made in other cases.

The lesson of this section is that although one can write down elementary perturbation expressions based upon the interaction between the fields, these often lead to infinite, uninterpretable results. This is why one needs much more sophisticated procedures for solving the equations for interacting fields, of the kinds to be discussed in subsequent chapters.

2.7 Holes and antiparticles

In many physical systems we find dense assemblies of fermions—for example, electrons in a metal or nucleons in nuclear matter. If the particles do not interact strongly, the exclusion principle works very effectively, and the ground state of the system may be described as a condensed Fermi gas. This is to say, the wave-vectors of all occupied states $|\mathbf{k}\rangle$ lie within a definite region in $\mathbf{k}$-space of energy $\mathscr{E}_F$. For N free particles each with two spin states, in a volume V, the Fermi surface is a sphere of radius

$$k_F = (3\pi^2 N/V)^{\frac{1}{3}}. \tag{2.67}$$

It is well known that the Fermi surface for electrons in a metal may be quite a complicated geometrical object; the electron states may still be classified as being 'above' or 'below' the Fermi surface according as $\mathscr{E}(\mathbf{k})$ is greater or less than $\mathscr{E}_F$.

The total energy spectrum of such a system is enormously complicated. But we are often interested in levels of low energy, near the ground state. It is convenient then to remove this large constant ground state energy, and to treat the condensed Fermi gas as a 'vacuum' of which these low levels are excitations. Although the fermions may be very numerous, the energy of most of them cannot change readily because of the exclusion principle—we assume, of course, that their number is conserved in all relevant interactions. It turns out to be much easier to describe the excited states as a rarefied gas of quasi-particles, although, as we shall see, the number of these is no longer conserved in typical physical phenomena.

Suppose that an assembly of independent electrons has Hamiltonian

$$H_0 = \sum_{\mathbf{k}} \mathscr{E}(\mathbf{k})\, b_{\mathbf{k}}^* b_{\mathbf{k}}, \tag{2.68}$$

and that all interactions conserve the particle number

$$N = \sum_{\mathbf{k}} b_{\mathbf{k}}^* b_{\mathbf{k}} = \sum_{k<k_F} 1. \tag{2.69}$$

The ground state will correspond to all levels being filled up to $\mathscr{E}(\mathbf{k}) = \mathscr{E}_F$, which we symbolize as the volume $k < k_F$ enclosed by the Fermi surface in **k**-space. We assume that the whole description is invariant under a transformation in which **k** goes to $-\mathbf{k}$, i.e.

$$\mathscr{E}(-\mathbf{k}) = \mathscr{E}(\mathbf{k}).$$

Let us define a new Hamiltonian operator whose expectation value is zero in this ground state. Using (2.69), we may write this

$$\begin{aligned}\tilde{H}_0 &= H_0 - \sum_{k<k_F} \mathscr{E}(\mathbf{k}) \\ &= \sum_{\mathbf{k}} \{\mathscr{E}(\mathbf{k}) - \mathscr{E}_F\} b_{\mathbf{k}}^* b_{\mathbf{k}} - \sum_{k<k_F} \{\mathscr{E}(\mathbf{k}) - \mathscr{E}_F\},\end{aligned} \tag{2.70}$$

which is just as good as H_0 for all attainable states of our system.

Now we introduce new annihilation and creation operators as follows:

$$\tilde{b}_{\mathbf{k}} = \begin{cases} b_{\mathbf{k}} \\ b_{-\mathbf{k}}^* \end{cases}; \quad \tilde{b}_{\mathbf{k}}^* = \begin{cases} b_{\mathbf{k}}^* \\ b_{-\mathbf{k}} \end{cases} \quad \text{for} \quad \begin{cases} k > k_F, \\ k < k_F. \end{cases} \tag{2.71}$$

In other words, for modes within the Fermi surface the roles of annihilation and creation operators are interchanged, and the wave-vector **k** is reversed. But because these are fermion operators, obeying anti-commutation relations, we still have

$$\{\tilde{b}_{\mathbf{k}}, \tilde{b}_{\mathbf{k}'}^*\} = \delta_{\mathbf{k}\mathbf{k}'}, \tag{2.72}$$

just as for our original operators $b_{\mathbf{k}}$, $b_{\mathbf{k}}^*$.

If now we substitute from (2.71) into (2.70), we get

$$\begin{aligned}\tilde{H}_0 &= \sum_{k<k_F} \{\mathscr{E}(\mathbf{k}) - \mathscr{E}_F\} (1 - \tilde{b}_{-\mathbf{k}}^* \tilde{b}_{-\mathbf{k}}) + \sum_{k>k_F} \{\mathscr{E}(\mathbf{k}) - \mathscr{E}_F\} \tilde{b}_{\mathbf{k}}^* \tilde{b}_{\mathbf{k}} - \sum_{k<k_F} \{\mathscr{E}(\mathbf{k}) - \mathscr{E}_F\} \\ &= \sum_{\mathbf{k}} \tilde{\mathscr{E}}(\mathbf{k}) \tilde{b}_{\mathbf{k}}^* b_{\mathbf{k}},\end{aligned} \tag{2.73}$$

where
$$\tilde{\mathscr{E}}(\mathbf{k}) = |\mathscr{E}(\mathbf{k}) - \mathscr{E}_F|, \tag{2.74}$$

whether **k** lies inside or outside the Fermi surface.

This Hamiltonian obviously describes a system whose 'vacuum' $|\tilde{0}\rangle$ is the ground state of H_0. The properties of (2.73) are exactly those of an assembly of fermions, all of positive energy, which may be created out of the vacuum by the operators $\tilde{b}_{\mathbf{k}}^*$. But near the ground state this gas of *quasi-particles* is rarefied, so that their collisions may be very infrequent, in strong contrast to the dense system from which we started.

But notice that we have to give up the conservation condition (2.69) which now says that

$$\begin{aligned} N-\sum_{\mathbf{k}} b^*_{\mathbf{k}} b_{\mathbf{k}} &= N - \sum_{k<k_F}(1-\tilde{b}^*_{-\mathbf{k}}\tilde{b}_{-\mathbf{k}}) - \sum_{k>k_F}\tilde{b}^*_{\mathbf{k}}\tilde{b}_{\mathbf{k}} \\ &= \sum_{k<k_F}\tilde{b}^*_{\mathbf{k}}\tilde{b}_{\mathbf{k}} - \sum_{k>k_F}\tilde{b}^*_{\mathbf{k}}\tilde{b}_{\mathbf{k}} \\ &= \tilde{N}_{k<k_F} - \tilde{N}_{k>k_F} \end{aligned} \tag{2.75}$$

must vanish. Thus we can only say that the number of quasi-particles excited above the Fermi level must always equal the number below $\mathscr{E}_F$.

What we are really saying is that for modes below the Fermi level the operator $\tilde{b}^*_{\mathbf{k}}$ destroys an electron in $-\mathbf{k}$, or, as we might put it, *creates a hole* in that mode. This requires the energy $-\{\mathscr{E}(\mathbf{k})-\mathscr{E}_F\}$ relative to $\mathscr{E}_F$, and this is positive. Moreover, the destruction of a real particle of momentum $-\mathbf{k}$ *increases* the momentum of the system by $-(-\mathbf{k})$. It is thus consistent to ascribe the momentum $\mathbf{k}$ to the '*hole-like*' *quasi-particle* created by this operator.

The processes that would be described by interaction terms such as (2.34) and (2.43) are now somewhat more complicated. Thus, for example, an operator which we should interpret as merely scattering a real particle from one state to another would split up into four terms:

$$\begin{aligned} H_I &= \sum_{\mathbf{k},\mathbf{k}'} \mathscr{V}(\mathbf{k}-\mathbf{k}') b^*_{\mathbf{k}} b_{\mathbf{k}'} \\ &= \sum_{k,k'<k_F} \mathscr{V}(\mathbf{k}'-\mathbf{k})\tilde{b}_{\mathbf{k}}\tilde{b}^*_{\mathbf{k}'} + \sum_{k,k'>k_F} \mathscr{V}(\mathbf{k}-\mathbf{k}')\tilde{b}^*_{\mathbf{k}}\tilde{b}_{\mathbf{k}'} \\ &\quad + \sum_{k'<k_F<k} \mathscr{V}(\mathbf{k}+\mathbf{k}')\tilde{b}^*_{\mathbf{k}}\tilde{b}^*_{\mathbf{k}'} + \sum_{k<k_F<k'} \mathscr{V}(-\mathbf{k}-\mathbf{k}')\tilde{b}_{\mathbf{k}}\tilde{b}_{\mathbf{k}'}. \end{aligned} \tag{2.76}$$

The first two would describe processes in which a quasi-particle is scattered into another excitation of the same sort, but the next describes a process in which an electron-hole pair is created out of the vacuum, and the final term will allow two excitations to be mutually annihilated if they are of opposite species.

The possible processes for an interaction *between* the particles, as in (2.43), are much more complicated. We might characterize them by a series of diagrams in which ingoing and outgoing quasi-particles (i.e. quasi-particles destroyed or created in the process) were indicated by arrows. Then the 'genuine particle' process described by fig. 1 (*a*) could become any one of the quasi-particle processes described by 1 (*b*)...(*e*).

For example, 1 (*d*) represents an electron being scattered as it knocks an electron-hole pair out of the vacuum. Of course for a real process we should need to have enough energy available.

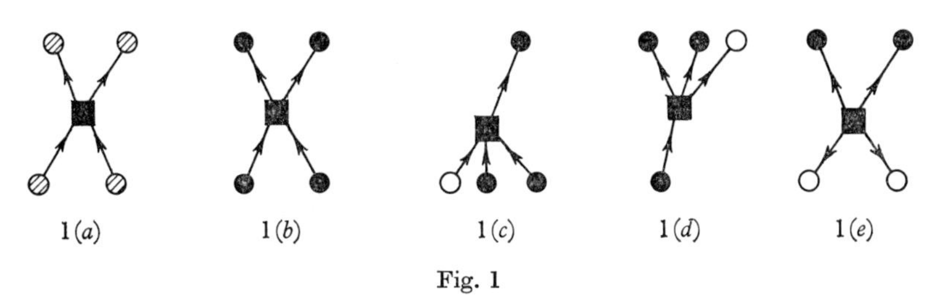

Fig. 1

The energy of a quasi-particle, as given by (2.74), may be expressed in the form

$$\begin{aligned}\tilde{\mathscr{E}}(\mathbf{k}) &= |\mathscr{E}(\mathbf{k}) - \mathscr{E}_F| \\ &\approx \left|(\mathbf{k} - \mathbf{k}_F)\cdot\frac{\partial\mathscr{E}(\mathbf{k})}{\partial\mathbf{k}}\right| \\ &\approx |k - k_F|\, v_F \qquad (2.77)\end{aligned}$$

for excitations near the Fermi surface. The velocity of an 'electron' quasi-particle is approximately the Fermi velocity in that neighbourhood. One can also see (though it takes careful thought to get it right) that a hole-like excitation created by $\tilde{b}^*_{\mathbf{k}}$ has the velocity

$$\mathbf{v}_{\mathbf{k}} = \frac{\partial\mathscr{E}(\mathbf{k})}{\partial\mathbf{k}} \qquad (2.78)$$

of the mode $\mathbf{k}$. This is because of the association of this operator with the destruction of a genuine particle in the mode $-\mathbf{k}$. It is worth noticing that our convention here is not the same as for hole states in the valence band of a semiconductor, where we do not reverse the wave-vector but where the gradient of $\mathscr{E}(\mathbf{k})$ is negative near the top of the band so that the velocity of a hole eventually comes out positive again. In both cases, however, if we had associated, say, a negative electrical charge with an electron state, then the 'hole-like' quasi-particles will carry a positive charge.

It need hardly be said that the above argument provides a model for a theory in which every particle has its *antiparticle*, of similar dynamical properties but opposite charge. The prototype of such theories is the Dirac hypothesis concerning the positron. It will be recalled that Dirac's relativistic wave equation for an electron (see §6.6) has two energy eigenvalues, $\pm\mathscr{E}(\mathbf{k})$ for each value of $\mathbf{k}$. If we assume that all the modes of negative energy are filled, and treat this as the vacuum $|\tilde{0}\rangle$, we can set up a theory in which all the observed excitations are quasi-particles of positive energy and behave like respectable dynamical objects. The concept of a 'Fermi surface' now shrinks, of course, to the point $\mathbf{k} = 0$, and we have to distinguish between the modes according as they are of positive or negative energy, but otherwise the formulation we have just given is applicable. Thus, for example, any interaction that can scatter real electrons from one another, as in fig. 1(*b*), must be capable of producing electron–positron pairs out of the vacuum, as in fig. 1(*d*). Indeed, by various algebraic devices sketched out in §§3.7, 6.8 and 6.10 we find that we can still use a single diagram such as 1(*a*) to describe all these different processes, provided that we allow an electron arrow 'going backward in time' to stand for a 'hole' travelling in the ordinary causal sense.

It will be noticed that a theory of charged fermions, with their antiparticles, is not quite the same in this analysis, as the theory of charged bosons put forward in §1.12.

CHAPTER 3

PERTURBATION THEORY

It were a dark and stormy night; and t'rain came down in loomps. *Cap'n said 'Tell us tale', and tale it ran as follows: 'It were a dark and stormy night; and t'rain...'.*

3.1 The Brillouin–Wigner series

It is very difficult to avoid a series expansion in the solution of problems in applied mathematics; in quantum mechanics the perturbation method is ubiquitous. The elementary theory of the first few terms of the time-independent and time-dependent series is well known. In this chapter we deal more systematically with the techniques for calculating the effect of a small extra potential or Hamiltonian H' acting on a system governed mainly by a more powerful Hamiltonian H_0.

Before embarking upon the general theory, it is instructive to derive a useful elementary formula, which is not widely known. Suppose for simplicity that our unperturbed Hamiltonian H_0 is an ordinary one-electron operator with an eigenfunction $|\Psi_0\rangle$ defined by

$$H_0|\Psi_0\rangle = \mathscr{E}_0|\Psi_0\rangle \tag{3.1}$$

and normalized to unity. When the perturbation H' is applied, we shall need to solve the equation

$$(\mathscr{E}-H_0)|\Psi\rangle = H'|\Psi\rangle. \tag{3.2}$$

Provided we make no condition on the normalization of the new eigenstate $|\Psi\rangle$, we may always split it into two terms

$$|\Psi\rangle = |\Psi_0\rangle + |\Phi\rangle, \tag{3.3}$$

where the 'change due to the perturbation' is to be orthogonal to the unperturbed state, i.e.

$$\langle\Psi_0|\Phi\rangle = 0. \tag{3.4}$$

This condition can be expressed elegantly by the formal introduction of *projection operators*. Think of the symbol

$$M \equiv |\Psi_0\rangle\langle\Psi_0|. \tag{3.5}$$

Suppose we put this in front of any ket, $|\Theta\rangle$ say. The symbolism forces us to write

$$\begin{aligned} M|\Theta\rangle &= |\Psi_0\rangle\langle\Psi_0|\Theta\rangle \\ &= (\langle\Psi_0|\Theta\rangle)\,|\Psi_0\rangle \end{aligned} \tag{3.6}$$

as if the effect of 'operating with M' were to calculate the inner product of $|\Psi_0\rangle$ with $|\Theta\rangle$ and then to multiply the ket $|\Psi_0\rangle$ by this number. If we think of $|\Psi_0\rangle$ as a unit vector in Hilbert space, then (3.6) is just the 'projection' of the vector $|\Theta\rangle$ in that direction.

Projection operators are not quite like physical observables; they are merely convenient algebraic scaffolding in the mathematics. But they have one property that makes them easily recognizable. Calculate

$$\begin{aligned} M^2|\Theta\rangle &= |\Psi_0\rangle\langle\Psi_0|\Psi_0\rangle\langle\Psi_0|\Theta\rangle \\ &= |\Psi_0\rangle\langle\Psi_0|\Theta\rangle \\ &= M|\Theta\rangle. \end{aligned} \tag{3.7}$$

This is true, whatever the vector $|\Theta\rangle$; hence the operator M is *idempotent*, i.e.

$$M^2 = M. \tag{3.8}$$

Now think of the 'complementary part' of $|\Theta\rangle$—what is left after we have taken out its component along $|\Psi_0\rangle$. This is obviously where we define the complementary projection operator

$$P = 1 - M. \tag{3.9}$$

It is easy to prove that this is also idempotent and that it projects any vector upon the manifold in Hilbert space that is orthogonal to $|\Psi_0\rangle$.

Returning to our perturbation problem, we see at once, from (3.3) and (3.4), that we can write

$$P|\Psi\rangle = |\Phi\rangle. \tag{3.10}$$

Applying the unperturbed Hamiltonian operator to this function we get

$$\begin{aligned} PH_0|\Phi\rangle &= H_0|\Phi\rangle - |\Psi_0\rangle\langle\Psi_0|H_0\Phi\rangle \\ &= H_0|\Phi\rangle - |\Psi_0\rangle\mathscr{E}_0\langle\Psi_0|\Phi\rangle \\ &= H_0|\Phi\rangle \\ &= H_0P|\Phi\rangle, \quad \text{etc.} \end{aligned} \tag{3.11}$$

showing that P commutes with H_0.

Now use this in (3.2), written in the form

$$(\mathscr{E} - H_0)|\Phi\rangle = H'|\Psi\rangle - (\mathscr{E} - \mathscr{E}_0)|\Psi_0\rangle; \tag{3.12}$$

projecting on both sides with P, we get

$$(\mathscr{E}-H_0)\,|\Phi\rangle = PH'\,|\Psi\rangle. \tag{3.13}$$

Again operating with M on (3.2), we get

$$\mathscr{E} = \mathscr{E}_0+\langle\Psi_0|\,H'\,|\Psi\rangle. \tag{3.14}$$

These equations are exact, but can only be solved approximately. Let us write (3.13) as follows:

$$|\Psi\rangle = |\Psi_0\rangle+(\mathscr{E}-H_0)^{-1}\,PH'\,|\Psi\rangle. \tag{3.15}$$

This implies that there is an operator $(\mathscr{E}-H_0)^{-1}$ that is the inverse of $(\mathscr{E}-H_0)$. To give meaning to such a symbol, suppose we were to express all our states in terms of the eigenstates of H_0, i.e.

$$|\Theta\rangle = \sum_n a_n\,|\Psi_n\rangle. \tag{3.16}$$

If now we define

$$(\mathscr{E}-H_0)^{-1}\,|\Theta\rangle \equiv \sum_n a_n \frac{1}{\mathscr{E}-\mathscr{E}_n}\,|\Psi_n\rangle, \tag{3.17}$$

we shall have constructed a state that automatically satisfies the inversion identity, i.e.

$$\begin{aligned}(\mathscr{E}-H_0)\,(\mathscr{E}-H_0)^{-1}\,|\Theta\rangle &= \sum_n a_n \frac{1}{\mathscr{E}-\mathscr{E}_n}\,(\mathscr{E}-H_0)\,|\Psi_n\rangle \\ &= \sum_n a_n \frac{1}{\mathscr{E}-\mathscr{E}_n}\,(\mathscr{E}-\mathscr{E}_n)\,|\Psi_n\rangle \\ &= |\Theta\rangle. \end{aligned} \tag{3.18}$$

Provided that $\mathscr{E}$ does not coincide with any eigenvalue $\mathscr{E}_n$ of H_0, this definition is unambiguous.

Now we solve (3.15) by iteration. Thus

$$\begin{aligned}|\Psi\rangle &= |\Psi_0\rangle+\frac{1}{\mathscr{E}-H_0}PH'\left(|\Psi_0\rangle+\frac{1}{\mathscr{E}-H_0}PH'\,|\Psi\rangle\right) \\ &= |\Psi_0\rangle+\frac{1}{\mathscr{E}-H_0}PH'\,|\Psi_0\rangle+\frac{1}{\mathscr{E}-H_0}PH'\frac{1}{\mathscr{E}-H_0}PH'\,|\Psi_0\rangle+\dots. \end{aligned} \tag{3.19}$$

This series is known as the *Brillouin–Wigner perturbation expansion.*

To make clear the significance of this formula, put (3.19) into the expression for the energy (3.14), and represent all states and operators in eigenstates of H_0, just as in (3.17). We get

$$\mathscr{E} = \mathscr{E}_0+\langle\Psi_0|\,H'\,|\Psi_0\rangle+\sum_{n\neq 0}\frac{|\langle\Psi_0|\,H'\,|\Psi_n\rangle|^2}{\mathscr{E}-\mathscr{E}_n}+\dots. \tag{3.20}$$

The projection operator automatically excludes certain matrix elements from the summations in each term of the series.

This formula is very similar to the *Rayleigh–Schrödinger* series of conventional perturbation theory, except that the *perturbed* energy $\mathscr{E}$ appears in each denominator instead of the unperturbed energy $\mathscr{E}_0$. It is thus an implicit equation to be solved for $\mathscr{E}$, which makes it more complicated to use.

But it has its advantages. Consider the very simple case where H' couples just two states of nearly equal energy. If the expectation value of H' is zero in each of these states, we get

$$\mathscr{E} = \mathscr{E}_0 + \frac{|H'_{01}|^2}{\mathscr{E} - \mathscr{E}_1}, \tag{3.21}$$

which has two roots. But this is the same as the solution of the determinantal equation for the degenerate or nearly degenerate Rayleigh–Schrödinger perturbation theory, i.e.

$$\begin{vmatrix} \mathscr{E} - \mathscr{E}_0 & H'_{01} \\ H'_{10} & \mathscr{E} - \mathscr{E}_1 \end{vmatrix} = 0. \tag{3.22}$$

This is not to say that the Brillouin–Wigner series can always be relied on in such cases, but it does show that the extra roots may sometimes have physical significance. It is, so to speak, a prototype of a self-consistent solution (cf. § 3.9) likely to be more accurate than the term of corresponding order in the Rayleigh–Schrödinger series. When tackling a complicated problem, one may often be guided by a preliminary calculation using this formula. There is also something to be learnt from the elementary applications of projection operators and inverse operators in the above derivation.

3.2 The Heisenberg representation

We now need a more sophisticated theory of time-dependent problems. In the elementary formalisms the time occurs as follows. Each dynamical variable is represented by an operator that does not depend on the time. But the state vector $|\psi(t)\rangle$ varies with time, being obedient to the time-dependent Schrödinger equation.

$$H\,|\psi(t)\rangle = i\hbar \frac{\partial}{\partial t}|\psi(t)\rangle. \tag{3.23}$$

This means that observable properties of the system will vary with time, because $|\psi(t)\rangle$ appears in all formulae for expectation values. This is the familiar *Schrödinger representation.*

Let us just calculate these quantities for some dynamical variable represented by the operator A. The expectation value of A at time t is given by

$$\mathscr{A}(t) = \langle\psi(t)|\, A\, |\psi(t)\rangle. \tag{3.24}$$

The time derivative of this expression can be worked out using the equation of motion (3.23). Bearing in mind the rules for Hermitian conjugation in the Dirac notation, we get

$$\begin{aligned} i\hbar\frac{\partial\mathscr{A}(t)}{\partial t} &= \left\{i\hbar\frac{\partial}{\partial t}\langle\psi(t)|\right\} A\,|\psi(t)\rangle + \langle\psi(t)|\, A\, i\hbar\frac{\partial}{\partial t}|\psi(t)\rangle \\ &= \langle\psi(t)|\, -HA\,|\psi(t)\rangle + \langle\psi(t)|\, AH\, |\psi(t)\rangle \\ &= \langle\psi(t)|\, [A, H]\, |\psi(t)\rangle: \end{aligned} \tag{3.25}$$

the rate of change of the expectation value of the operator is the expectation value of the commutator of A with the Hamiltonian.

This is rather nice. It proves, for example, that any operator that commutes with the Hamiltonian is a 'constant of the motion'—a principle whose general significance is discussed in §7.8. It also suggests that there may be a sense in which one could define a time-varying operator satisfying an equation

$$i\hbar\frac{\partial A(t)}{\partial t} = [A(t), H] \tag{3.26}$$

independently, so to speak, of the state function used to calculate an expectation value on each side.

To define just such an operator we proceed as follows. Suppose we think of the time variation of the Schrödinger state function $|\psi_S(t)\rangle$ as being represented by the continuous action of an operator, i.e.

$$|\psi_S(t)\rangle = U(t)\,|\psi(0)\rangle, \tag{3.27}$$

where $|\psi(0)\rangle$ is a standard fixed state, as at time zero. The operator must satisfy two types of condition.

In the first place, to preserve the normalization properties of our state function as time proceeds, it must be *unitary*, i.e. if

$$1 = \langle\psi_S(t)\,|\,\psi_S(t)\rangle = \langle\psi(0)|\, U^*(t)\, U(t)\,|\psi(0)\rangle, \tag{3.28}$$

then
$$U^*(t)\, U(t) = U(t)\, U^*(t) = 1,$$

or
$$U^*(t) = U^{-1}(t). \tag{3.29}$$

Notice that this is quite a different type of operator from a dynamical variable, say, which is Hermitian. $U(t)$ is a *transformation operator*: its job is to rotate axes in Hilbert space in such a way as to keep the real physical quantities working.

But this *time evolution operator* must be consistent with the equation of motion (3.23), i.e.

$$HU(t)\,|\psi(0)\rangle = i\hbar\frac{\partial}{\partial t}U(t)\,|\psi(0)\rangle, \tag{3.30}$$

which suggests the equation of motion for $U(t)$—

$$i\hbar\frac{\partial U(t)}{\partial t} = HU(t), \tag{3.31}$$

and the conjugate equation

$$-i\hbar\frac{\partial U^*(t)}{\partial t} = U^*(t)H. \tag{3.32}$$

These equations can be solved quite easily. By the elementary theory of differential equations, we have

$$U(t) = e^{-(i/\hbar)Ht}, \tag{3.33}$$

bearing in mind that we must have $U(0) = 1$. The fact that H is an operator presents no difficulty. The meaning of the exponential function is simply its series expansion.

Our Schrödinger equation (3.23) has thus been solved (in the abstract) in the form (3.27) and (3.33). We shall be using equations like these a great deal in this chapter. But for the moment let us do something else. For any dynamical variable with 'Schrödinger' operator A_S define a new operator that depends on time by the relation

$$A_H(t) = U_S{}^*(t)\,A\,U(t). \tag{3.34}$$

This operator—the *Heisenberg representative* of the variable A—has the same expectation value in the state $|\psi(0)\rangle$ as A_S has in the state $|\psi_S(t)\rangle$:

$$\begin{aligned}\langle\psi(0)|\,A_H(t)\,|\psi(0)\rangle &= \langle\psi(0)|\,U^*(t)\,A_S\,U(t)\,|\psi(0)\rangle \\ &= \langle\psi_S(t)|\,A_S\,|\psi_S(t)\rangle.\end{aligned} \tag{3.35}$$

The equation of motion of $A_H(t)$ can also be derived from (3.31) and (3.32):

$$\begin{aligned} i\hbar\frac{\partial A_H(t)}{\partial t} &= i\hbar\frac{\partial U^*(t)}{\partial t}A_S\,U(t) + U^*(t)\,A_S i\hbar\frac{\partial U(t)}{\partial t} \\ &= -U^*(t)\,HA_S\,U(t) + U^*(t)\,A_S HU(t) \\ &= -HU^*(t)\,A_S\,U(t) + U^*(t)\,A_S\,U(t)\,H \\ &= [A_H(t), H] \end{aligned} \tag{3.36}$$

(we have used (3.33) to commute $U(t)$ with H). This is just the same as our conjectural relation (3.26). The operator $A_H(t)$ is the time-varying operator whose expectation value in a standard state $|\psi(0)\rangle$ is the value of the physical quantity A measured on the system at the time t.

The whole of quantum mechanics can thus be transformed into a theory in which the 'state' of the system is constant—a dummy background for all the dynamics—whilst the operators themselves vary with time. This is called the *Heisenberg representation.* It has the advantage of throwing the whole of the physics on to the operators, whose equation of motion (3.36) is quite similar to the ordinary classical equations of Hamiltonian dynamics. Thus, for example, consider the one-particle momentum operator

$$p = \frac{\hbar}{\mathrm{i}}\frac{\partial}{\partial q} \tag{3.37}$$

in the Schrödinger representation. In the Heisenberg representation this operator satisfies the equation of motion

$$\begin{aligned}\frac{\partial p}{\partial t} &= -\frac{\mathrm{i}}{\hbar}[p(t), H] \\ &= -\left[\frac{\partial}{\partial q}H - H\frac{\partial}{\partial q}\right] \\ &= -\frac{\partial H}{\partial q}\end{aligned} \tag{3.38}$$

which is one of the canonical equations of Hamilton. This could be the starting point for investigations of the correspondence principle, and other general relations between classical and quantum mechanics.

To see how the time-variation actually appears in the theory, let us work out the case of a simple harmonic oscillator, with Hamiltonian

$$H = \tfrac{1}{2}\hbar\omega(a^*a + aa^*). \tag{3.39}$$

An annihilation operator, in the Heisenberg representation, then must satisfy the equation

$$\begin{aligned}\mathrm{i}\hbar\frac{\partial a(t)}{\partial t} &= [a(t), H] \\ &= \hbar\omega a(t)\end{aligned} \tag{3.40}$$

by elementary manipulations of the commutators for a and a^*. This means

$$a(t) = \mathrm{e}^{-\mathrm{i}\omega t}\, a(0), \tag{3.41}$$

and similarly

$$a^*(t) = \mathrm{e}^{\mathrm{i}\omega t}\, a^*(0).$$

Thus, an annihilation operator carries a negative frequency factor, consistent with the effect of dropping one unit $\hbar\omega$ in energy. It is worth noting that the transformation (2.71) to 'hole' or 'antiparticle' states, $\tilde{b}^*_{\mathbf{k}} \to \tilde{b}_{-\mathbf{k}}$, has the effect of reversing the sign of the 'frequency' of the state, as well as reversing the wave-vector. This is consistent with the relabelling of energies. The operator for a positron is just equivalent to the operator for an electron 'travelling backwards in time'—a principle of some importance in the general theory (see §§ 3.7 and 6.8).

Another advantage of the Heisenberg picture is that it allows one to isolate the interesting parts of a problem and leave out any constant effects. Thus, suppose we had added an arbitrary constant energy E_0 to H—for example, by putting the atom into some region of high electrostatic potential. In the Schrödinger representation all the state vectors would be changed; we should have to write

$$|\psi'_S(t)\rangle = e^{-(i/\hbar)E_0 t}|\psi_S(t)\rangle, \quad \text{etc.} \tag{3.42}$$

But in the Heisenberg representation this has no effect. Thus

$$\begin{aligned} A'_H(t) &= e^{(i/\hbar)(H_0+E_0)t} A_S e^{-(i/\hbar)(H+E_0)t} \\ &= A_H(t), \end{aligned} \tag{3.43}$$

because the number $\exp\{iE_0 t/\hbar\}$ simply commutes and cancels through. The Heisenberg equations are invariant to such arbitrary changes of phase of all the states. Perhaps this argument will be helpful to those who have worried about the meaning of the frequency-factor attached to wave functions in the time-dependent Schrödinger representation. This factor only has meaning when measured relative to the frequency of another state vector in the same representation.

As is well known, the Heisenberg formulation of quantum mechanics was actually discovered at nearly the same time as the Schrödinger formulation. Unfortunately, it is extremely difficult to solve problems in the Heisenberg language, for all its abstract elegance. In the case of the hydrogen atom, for example, it is much easier to use the familiar theory of partial differential equations than to struggle with diagonalizing the corresponding very complicated Hamiltonian matrix.

3.3 Interaction representation

Nevertheless, it is convenient in perturbation problems to go some way towards the Heisenberg picture. In the standard treatments of time-dependent perturbation theory all states of the system are expressed in terms of the eigenstates of the unperturbed Hamiltonian, including their intrinsic time factors. Thus we might write

$$|\psi_S(t)\rangle = \sum_n a_n(t)\, e^{-(i/\hbar)\mathscr{E}_n t}\, |n\rangle, \tag{3.44}$$

where $|n\rangle$ is the solution of a time-*independent* Schrödinger equation

$$H_0\,|n\rangle = \mathscr{E}_n\,|n\rangle. \tag{3.45}$$

The essence of the problem is then to find the time variation of the coefficients $a_n(t)$ which tell about rates of transition between levels, etc.

We get similar results, more elegantly, by removing this time factor from each basis state and putting it into the definition of the operator. Use the unperturbed Hamiltonian to transform to a new representation of each state vector,

$$|\psi_I(t)\rangle = e^{(i/\hbar)H_0 t}\,|\psi_S(t)\rangle, \tag{3.46}$$

where the perturbation becomes

$$H'_I(t) = e^{(i/\hbar)H_0 t}\, H'_S\, e^{-(i/\hbar)H_0 t}. \tag{3.47}$$

In the *interaction representation* both states and operators may depend on time. But the time variation of the state functions has genuine physical significance. Applying (3.46) to (3.44), say, we get

$$|\psi_I(t)\rangle = \sum_n a_n(t)\,|n\rangle, \tag{3.48}$$

showing that this depends on the mixing coefficients and matrix elements for transitions between states of the free fields, not on any intrinsic time variation of each state.

On the other hand, the perturbing Hamiltonian, which is constant in the Schrödinger picture, has acquired an intrinsic time variation. From (3.45) and (3.47) we get

$$\langle n|\,H'_I(t)\,|m\rangle = e^{i(\mathscr{E}_n - \mathscr{E}_m)t/\hbar}\,\langle n|\,H'_S\,|m\rangle. \tag{3.49}$$

Thus, the mnth matrix element oscillates at a frequency corresponding to the difference of the intrinsic frequencies of the free-field modes it couples. The properties of the operators are summed up in their equation of motion; for any operator,

$$i\hbar\frac{\partial A_I(t)}{\partial t} = [A_I(t), H_0], \tag{3.50}$$

which is easily derived by analogy with (3.36). Thus, every operator behaves as it would in the Heisenberg representation for a *non-interacting* system.

The physics is contained in the equation of motion

$$H'_I(t)\,|\psi_I(t)\rangle = i\hbar\frac{\partial\,|\psi_I(t)\rangle}{\partial t}, \tag{3.51}$$

which is easily derived from (3.46), (3.47) and, of course (3.23). Thus, the time variation of the state vector in the interaction representation is governed by a 'Schrödinger equation' in which the perturbing Hamiltonian alone appears. This is a very convenient starting point for further investigations.

It is interesting to note that we implicitly use the interaction representation in all ordinary problems of quantum mechanics when we ignore 'the rest of the world' and its Hamiltonian and isolate, shall we say, a single electron moving in the field of a point charge. All time factors belonging to the huge but constant energy that has been ignored are automatically eliminated by this device.

3.4 Time–integral expansion series

In the interaction picture, which will be used henceforth without special labelling, we have to solve the equation of motion (3.51). Let us represent the solution by means of a unitary operator, just as in (3.30). The state function at some general time t evolves out of the state function at some standard time t_1, as if

$$|\psi(t)\rangle = U(t,t_1)\,|\psi(t_1)\rangle. \tag{3.52}$$

The evolution operator must satisfy an equation like (3.31):†

$$i\frac{\partial U(t,t_1)}{\partial t} = H'(t)\,U(t,t_1). \tag{3.53}$$

If H' had been independent of t, this would have had the solution (3.33):

$$U(t,t_1) = e^{-i(t-t_1)H'}. \tag{3.54}$$

Alternatively, if H' were a 'c-number' (i.e. an ordinary scalar function, not an operator) we should have the standard integral of the differential equation:

$$U(t,t_1) = \exp\left\{-i\int_{t_1}^{t} H'(t)\,dt\right\}. \tag{3.55}$$

† From this point we shall take $\hbar = 1$ and drop it from the equations. In a subject like this it is not always easy to decide how far one should carry such symbols which help give dimensional sense to the mathematical expressions as one is learning one's way through them. Similarly, it is convenient to keep 'the volume of the box' in expressions involving Fourier transformation, until one has learned to keep track of such factors for oneself.

Unfortunately, we cannot use this, because it implies that we have a convention for giving a meaning to $H'(t) \,.\, H'(t')$ say, where t and t' are different times. These operators do not necessarily commute with one another, so that (3.55) is not necessarily true.

It is not difficult, however, to construct a formal series solution to (3.53). First of all, integrate this equation as it stands, putting $U(t_1, t_1) = 1$ as a natural initial condition. We get

$$U(t_2, t_1) = 1 - \mathrm{i}\int_{t_1}^{t_2} H'(t) \,.\, U(t, t_1)\,\mathrm{d}t. \tag{3.56}$$

Now put the right-hand side back under the integral sign as a formula for $U(t, t_1)$ and iterate: this gives

$$U(t_2, t_1) = 1 - \mathrm{i}\int_{t_1}^{t_2} H'(t)\,\mathrm{d}t + (-\mathrm{i})^2 \int_{t_1}^{t_2} H'(t)\,\mathrm{d}t \int_{t_1}^{t} H'(t')\,\mathrm{d}t'$$

$$+ (-\mathrm{i})^n \int_{t_1}^{t_2} \mathrm{d}t \int_{t_1}^{t} \mathrm{d}t' \ldots \int_{t_1}^{t^{(n)}} \mathrm{d}t^{(n)}\, H'(t) \,.\, H'(t') \ldots H'(t^{(n)}). \tag{3.57}$$

This series is quite explicit as to the order of the operators, and is exact. It is obvious that the nth term is of order n in powers of the perturbation. The basic problem of perturbation theory is to evaluate the successive terms, and, if possible, to sum the whole series analytically.

To see the connection with the elementary Rayleigh–Schrödinger series let us suppose that the perturbing Hamiltonian of a simple one-electron system is just an extra small potential $\mathscr{V}(\mathbf{r})$. Now to deal with this we need a little trick. The trouble is that if this potential is always present then the system does not 'evolve'; it just stays in an eigenstate of the total Hamiltonian $H_0 + \mathscr{V}$.

We must pretend that the perturbing potential was initially absent, and then was slowly switched on. In the Schrödinger representation we suppose that the perturbation is

$$H'_S(t) = \mathrm{e}^{\alpha t}\,\mathscr{V}, \tag{3.58}$$

where α is some very small positive constant. This means that at time $t_1 = -\infty$ the system must have been in one of the eigenstates $|n\rangle$ of the unperturbed Hamiltonian, and has gradually evolved so that at time $t = 0$ it is in a state of the full Hamiltonian $H_0 + \mathscr{V}$. Thus, the operator $U(0, -\infty)$ applied to $|n\rangle$, should give us the corresponding state of the perturbed system.

We calculate the first-order approximation from the series (3.57).

Remembering the transformation (3.47) to the interaction representation, we have

$$U(0, -\infty) \approx 1 - \mathrm{i}\int_{-\infty}^{0} \mathrm{e}^{\mathrm{i}H_0 t}\mathrm{e}^{\alpha t}\mathscr{V}\,\mathrm{e}^{-\mathrm{i}H_0 t}\,\mathrm{d}t. \tag{3.59}$$

This is rather abstract. But let it operate on our state $|n\rangle$. By a succession of matrix multiplications, etc. we get the following:

$$\begin{aligned} U(0, -\infty)\,|n\rangle &\approx |n\rangle - \mathrm{i}\int_{-\infty}^{0} \mathrm{d}t\,\mathrm{e}^{\mathrm{i}H_0 t}\mathrm{e}^{\alpha t}\mathscr{V}\,\mathrm{e}^{-\mathrm{i}\mathscr{E}_n t}\,|n\rangle \\ &\approx |n\rangle - \mathrm{i}\int_{-\infty}^{0} \mathrm{d}t\,\mathrm{e}^{\alpha t}\,\mathrm{e}^{-\mathrm{i}\mathscr{E}_n t}\,\mathrm{e}^{\mathrm{i}H_0 t}\sum_m \mathscr{V}_{nm}\,|m\rangle \\ &\approx |n\rangle - \mathrm{i}\int_{-\infty}^{0} \mathrm{d}t\,\mathrm{e}^{\alpha t}\,\mathrm{e}^{-\mathrm{i}\mathscr{E}_n t}\sum_m \mathscr{V}_{nm}\,\mathrm{e}^{\mathrm{i}\mathscr{E}_m t}\,|m\rangle \\ &\approx |n\rangle - \mathrm{i}\sum_m\int_{-\infty}^{0} \mathrm{e}^{\mathrm{i}(\mathscr{E}_m - \mathscr{E}_n - \mathrm{i}\alpha)t}\,\mathrm{d}t\,\mathscr{V}_{nm}\,|m\rangle \\ &\approx |n\rangle - \sum_m \frac{\mathscr{V}_{nm}}{\mathscr{E}_m - \mathscr{E}_n - \mathrm{i}\alpha}\,|m\rangle. \end{aligned} \tag{3.60}$$

For simplicity I suppose that $\mathscr{V}_{nn}$ is zero so that there are no singular terms. We may now drop the small quantity α (whose function was mainly to guarantee the convergence of the integral) and recognize the familiar series of time-independent perturbation theory.

The device (3.58) is quite useful in practice, even for very sophisticated problems, although one has to go to some trouble to justify it mathematically. This is, of course, the limiting case of an *adiabatic* perturbation.

3.5 *S*-matrix

Many problems deal with scattering events, which cannot be probed in detail. All we can then state are initial conditions, at some time t_1 far in the past, when the particles had not 'seen' each other, and final conditions a long time t_2 after the interaction, when they are again far apart. We then define the *scattering matrix*,

$$S \equiv U(\infty, -\infty), \tag{3.61}$$

for which these initial and final times have been pushed away to infinity, so that they will not appear explicitly in any expression derived from (3.57). A little artistry is needed to take these limits so that the integrals converge properly.

The S-matrix is unitary; but it often contains, so to speak, a great amount of the unscattered states. The *transition matrix*

$$\mathscr{T} = 1 - S \tag{3.62}$$

represents the part of the S-matrix that is not mere identity; the matrix elements of $\mathscr{T}$ are probability amplitudes between initial and final states of the systems. Thus, we may write, symbolically,

$$P_{a\to b} = |\langle a|\,\mathscr{T}\,|b\rangle|^2\,\delta(\mathscr{E}_a - \mathscr{E}_b), \tag{3.63}$$

for the probability of transition between two states $|a\rangle$ and $|b\rangle$, of energies $\mathscr{E}_a$ and $\mathscr{E}_b$. These two relations, (3.62) and (3.63) link the theory of the S-matrix with more familiar theories of scattering. For example, in the Born approximation, a potential $\mathscr{V}$ gives rise to transitions for which

$$\langle a|\,\mathscr{T}\,|b\rangle \approx \langle a|\,\mathscr{V}\,|b\rangle. \tag{3.64}$$

These relations will be discussed further in §4.14.

Although we have arrived at the S-matrix through the theory of the interaction representation, the idea is of more general validity. In §3.1, for example, we could have said that the perturbed state $|\Psi\rangle$ is related to the unperturbed state $|\Psi_0\rangle$ by the unitary transformation

$$|\Psi\rangle = S\,|\Psi_0\rangle. \tag{3.65}$$

The Brillouin–Wigner theory, as summed up in (3.15), is then equivalent to the integral equation

$$S = 1 + \frac{1}{\mathscr{E} - H_0}PH'S. \tag{3.66}$$

Another way of looking at time-independent perturbation theory is to treat S as a canonical transformation matrix which diagonalizes the Hamiltonian. Thus, we may deduce from (3.2) and (3.65) that $|\Psi_0\rangle$ is an eigenstate of the transformed Hamiltonian

$$\tilde{H} = S^{-1}(H_0 + H')\,S, \tag{3.67}$$

which must therefore be diagonal in a representation in terms of the eigenstates $|\Psi_n\rangle$ of H_0.

Suppose we write the unitary matrix S in the form

$$S = \mathrm{e}^{\mathrm{i}W}, \tag{3.68}$$

where W is an operator to be discovered. This expression means, of course, the usual exponential series, which gives, when put in (3.67), the following expansion:

$$\tilde{H} = H_0 + H' + \mathrm{i}[H_0, W] + \mathrm{i}[H', W] + \tfrac{1}{2}\mathrm{i}^2[[H_0, W], W] + \dots \quad (3.69)$$

Now let us choose W so that

$$H' + \mathrm{i}[H_0, W] = 0. \quad (3.70)$$

This removes from $\tilde{H}$ all terms that are linear in the perturbation. To the order of terms quadratic in H' we shall then get

$$\tilde{H} \approx H_0 + \tfrac{1}{2}\mathrm{i}[H', W], \quad (3.71)$$

which is a useful approximation to the required diagonal operator.

In elementary cases this takes us back to the ordinary Rayleigh–Schrödinger series. In our basic representation, (3.70) becomes

$$\langle \Psi_n | H' | \Psi_m \rangle + \mathrm{i}\langle \Psi_n | H_0 W - W H_0 | \Psi_m \rangle = 0, \quad (3.72)$$

which has the simple solution

$$\mathrm{i}\langle \Psi_n | W | \Psi_m \rangle = \frac{\langle \Psi_n | H' | \Psi_m \rangle}{\mathscr{E}_m - \mathscr{E}_n} \quad (3.73)$$

for the elements of the transformation matrix W (and hence, from (3.68), of the matrix S). Putting these back into (3.71) gives an operator whose diagonal elements are the usual expressions for the energy in second-order perturbation theory.

This technique has some advantages when we want only to diagonalize part of the Hamiltonian. For example, let us eliminate the term in the fermion–boson interaction (2.47) that creates and annihilates single bosons, so that we can concentrate on the properties of the fermion field. The trick then is to evaluate $\tilde{H}$ only in the vacuum state for bosons, so that only single-boson excitations out of, and back to, the vacuum need be considered. These excitations imply, of course, concomitant transitions of the fermion states, but instead of writing these all down explicitly let us simply keep the appropriate fermion annihilation and creation operators in evidence. In other words, the operator W is represented as an *operator* on the fermion states, but as a *matrix* between certain boson levels. By analogy with (3.73) we get

$$\mathrm{i}\langle 1_{\mathbf{q}} | W | 0 \rangle = -\sum_{\mathbf{k}} \frac{F(q)\, b^*_{\mathbf{k}-\mathbf{q}} b_{\mathbf{k}}}{\mathscr{E}(\mathbf{k}) - \mathscr{E}(\mathbf{k}-\mathbf{q}) - \omega_{\mathbf{q}}} \quad (3.74)$$

and

$$\mathrm{i}\langle 0 | W | 1_{\mathbf{q}} \rangle = \sum_{\mathbf{k}'} \frac{F(q)\, b^*_{\mathbf{k}'+\mathbf{q}} b_{\mathbf{k}'}}{\mathscr{E}(\mathbf{k}') + \omega_{\mathbf{q}} - \mathscr{E}(\mathbf{k}'+\mathbf{q})}. \quad (3.75)$$

Putting these expressions into (3.71), we get an addition to the effective (unperturbed) Hamiltonian of the form

$$
\begin{aligned}
\tfrac{1}{2}\mathrm{i}[H', W] &= \tfrac{1}{2}\mathrm{i}\langle 0|\, H'W - WH' \,|0\rangle \\
&= \tfrac{1}{2}\mathrm{i} \sum_{\mathbf{q}} \{\langle 0|\, H' \,|1_{\mathbf{q}}\rangle \langle 1_{\mathbf{q}}|\, W \,|0\rangle - \langle 0|\, W \,|1_{\mathbf{q}}\rangle \langle 1_{\mathbf{q}}|\, H' \,|0\rangle\} \\
&= \tfrac{1}{2} \sum_{\mathbf{k},\mathbf{k}',\mathbf{q}} |F(q)|^2 b^*_{\mathbf{k}'+\mathbf{q}} b_{\mathbf{k}'} b^*_{\mathbf{k}-\mathbf{q}} b_{\mathbf{k}} \\
&\quad \times \left\{ \frac{1}{\mathscr{E}(\mathbf{k}) - \mathscr{E}(\mathbf{k}-\mathbf{q}) - \omega_{\mathbf{q}}} - \frac{1}{\mathscr{E}(\mathbf{k}') + \omega_{\mathbf{q}} - \mathscr{E}(\mathbf{k}'+\mathbf{q})} \right\}. \qquad (3.76)
\end{aligned}
$$

By separating the sums in $\mathbf{q}$ and $-\mathbf{q}$ (remembering that $\omega_{\mathbf{q}} = \omega_{-\mathbf{q}}$) and relabelling $\mathbf{k}$ and $\mathbf{k}'$, we get eventually an operator

$$
H_{el.el} = \sum_{\mathbf{k},\mathbf{k}',\mathbf{q}} \frac{\omega_{\mathbf{q}} |F(q)|^2}{\{\mathscr{E}(\mathbf{k}) - \mathscr{E}(\mathbf{k}-\mathbf{q})\}^2 - \omega_{\mathbf{q}}^2} b^*_{\mathbf{k}'+\mathbf{q}} b_{\mathbf{k}'} b^*_{\mathbf{k}-\mathbf{q}} b_{\mathbf{k}}. \qquad (3.77)
$$

Thus, *from the point of view of the fermions*, the perturbation H' acts as if there were an interaction between them, due to the exchange of bosons—an interaction that will be attractive or repulsive according to the range of values of momenta involved. This is just the effective interaction between electrons, due to the exchange of virtual phonons, that gives rise to superconductivity. The Yukawa force—the interaction between nucleons by exchange of pions—could also be calculated in this way; the derivation in § 1.11 is simpler in that the change of energy of the pion field (i.e. the *recoil* effect) was neglected in the energy denominators.

Of course this reduction is only approximate; there will be other terms in the transformed Hamiltonian $\tilde{H}$ corresponding to residual fermion–boson scattering, etc. Sometimes these terms may be removed by yet another canonical transformation, and so on. This method is not general and systematic, but is often used as a means of removing awkward terms from a Hamiltonian without going to all the labour of working out the S-matrix expansion (3.57), say.

3.6 S-matrix expansion: algebraic theory

The explicit series expansion for the S-matrix in the interaction representation, as defined by (3.57) and (3.61), has the following term of nth order in the perturbation H':

$$
S_n = (-\mathrm{i})^n \int_{-\infty}^{\infty} \mathrm{d}t_1 \int_{-\infty}^{t_1} \mathrm{d}t_2 \ldots \int_{-\infty}^{t_{n-1}} \mathrm{d}t_n \{H'(t_1)\, H'(t_2) \ldots H'(t_n)\}. \quad (3.78)
$$

This expression is awkward because each integration involves the upper limit of the previous one, so that all are linked together, and cannot be unravelled except in the official order.

The corresponding term in the expansion of a 'non-operator' exponential function, such as (3.55) may be written

$$S_n' = (-\mathrm{i})^n \frac{1}{n!}\left\{\int_{-\infty}^{\infty} H'(t)\,\mathrm{d}t\right\}^n$$

$$= (-\mathrm{i})^n \frac{1}{n!}\int_{-\infty}^{\infty}\mathrm{d}t_1\int_{-\infty}^{\infty}\mathrm{d}t_2\ldots\int_{-\infty}^{\infty}\mathrm{d}t_n\{H'(t_1)\,H'(t_2)\ldots H'(t_n)\}. \quad (3.79)$$

The difference between (3.78) and (3.79) is that the region of integration in the n-dimensional space of the coordinates $\mathrm{d}t_1\,\mathrm{d}t_2\ldots\mathrm{d}t_n$ is limited to a polyhedron in which

$$-\infty < t_n \leqslant t_{n-1} \leqslant t_{n-2} \leqslant \ldots \leqslant t_2 \leqslant t_1 < \infty. \quad (3.80)$$

But we would get the same answer if we permuted the labels of these variables in any of $n!$ ways—i.e. if we had evaluated the appropriate product over any other of the $n!$ similar polyhedra into which this 'space' can be dissected. Adding up all these different ways of calculating S_n, and dividing by $n!$, we get an answer very similar to (3.79),

$$S_n = (-\mathrm{i})^n \frac{1}{n!}\int_{-\infty}^{\infty}\mathrm{d}t_1\int_{-\infty}^{\infty}\mathrm{d}t_2\ldots\int_{-\infty}^{\infty}\mathrm{d}t_n P\{H'(t_1)\,H'(t_2)\ldots H'(t_n)\}. \quad (3.81)$$

Each integral is now over all values of each co-ordinate $t_1 \ldots t_n$, but the product to be integrated must now be 'time-ordered', as indicated by the symbol P. This means that if we are in a region where, say

$$-\infty < t_k < t_l < \ldots < t_m < \ldots < t_n < t_q < \infty, \quad (3.82)$$

then the integrand shall be taken to be

$$H'(t_q)\,H'(t_n)\ldots H'(t_m)\ldots H'(t_l)\,H'(t_k). \quad (3.83)$$

This ordering is important, of course, whenever the operators $H'(t_1)$ and $H'(t_2)$ do not commute with one another—otherwise (3.81) would be identical with (3.79).

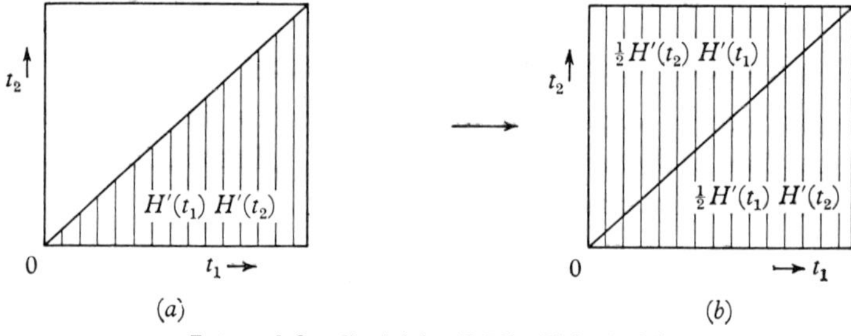

Integral for S_2: (a) in (3.78), (b) in (3.81).

We may make (3.81) a little more general by recalling that each $H'(t)$ is itself an integral of a Hamiltonian density over ordinary space $\mathbf{r}$. Thus, the general form of (3.81) is

$$S_n = \frac{(-\mathrm{i})^n}{n!}\int\cdots\int \mathrm{d}^4x_1\,\mathrm{d}^4x_2\ldots\mathrm{d}^4x_n\,P\{\mathscr{H}'_I(x_1)\,\mathscr{H}'_I(x_2)\ldots\mathscr{H}'_I(x_n)\} \quad (3.84)$$

when we use the symbol x_i to refer to three space co-ordinates $\mathbf{r}_i$ and a time co-ordinate t_i of some point in a 'space-time' continuum. This notation is used here for brevity, but we do not mean that our expressions are relativistically invariant nor do we allow space-like and time-like co-ordinates to be transformed into one another until we reach § 6.4.

The essence of our problem is to evaluate various matrix elements of S, or of its terms. Thus, the calculation of the probability of some process carrying the system from an initial state $|\Psi_i\rangle$ to a final state $|\Psi_f\rangle$ requires knowledge of matrix elements such as

$$\langle\Psi_f|\,S\,|\Psi_i\rangle = \sum_n \langle\Psi_f|\,S_n\,|\Psi_i\rangle. \quad (3.85)$$

These states are often quite simple—e.g. there may be only one or two real particles present initially, and these are just allowed to interact, separate and go into other well-defined states. This means that we only need to consider intermediate states to which these are coupled; not all the factors in the product of interaction Hamiltonians are capable of acting.

For example, consider the second-order term arising from the fermion–boson interaction of § 2.7. The working parts will be products of field operators:

$$\begin{aligned} S_2 &\sim \int \mathrm{d}^4x_1 \int \mathrm{d}^4x_2\,P\{\psi^*(x_1)\,\phi(x_1)\,\psi(x_1)\,\psi^*(x_2)\,\phi(x_2)\,\psi(x_2)\} \\ &\sim \sum_{\mathbf{k}_1\,\mathbf{k}_2,\,\mathbf{k}_3,\,\mathbf{k}_4,\,\mathbf{q},\,\mathbf{q}'} \Bigg[\iint \mathrm{e}^{\mathrm{i}(\mathbf{k}_1-\mathbf{q}-\mathbf{k}_2)\cdot\mathbf{r}_1}\,\mathrm{e}^{\mathrm{i}(\mathbf{k}_3-\mathbf{q}'-\mathbf{k}_4)\cdot\mathbf{r}_2}\,\mathbf{d}^3\mathbf{r}_1\,\mathbf{d}^3\mathbf{r}_2 \\ &\qquad \times \iint \mathrm{e}^{\mathrm{i}(\mathscr{E}_1\pm\omega-\mathscr{E}_2)t_1}\,\mathrm{e}^{\mathrm{i}(\mathscr{E}_3\pm\omega'-\mathscr{E}_4)t_2} \\ &\qquad \times P\{b_1^*(a_\mathbf{q}-a^*_{-\mathbf{q}})\,b_2\,b_3^*(a_{\mathbf{q}'}-a^*_{-\mathbf{q}'})\,b_4\}\,\mathrm{d}t_1\,\mathrm{d}t_2\Bigg], \end{aligned} \quad (3.86)$$

when transformed into free-field representations, with b_1^* creating an electron of momentum $\mathbf{k}_1$ and energy $\mathscr{E}_1$ at point $\mathbf{r}_1$, time t_1, etc.

Now suppose this matrix acts on a state containing just one electron. For some combinations of time variables—i.e. when $t_2 < t_1$—the order of operators in the P-product is as written, and we can allow b_4 to destroy this electron, and b_3^* to create another, which then only b_2 could destroy. This contribution links together b_3^* and b_2, which must therefore correspond to an electron of the same momentum, i.e. $\mathbf{k}_2 = \mathbf{k}_3$.

But in that case the anticommutator of b_3^* and b_2 is not zero. In other words, we can generate this condition by exchanging the order of these two operators and using (2.19), i.e.

$$\begin{aligned} b_1^*(a_{\mathbf{q}} - a_{-\mathbf{q}}^*)\, b_2 b_3^*(a_{\mathbf{q}'} - a_{-\mathbf{q}'}^*)\, b_4 \\ = -b_1^*(a_{\mathbf{q}} - a_{-\mathbf{q}}^*)\, b_3^* b_2 (a_{\mathbf{q}'} - a_{-\mathbf{q}'}^*)\, b_4 \\ + b_1^*(a_{\mathbf{q}} - a_{-\mathbf{q}}^*)\,(a_{\mathbf{q}'} - a_{-\mathbf{q}'}^*)\, b_4\, \delta(\mathbf{k}_2 - \mathbf{k}_3). \end{aligned} \tag{3.87}$$

The first term on the right gives zero when applied to our initial state, where there is not a second electron to destroy. Thus, the contribution from the 'allowed intermediate state' is generated algebraically by the non-vanishing anticommutator in the second term.

Even this term will not contribute unless the phonon operators are correctly related—e.g. so that the phonon created by $a_{-\mathbf{q}'}^*$ can be destroyed by $a_{\mathbf{q}}$. This is related to the commutation relations of these operators. Then, when $t_1 < t_2$, the order of the operators at x_1 and x_2 must be interchanged, and there will be different types of intermediate state to investigate—in this case, the anticommutator of b_4 and b_1^* will be the important effect.

The job, then, is to show up explicitly all possible pairs of non-vanishing commutators and anticommutators. What we need to do is to reduce the product to *normal form*, where all creation operators are on the left and all annihilators are on the right. As shown in (3.87), this product has zero expectation value in the vacuum, and in many simple excited states, so all we need consider is the residual parts due to non-vanishing commutators or anticommutators.

To this end, we *define* the operation

$$N(ABC \ldots XYZ) \equiv (-1)^P LMN \ldots QRS, \tag{3.88}$$

where $LMN \ldots QRS$ is the set of operators $ABC \ldots XYZ$ rearranged so that all creation operators precede all annihilation operators. The sign factor merely counts the number of times that fermion operators have had to be exchanged in the rearrangement. Thus, for example,

$$N\{\psi^*(x)\,\psi(x')\} = \psi^*(x)\,\psi(x'),$$

but

$$N\{\psi(x)\,\psi^*(x')\} = -\psi^*(x')\,\psi(x). \tag{3.89}$$

For bosons the field operators ϕ and π of § 1.9 each contain both annihilation and creation operators, i.e. we have to split them up into

$$\phi(x) = \sum_{\mathbf{q}} (2\omega_{\mathbf{q}})^{-\frac{1}{2}} (a^*_{\mathbf{q}} - a_{-\mathbf{q}})\, e^{i\mathbf{q}\cdot\mathbf{r}} = \phi^+(x) - \phi^-(x), \quad \text{etc.} \tag{3.90}$$

The *normal product* now takes the form

$$N\{\phi(x)\,\phi(x')\} = \phi^+(x)\,\phi^+(x') \\ - \phi^+(x')\,\phi^-(x) - \phi^+(x)\,\phi^-(x') + \phi^-(x)\,\phi^-(x'), \tag{3.91}$$

which is not so easily represented in terms of ϕ and π.

Basically, the normal product is constructed as if all boson operators always commute, and all fermion operators always anticommute. But of course this is not true. In the Schrödinger representation,

$$\psi(\mathbf{r})\,\psi^*(\mathbf{r}') = -\psi^*(\mathbf{r}')\,\psi(\mathbf{r}) + \delta(\mathbf{r}-\mathbf{r}'). \tag{3.92}$$

Thus, in the interaction representation, with energy/time factors (e.g. (3.41)) for all field variables, we have

$$\psi(x)\,\psi^*(x') = N\{\psi(x)\,\psi^*(x')\} + \delta(\mathbf{r}-\mathbf{r}')\, e^{i(\mathscr{E}'t' - \mathscr{E}t)}. \tag{3.93}$$

Unfortunately we cannot yet apply this normal product ordering to the terms in the S-matrix, because of the time ordering rule. In (3.81) this was expressed through a symbol P, which told us to put the operators in the order of increasing times from right to left. We shall replace this by another symbol—*Wick's chronological T-operator*—that has the same effect, but also puts in a factor (-1) for every time that two fermion operators pass each other in the rearrangement process. This does not make any difference in practice, because fermion operators always occur in pairs in each interaction Hamiltonian. To be quite explicit, we now write, in place of (3.84)

$$S_n = (-i)^n \frac{1}{n!} \iint \dots \int d^4x_1\, d^4x_2 \dots d^4x_n\, T\{\mathscr{H}'(x_1)\,\mathscr{H}'(x_2) \dots \mathscr{H}'(x_n)\}, \tag{3.94}$$

where

$$T\{A_1(x_1)\,A_2(x_2) \dots A_n(x_n)\} \\ \equiv (-1)^P A_j(x_j)\,A_k(x_k) \dots A_m(x_m), \tag{3.95}$$

with $t_m < \dots < t_k < t_j$ and P counting the number of permutations of fermion operators to get to this chronologically ordered product. This procedure is thus similar to (3.88) in that it seems to assume that all boson operators commute and all fermion operators anticommute.

The purpose of these sign factors is simply to put boson and fermion operators on the same footing in the next stage of the argument.

We come now to the key point. The normal product of any set of operators has zero expectation value in the vacuum state. Let us define the *contraction* or *chronological pairing* of two operators as follows:

$$\overbrace{A_1(x_1)\,A_2}(x_2) = T\{A_1(x_1)\,A_2(x_2)\} - N\{A_1(x_1)\,A_2(x_2)\}. \tag{3.96}$$

Then the expectation value of the T-product in the vacuum state will be just this quantity, i.e.

$$\overbrace{A_1(x_1)\,A_2}(x_2) = \langle 0|\,T\{A_1(x_1)\,A_2(x_2)\}\,|0\rangle. \tag{3.97}$$

Basically this is just a function that picks out cases of non-vanishing commutators or anticommutators between the operators in the product. It is, in fact, a c-number—a function that is independent of the states on which it acts—so we can evaluate it directly from (3.97).

Consider, for example, the pairing of fermion field operators:

$$\overbrace{\psi^*(x)\,\psi}(x') = \langle 0|\,T\{\psi^*(x)\,\psi(x')\}\,|0\rangle. \tag{3.98}$$

Then (i) if $t > t'$, the chronological product is the same as the ordinary product, and vanishes:

$$\overbrace{\psi^*(x)\,\psi}(x') = 0 \quad \text{for} \quad t > t'; \tag{3.99}$$

(ii) if $t < t'$, we can use the commutation relation in (3.93), and get

$$\begin{aligned}\overbrace{\psi^*(x)\,\psi}(x') &= \langle 0|\,-\psi(x')\,\psi^*(x)\,|0\rangle \\ &= \langle 0|\,\psi^*(x)\,\psi(x') - \delta(\mathbf{r}-\mathbf{r}')\,e^{i(\mathscr{E}t-\mathscr{E}'t')}\,|0\rangle \\ &= -\,\delta(\mathbf{r}-\mathbf{r}')\,e^{i(\mathscr{E}t-\mathscr{E}'t')} \quad \text{for} \quad t < t'.\end{aligned} \tag{3.100}$$

This result is slightly spurious, in that the time factors for each field operator have been simplified. The proper formula will be given in due course. The main point to notice is that the pairing is a 'causal propagator': it is zero unless the field to be destroyed at time t' already exists, having been created at a previous time t. It is this characteristic that marks the distinction between a pairing and an ordinary delta function commutator or anticommutator.

All we need now is a general procedure for reducing any more complicated T-product to N-products and pairings. This comes from *Wick's theorem* which states that *any chronological product is equal to the sum of all possible normal products that can be formed with all possible pairings.* What this means may be seen from the first few cases. Thus, by the definition (3.96),

$$T[AB] = N[AB] + \overline{AB}$$
$$= N[AB] + N[\overline{AB}], \tag{3.101}$$

by an extension of notation. Again we have

$$T[ABC] = N[ABC] + N[\overline{AB}C] + N[\overline{A\,B\,C}] + N[A\overline{BC}] \tag{3.102}$$

(where in $N[\overline{A\,B\,C}]$ the pairing is between A and C) and

$$\begin{aligned} T[ABCD] = {} & N[ABCD] + N[\overline{AB}CD] + N[\overline{A\,B\,C}D] \\ & + N[\overline{A\,B\,C\,D}] + N[A\overline{BC}D] + N[A\overline{B\,C\,D}] \\ & + N[AB\overline{CD}] + N[\overline{AB}\,\overline{CD}] + \\ & + N[\overline{A\overline{BC}D}] + N[A^{\bullet}B^{\circ}C^{\bullet}D^{\circ}], \end{aligned} \tag{3.103}$$

and so on. Here we use the notation

$$N[AB\overline{CD}] \equiv \eta\overline{CD}N[AB], \tag{3.104}$$

where η is the sign factor for the permutation of fermion operators in going from $ABCD$ to $CDAB$.

The proof of this theorem is given in all the standard books, usually by induction. The basic argument is as follows. To get an N-product from a T-product we need to interchange the order of certain creation operators with certain annihilation operators. Each time we do this, we add a term with the corresponding pairing on the right. This produces pairings for those operators whose order in the T-product is not already normal. But the value of the pairing of any two operators that are already normally ordered in T is zero (e.g. it might be two factors $\psi^*(x)\,\psi^*(x')$), so that we can include all these pairings as well without spoiling the result. Thus, we count formally on the right, *all* possible pairings of *all* operators in the product.

We thus have a systematic algebraic machine for evaluating the terms in the S-matrix. We apply Wick's theorem to the expression (3.94), and obtain a number of different contributions corresponding to the different possible pairings. Each contribution will be a multiple integral over space and time, but the integrand will contain only two types of factor (i) pairings of field operators, which are explicit

c-number functions such as (3.100), (ii) normal products of field operators, whose expectation values or other matrix elements between initial and final states of the system can be written down by inspection—and are often identically zero. Although this reduction process is stated somewhat abstractly, it is no more than a formalization of the procedures that were invoked when we inspected the elementary example (3.86).

3.7 Diagrammatic representation

This algebraic theory, although quite precise, is rather heavy work. Fortunately, the properties of the operators that can actually occur in a term in the S-matrix allow of a very elegant topological representation.

First let us represent Wick's theorem for some arbitrary product of operators $T[ABCDE]$ by a series of diagrams in which each operator is represented by a point in space, and each pairing by a line. Then we have a lot of cases to consider, such as

A B / C D E + A B / C D E + A B / C D E + ··· + A B / C D E + A B / C D E + ···

$N[ABCDE]$ $N[ABCDE]$ $N[ABCDE]$ $N[ABCDE]$ $N[ABCDE]$

But not all these diagrams are relevant, for in many cases they correspond to pairs that vanish. What are the rules for avoiding such cases?

(*a*) Each operator carries a space-time label x_i, say—and there are only n such labels in S_n. Pairings between operators with the same label are not distinct, for they correspond, eventually to finding the commutator or anticommutator of an operator with itself. Thus, we can separate the various points on the diagram into n groups, and need only consider pairings that link operators in different groups. This reduces considerably the number of relevant terms.

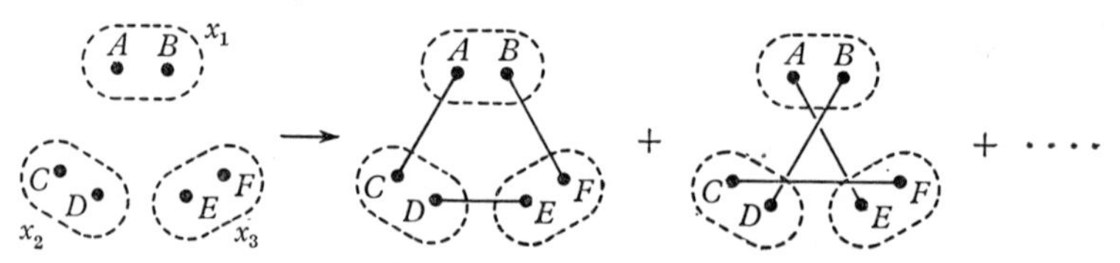

(*b*) Each type of field pairs only with itself—bosons with bosons, fermions with fermions, etc. Each point on the diagram ought therefore to carry a label telling us its type. This could be indicated by the type of line that it can terminate—for example, we represent fermion pairings by solid lines and boson pairings by dotted lines.

(*c*) But the pairing of two fermion creation operators, say, is zero. We need only consider lines joining $\psi^*(x_i)$ with $\psi(x_j)$. To indicate this condition on the diagram, let us put an arrow outwards on the line leaving $\psi^*(x_i)$ and an arrow inwards on the line ending on $\psi(x_j)$. Such an arrow is not needed on a boson pairing, because the boson field operator ϕ contains both annihilation and creation operators (cf. (3.90)), so that $\phi(x_i)$ and $\phi(x_j)$ can always be linked.

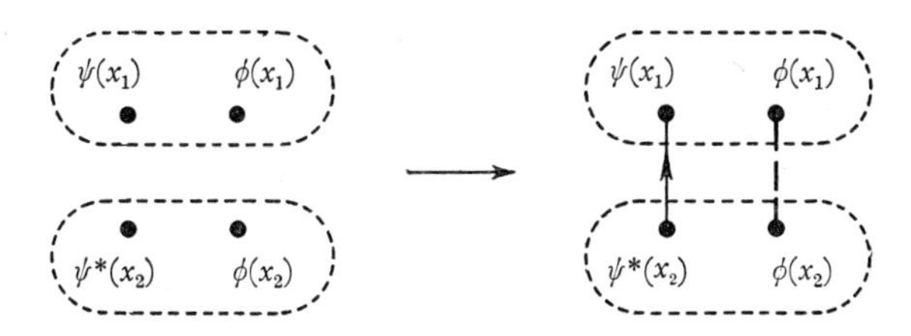

(*d*) Each operator must either be paired with another, or it must have the possibility of acting on the initial state of the system, or of contributing to the final state. Thus, all points in the diagram must be connected to other points, or else connected to 'external' points corresponding to these initial and final states. For example, if the initial and final states each contain one electron but no phonons, there is an electron line in, and an electron line out, available to be connected to ψ and ψ^*, but all phonon operators must be connected internally to one another: e.g.

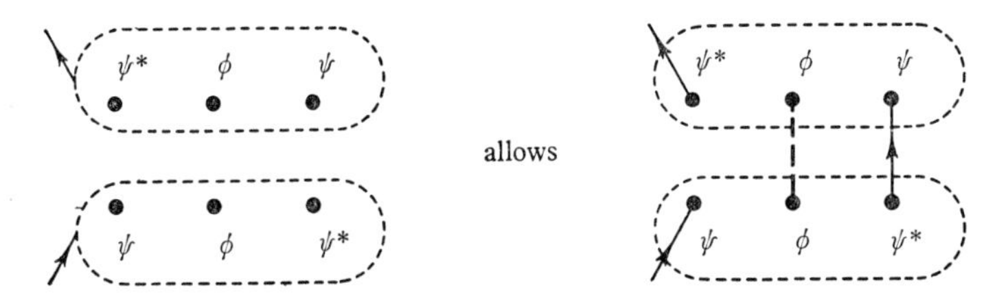

(*e*) The *n*th order term in the S-matrix is a T-product of n identical perturbation Hamiltonians, one at each of n different points in space-time. But each Hamiltonian contains the same product of field operators—for example, in (2.47) the product $\psi^*(x)\,\phi(x)\,\psi(x)$. Thus, every group in the diagram has the same types of lines in and out. We can therefore combine all the operators in this group into a *vertex*—a single point which must be connected to the rest of the diagram by the proper combination of arrowed and dotted lines. Thus

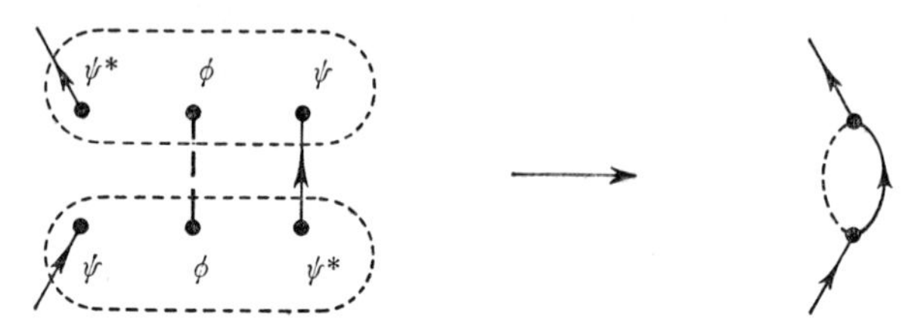

The operators present at each vertex speak for themselves, and need not be specifically labelled.

(*f*) Since each vertex belongs to a point in space-time, we can actually give it co-ordinates, on the paper, that represent its position. Thus, we can think of the time axis as being the co-ordinate direction vertically up the page, and then we let all three space co-ordinates $\mathbf{r}$ of the event be represented by a horizontal axis. Our diagram now has a very direct physical interpretation. We think of an electron emitting a phonon at the point $\mathbf{r}_1$, at time t_1, and then travelling to $\mathbf{r}_2$ where the phonon is reabsorbed at time t_2. Of course the lines on this diagram do not really represent the actual paths of actual particles in space and time (for example, here we deliberately curve them, so as to draw them separately and show up the connectivity of the diagram) but for many purposes it is quite proper to think of them in that way. Strictly speaking a *Feynman diagram* is only a topological representation of an algebraic expression, but it has the power of depicting recognizable physical processes.

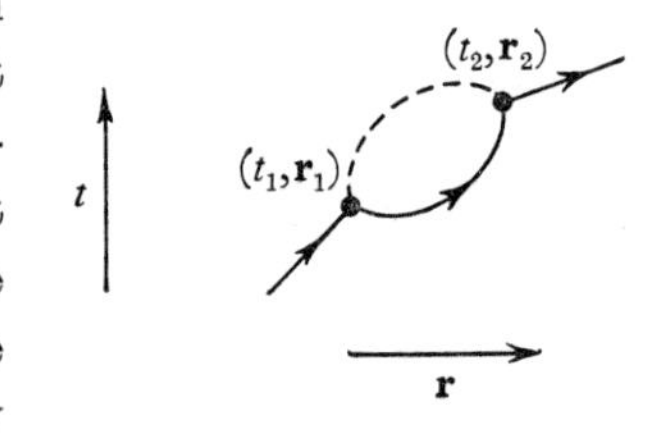

For example, let us consider the second-order diagrams that can be constructed out of our standard fermion–boson interaction (2.47). Each vertex must have a fermion line in and out, and a single boson line. Then we can make up the following pictures.

Compton scattering. An electron absorbs a photon at 1 and re-emits it, in a different direction, at 2.

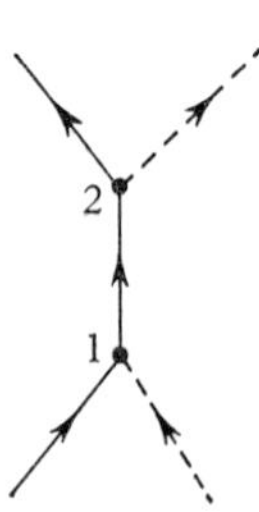

In this diagram we have had to put arrows on the external photon lines: the initial state already has a photon to be absorbed, and we must emit one into the final state. In fact, the Compton effect also includes a contribution from a similar but topologically distinct diagram in which the emission of the final photon occurs before the absorption of the initial one. These two contributions to the matrix element have to be evaluated separately.

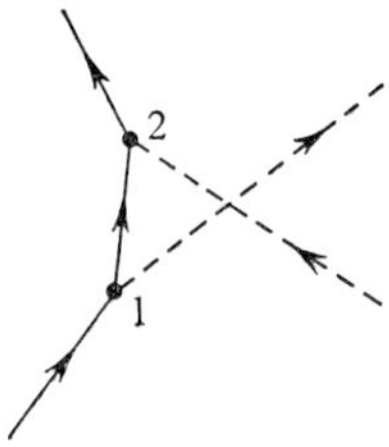

Cerenkov effect. This diagram is very similar to the first two, except that both photons are emitted. As we saw in (2.61), such processes are often forbidden by the rules for the overall conservation of energy and momentum, but for electrons and phonons they ought to be included in an exact theory of electrical resistivity, say, or of the mobility of a polaron (§ 2.7). Notice that the intermediate state here may be virtual, so that this is not the same as two successive real processes in which photons are produced.

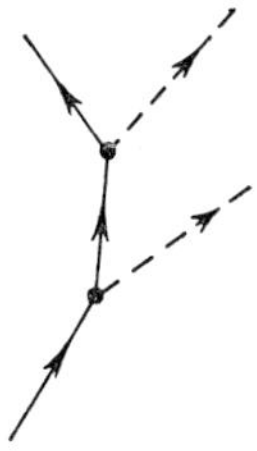

Self-energy of fermion. If now we let the emitted boson be reabsorbed, we describe a process of the kind calculated in (2.63)—the effect of the emission and reabsorption of virtual phonons on the energy of a free electron. This gives rise, for example, to the effective-mass correction for the polaron.

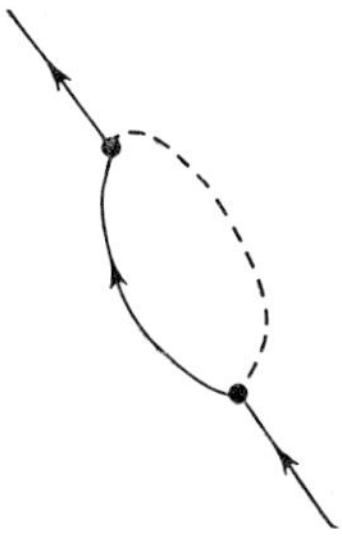

Fermion interaction by boson exchange. This is the sort of process that we have already discussed in §§ 1.11 and 3.5. Notice that we do not need to put an arrow on the boson line. If vertex 1 is before vertex 2, then the boson must have been propagated from 1 to 2; if 2 precedes 1, then the boson has to go the other way. But this is automatically taken care of in the boson pairing function; the appropriate term is put into the integrand according as $t_1 < t_2$ or $t_1 > t_2$.

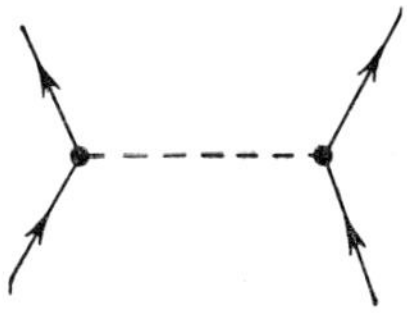

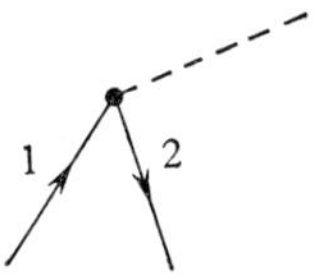

In these processes, a fermion is already supposed to be present. But suppose, on our sketching diagrams, we had produced a vertex like this. The formal requirements of diagram theory would be satisfied, in that these are fermion lines in and out; but what does it mean physically? The electron from 1 has evidently been destroyed at the vertex, and does not proceed beyond it (since we treat the vertical direction as the time axis). This can only happen if a 'hole' had also reached this vertex, for then the pair would mutually annihilate one another. Thus, a line like this—*a fermion apparently travelling backwards in time*—must correspond to the ordinary forward propagation of a hole from 2 into the vertex.

It is a peculiar and powerful property of the diagrammatic technique that this interpretation, forced upon us by the physics, is in fact correct. We have already noted, in §3.2, that the fermion–antifermion transformation (§2.8) does have the effect of apparently reversing the direction of time. One can check the argument through, and show that the contribution to the S-matrix of processes involving holes is given by diagrams with propagators of this kind, and that the rules for converting a diagram into an algebraic expression do not have to be specially modified. But it must be remembered that sometimes hole states are not really permitted in the system—for example, in the polaron problem, where the actual number of electrons present is rigorously conserved, and the energy required to create an 'antiparticle' is far larger than needs to be considered in the problem.

We now have several other second-order processes to consider. Thus, for example:

Self-energy of boson. This is the correction to the energy of a boson due to the excitation of virtual fermion–antifermion pairs. In quantum electrodynamics this would influence the propagation of light: in solid state physics this diagram describes the 'renormalization of the velocity of sound' due to interaction with the electrons in a metal.

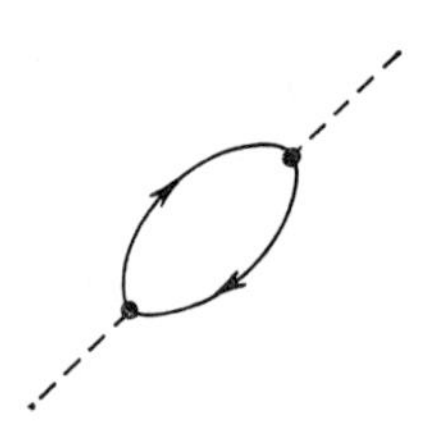

Vacuum polarization. Even in the vacuum, we may have the excitation of virtual bosons and fermion pairs, and their mutual annihilation. These processes are formally serious, because they seem to make an infinite correction to the energy of the ground state of even empty space—but can be eliminated, as will be shown in §3.9.

Particle–antiparticle interaction. This is, of course, very similar to the boson-exchange interaction between fermions (i). But notice that we can now draw another diagram (ii) having the same final consequences, in which the initial electron and positron mutually annihilate, and the virtual photon produced recreates the pair at a later time. These two diagrams are not equivalent, and must be added to give the full matrix element of the process.

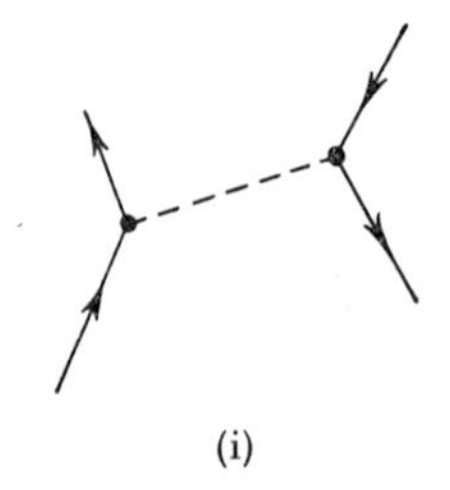

(i)

From this discussion it almost looks as if we could draw the Feynman diagram for any physical event—a reaction between fundamental particles, shall we say—by transcribing the cloud-chamber or bubble-chamber photograph of the actual phenomenon. But this would tell us only the external, real lines. There are infinitely many diagrams, corresponding to the infinite set of terms in the expansion of the S-matrix, to be constructed by the insertion of 'virtual', internal lines and vertices between the observed initial and final states. The essential problem of perturbation theory is to draw all such diagrams, and to evaluate, algebraically, their contribution to the over-all transition probability or energy.

(ii)

Consider, for example, the correction to the energy of a single fermion—the 'effective mass' calculation of (2.63). The next term in the series will be S_4, to which the following diagrams ((i)–(iv)) contribute:

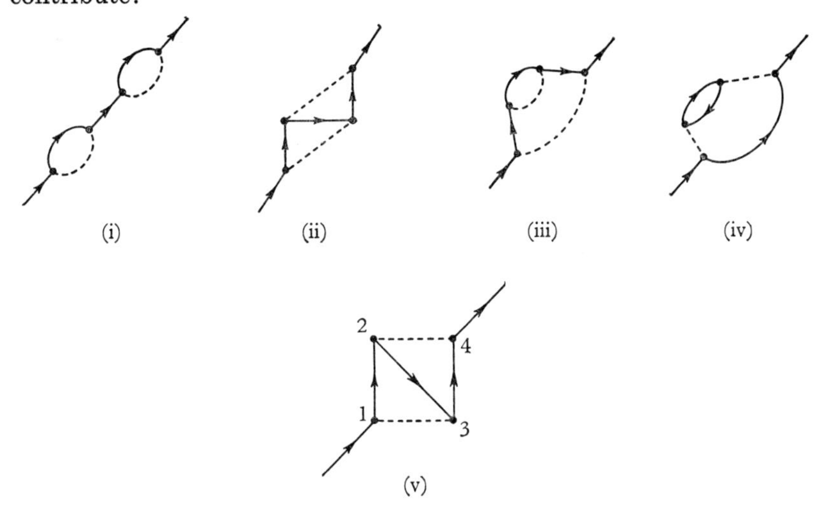

(i) (ii) (iii) (iv)

(v)

Are there any more? We might look at a process such as (v) in which a hole seems to be involved. But topologically, this is just the same as (ii). The fact that vertex 3 is now pushed back to precede vertex 2 does not alter the calculation; it is automatically taken account of in the definition of the pairing functions that occur in the integral. Here we see the power of the Feynman principle by which the time-ordering of vertices can be transformed into a convention for introducing antiparticle states when required.

3.8 Momentum representation

Having constructed all the diagrams of a given order, we need rules for writing down the algebraic formulae for the matrix elements to which they refer. These formulae are just multiple integrals over space and time of products of 'pairings', or of matrix elements of normal products. Let us now investigate these functions in more detail.

As we have seen, a pairing function corresponds to a line in a diagram, as if a particle had been propagated from one vertex to the other. We therefore define the *fermion propagator*,

$$G_0(x-x') \equiv \mathrm{i}\overline{\psi^*(x)\,\psi(x')}, \qquad (3.105)$$

and the corresponding *boson propagator*,

$$D_0(x-x') \equiv \overline{\phi(x)\,\phi(x')}. \qquad (3.106)$$

A diagram such as the second-order contribution to vacuum polarization (p. 78) corresponds to a double four-dimensional integral

$$S_2 = \frac{1}{2!}\iint G_0(x-x')\,G_0(x'-x)\,D_0(x-x')\,\mathrm{d}^4x\,\mathrm{d}^4x'. \qquad (3.107)$$

More complicated diagrams, such as 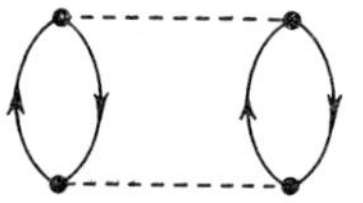 correspond to much more complicated integrals, with the various variables of integration linked by propagators in a most unpleasant tangle. But for problems involving the interactions between free fields these integrals may be somewhat simplified by going to a *momentum representation.* That is to say, we use the fact that the unperturbed Hamiltonian for each type of excitation has momentum as a good quantum number, so that in the interaction representation (cf. (2.22) and (3.41)) we have

$$\psi(x) = \sum_{\mathbf{k}} \mathrm{e}^{\mathrm{i}\mathbf{k}\cdot\mathbf{r}}\,\mathrm{e}^{-\mathrm{i}\mathscr{E}(\mathbf{k})t}\,b_{\mathbf{k}}, \quad \text{etc.} \qquad (3.108)$$

Notice that this representation would be quite inappropriate if we were dealing with a system with rotational symmetry, such as electron states in an atom, although the general theory of the S-matrix would hold up to this point.

As we have already shown in (3.99), the fermion propagator vanishes if the moment of annihilation precedes the moment of creation, i.e.

$$G_0(x-x') = 0 \quad \text{if} \quad t > t'. \qquad (3.109)$$

Otherwise, in the representation (3.108), we have from (3.105)

$$G_0(x-x') = -\mathrm{i}\langle|\sum_{\mathbf{k},\mathbf{k}'} \mathrm{e}^{-\mathrm{i}(\mathbf{k}\cdot\mathbf{r}-\mathbf{k}'\cdot\mathbf{r}')}\,\mathrm{e}^{\mathrm{i}\{\mathscr{E}(\mathbf{k})t-\mathscr{E}(\mathbf{k}')t'\}}\, b_{\mathbf{k}}^{*}\, b_{\mathbf{k}'}|\rangle$$
$$= -\mathrm{i}\sum_{\mathbf{k}} \mathrm{e}^{-\mathrm{i}\mathbf{k}\cdot(\mathbf{r}-\mathbf{r}')}\,\mathrm{e}^{\mathrm{i}\mathscr{E}(\mathbf{k})(t-t')} \quad \text{if} \quad t<t', \tag{3.110}$$

using the commutation relation (2.19).

At this point, we need an analytical expression for a function with the following properties:

$$G(\mathscr{E}) = \begin{cases} -\mathrm{i}\,\mathrm{e}^{-\mathrm{i}\mathscr{E}T} & \text{for} \quad T>0, \\ 0 & \text{for} \quad T<0. \end{cases} \tag{3.111}$$

This can be constructed by contour integration. Consider

$$G(\mathscr{E}) = \frac{1}{2\pi}\int_{-\infty}^{\infty} \frac{\mathrm{e}^{-\mathrm{i}\Omega T}}{\Omega-\mathscr{E}+\mathrm{i}\delta}\,\mathrm{d}\Omega, \tag{3.112}$$

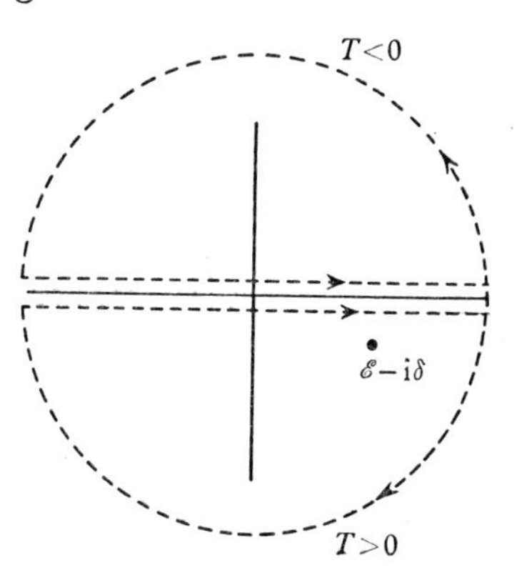

where the integral is along the real axis in the plane of the complex variable Ω, and where δ is an infinitesimal positive real number. Now if $T>0$, we can complete the contour in the lower half plane, enclosing the pole at $\mathscr{E}-\mathrm{i}\delta$ and hence giving rise to the residue of (3.111). On the other hand, when $T<0$ we must complete the contour in the upper half plane, where the imaginary part of Ω is negative and the result is zero. This proves the equivalence of (3.111) and (3.112).

When (3.112) is put into (3.109) and (3.110), it gives the following formula for the fermion propagator:

$$G_0(x-x') = \frac{1}{(2\pi)^4}\iint \frac{\mathrm{e}^{-\mathrm{i}(\mathbf{k}\cdot\mathbf{r}-\Omega t)}\,\mathrm{e}^{\mathrm{i}(\mathbf{k}\cdot\mathbf{r}'-\Omega t')}}{\Omega-\mathscr{E}(\mathbf{k})+\mathrm{i}\delta}\,\mathbf{d}^3\mathbf{k}\,\mathrm{d}\Omega. \tag{3.113}$$

The small quantity δ here is simply a convergence factor; if the fermion line had been a 'hole'-like or antiparticle state, then we should only have to change the sign of δ to keep the books straight.

We have to substitute (3.113) for each fermion propagator in an integral such as (3.107)—for each fermion line in the relevant diagram. The result of such substitutions is to produce an integrand with the following factors.

(i) Each fermion line is assigned a 'momentum' $\mathbf{k}$ and an 'energy' Ω.

(ii) The vertex at $(\mathbf{r},t)$ acquires a factor $e^{i(\mathbf{k}\cdot\mathbf{r}-\Omega t)}$ for each propagator that *enters* it with 'momentum' $\mathbf{k}$ and 'energy' Ω. It also acquires a factor $e^{-i(\mathbf{k}'\cdot\mathbf{r}-\Omega' t)}$ for each line that *leaves* it with momentum $\mathbf{k}'$ and energy Ω'.

(iii) The fermion line now contributes a four-fold integral over its 'momentum' and 'energy':

$$\frac{1}{(2\pi)^4}\iint \frac{d^3\mathbf{k}\, d\Omega}{\Omega - \mathscr{E}(\mathbf{k}) + i\delta}. \tag{3.114}$$

The theory of the boson propagator (3.106) is rather similar, though a little more complicated, and can be worked out from (3.41) and (3.90). In effect, it is the sum of two distinct propagators, corresponding to the boson going one way or another between the vertices, as determined by their time ordering. The boson line acquires a momentum $\mathbf{q}$ (whose sense we pencil in on the diagram, to avoid ambiguity), and we integrate

$$\frac{1}{2\omega_{\mathbf{q}}}\left[\frac{1}{\omega-\omega_{\mathbf{q}}+i\delta}-\frac{1}{\omega+\omega_{\mathbf{q}}-i\delta}\right] = \frac{1}{\omega^2-\omega_{\mathbf{q}}^2+i\delta} \tag{3.115}$$

over all $\mathbf{q}$ and ω. The factor $1/2\omega_{\mathbf{q}}$ comes from $(2\omega_{\mathbf{q}})^{-\frac{1}{2}}$ in (3.90), whilst the two different energy denominators stem from the two alternative processes of emission or absorption of the boson, and hence the increase or decrease of energy along the direction of the vector. As will be shown in §6.4, this formula has a simple relativistic generalization.

There may, of course, be further factors in the integrand, coming from the interaction Hamiltonian. Thus, the boson–fermion interaction (2.47) introduces, at each vertex, a form factor $F(q)$. This can easily be incorporated into the integrations as an additional factor $|F(\mathbf{q})|^2$ in the boson propagator (3.115).

At each vertex we now have only a product of simple periodic functions, in space and time, for the particles that interact there. But we are now instructed to integrate, without let or hindrance, over the space-time co-ordinates of each vertex. This can be done immediately, and gives us, say,

$$\iint e^{i(\mathbf{k}-\mathbf{k}'+\mathbf{q})\cdot\mathbf{r}}\, e^{-i(\Omega-\Omega'+\omega)t}\, d^3\mathbf{r}\, dt = \delta(\mathbf{k}-\mathbf{k}'+\mathbf{q}).\delta(\Omega-\Omega'+\omega). \tag{3.116}$$

In other words, *momentum and 'energy' are conserved at each vertex.*

The evaluation of the contribution of a diagram now proceeds as follows.

(i) Associate a momentum $\mathbf{k}$ and 'energy' Ω, or momentum $\mathbf{q}$ and 'energy' ω, with each internal fermion or boson line.

(ii) Assign these so that momentum and 'energy' are conserved at each vertex—i.e. reduce the number of independent variables of this type to a minimum consistent with the given parameters of the external lines.

(iii) Put the appropriate propagator factor, (3.114), (3.115), etc., on each line, and integrate the product over all independent momenta and 'energies'.

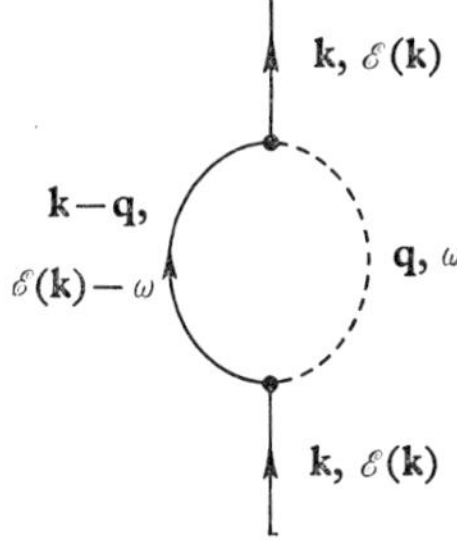

Consider, for example, the diagram for the correction to the self-energy of an electron by the emission and re-absorption of a phonon. We suppose that this electron has initially momentum $\mathbf{k}$ and energy $\mathscr{E}(\mathbf{k})$. The momentum $\mathbf{q}$ and 'energy' ω of the phonon is arbitrary, but then the parameters of the electron in the intermediate state are fixed. The matrix element must contain the following integral

$$I = \iint \frac{|F(\mathbf{q})|^2}{\omega^2-\omega_{\mathbf{q}}^2+\mathrm{i}\delta} \frac{1}{\mathscr{E}(\mathbf{k})-\omega-\mathscr{E}(\mathbf{k}-\mathbf{q})+\mathrm{i}\delta} \mathbf{d}^3\mathbf{q}\, \mathrm{d}\omega. \quad (3.117)$$

It is interesting to compare this expression with (2.63). If we integrate over ω, we can make the poles of the phonon propagator substitute either $\omega = \omega_{\mathbf{q}}$ or $\omega = -\omega_{\mathbf{q}}$ in the other energy denominator. The first case is just the second-order perturbation term of (2.63). What does the second term mean? Evidently this comes from a phonon travelling in the opposite direction, as in 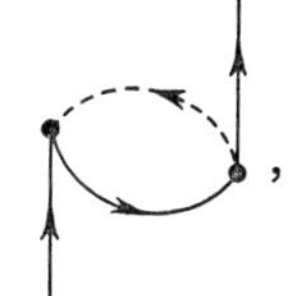,

where the process would now be described as the creation of an electron 'hole' pair followed by annihilation of the initial electron. This possibility was not envisaged in the physics of (2.63)—and would not be important in practice because of the large energy (the width of the band gap, say, in a semiconductor) required to produce the hole. Nevertheless, if we took care to define $\mathscr{E}(\mathbf{k}-\mathbf{q})$ properly for such a state this process would be correctly described and included in (3.117).

In the present chapter the functions (3.114) and (3.115) appear out of the algebra. As we shall show in §4.6, a propagator also has a deeper significance as a *Green function* of the corresponding field. Indeed, such functions take over most of the parts played by wave functions in the elementary theory.

Of course, in any proper calculation there must be further rules for the factors introduced by external lines. As we have seen, these derive from those terms in the Wick expansion that contain non-vanishing matrix elements of normal products of the field operators. In any particular case, it is not difficult to identify these matrix elements and to include them in the expression for the term in the S-matrix. For example, an incoming boson line would carry a factor $F(\mathbf{q})\,(2\omega_{\mathbf{q}})^{-\frac{1}{2}}$ into a fermion–boson interaction vertex and so on. But we leave this topic for its detailed discussion in the proper books.

Another point to notice is that the factor $1/n!$ in (3.84), say, can be dropped. The point is that when we transform a diagram into the momentum representation we have tacitly assumed a particular assignment of space-time labels, $(x_1, x_2, \ldots, x_n)$ to the n vertices. This is only one of $n!$ equivalent expressions, corresponding to the $n!$ ways of labelling these n points. Thus, the integral in (3.84) should be over all these diagrams—topologically equivalent except for these labels. But each of these figures contributes exactly the same amount to the total, so that the initial factor $1/n!$ is exactly cancelled in the final sum.

It may be objected that sometimes we draw a diagram for which the Pauli principle could be violated. It is hard to see why this should happen, because we have ensured proper antisymmetry of all fermion

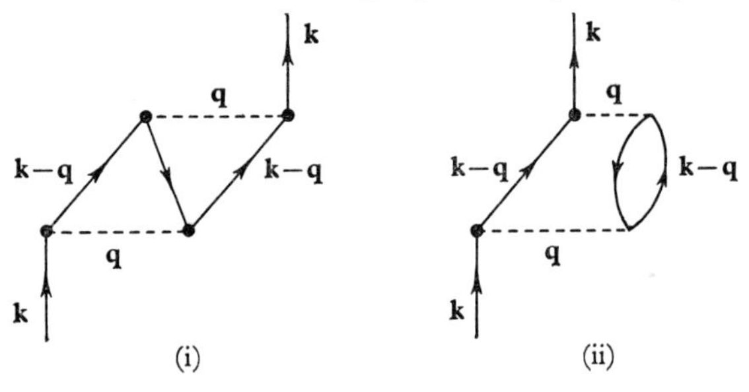

states by using the anticommuting operators $b_{\mathbf{k}}$, $b^*_{\mathbf{k}'}$, etc. Nevertheless, what are we to make of the process (i) where two bosons are interchanged, in the special case where these two have the same momentum, so that the fermion state of momentum $\mathbf{k}-\mathbf{q}$ seems to be occupied by two particles simultaneously? The answer is that the S-matrix contains another term which exactly cancels this contribution. Because of the way that we have carved up the T-product into normal products and pairings, this may occur in a topologically quite distinct diagram (ii).

In the discussion of diagrams, we have concentrated on examples of the effect of a fermion–boson interaction. There is no difficulty in setting up similar diagrams corresponding to other types of interaction. For example, a 'two-particle interaction' such as (2.42) or (2.43) gives rise to diagrams in which fermion lines are joined by, say, wavy lines corresponding to the interaction potential $\mathscr{V}(\mathbf{r}'-\mathbf{r})$ or its Fourier transform, $\mathscr{V}(\mathbf{k}''-\mathbf{k}''')$. The topology of such diagrams is similar to that

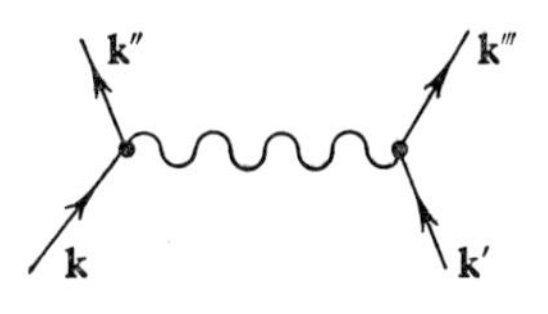

of the fermion–boson diagrams, except that there are no external wavy lines, and we can include 'loop' terms like

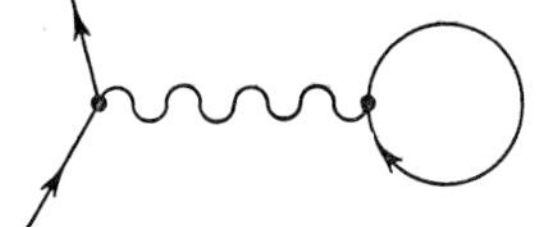

which actually give zero contribution in the fermion–boson case. In these diagrams the fermion propagator contributes as in (3.114), but we have to assign a momentum transfer $\mathbf{q}$ to the wavy line, and introduce a factor $\mathscr{V}(\mathbf{q})$ in the matrix element.

The essential point is that the diagrammatic technique can be applied quite generally to any perturbation problem as a systematic procedure for generating the successive terms in the perturbation series. The structure of the diagrams, the types of line that they may contain, the rules for the lines that may occur at each vertex, sign conventions, etc. depend upon the particular system, or upon the choice of representation. But some general features persist. The conservation of momentum at each vertex is always assumed, and also a rule about adding energies. These rules reflect the most elegant feature of the theory—the elimination of all space and time co-ordinates for the interaction events.

But it must be emphasized that a Feynman diagram represents only a succession of *virtual* processes. In the assignment of the variable Ω or ω to a fermion or boson propagator we have called it the 'energy', because it has this physical dimension. But the propagator (3.114) or (3.115) contains differences between this variable and the true, physically realizable energy, $\mathscr{E}(\mathbf{k})$ or $\omega_{\mathbf{q}}$, of a free particle of that momentum. This real energy is not conserved at each vertex, although, of course, it must be conserved over the diagram as a whole, from incoming to outgoing external lines. We recognize, of course, the typical energy denominator of elementary perturbation theory—the difference between the energy of the initial state, say, and of the intermediate state that is supposed to exist, briefly, before it decays into the final state of the particle.

3.9 The physical vacuum

The diagrammatic technique is a machine for constructing, systematically, all the terms of given order in a perturbation series. But a term-by-term approach does not always help, even if we could find the energy and skill to draw all the diagrams of, say, the 16th order and evaluate the integrals which they define.

In the first place, a finite number of finite terms does not give rise to essentially new physical phenomena; if each term is continuous and non-singular, then the sum will not have any singularities corresponding, perhaps, to a phase transition in a many-body problem, or to a bound state.

On the other hand, if an early term in the series is infinite—and, as we saw in §§ 1.11 and 2.7, this can happen in quite elementary field-theoretical systems—we cannot use any procedure for eliminating or neutralizing it unless we can show that the infinities in all higher terms are also removed by the same trick.

The peculiar and striking advantage of the diagrammatic representation of a perturbation series is that it allows us to *reduce* and partially sum the infinite series by elementary topological arguments, without recourse to algebra. As we shall see (for example, in §§ 5.5 and 5.9) this power is really more valuable than the algorithmic capacity for producing formulae for various specific terms.

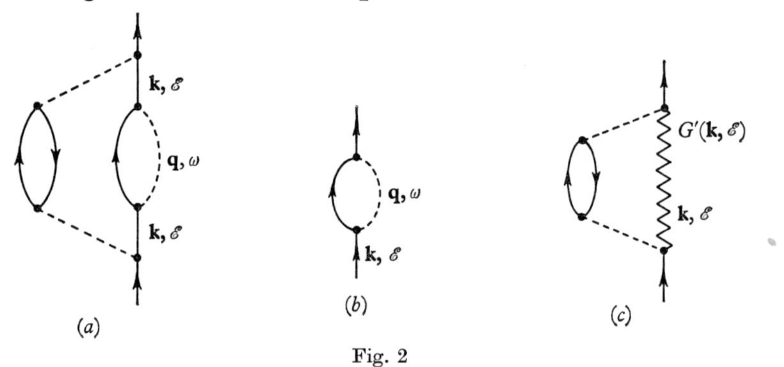

Fig. 2

Suppose, for example, that we have to evaluate a complicated diagram such as fig. 2 (*a*), and that we had assigned some of the momenta and 'energy' parameters as shown. Then we might begin the calculation by integrating over the internal boson line $\mathbf{q}$, before tackling the other factors. But this integration will involve only the propagators in this sub-diagram 2 (*b*)—just as in (3.117). The result will be a function

$G'(\mathbf{k}, \mathscr{E})$ say, of the supposed momentum and 'energy' of the incoming and outgoing electron.

This means that we may calculate the contribution of the big diagram as if it were of the form 2 (c), where a zig-zag line is drawn in place of the sub-diagram 2 (b). All we need to do is to associate with the zig-zag line the *modified propagator* $G'(\mathbf{k}, \mathscr{E})$ instead of the usual free-particle propagator.

But of course there are lots of other possible sub-diagrams that might occur in place of an ordinary propagator. For example, we might have

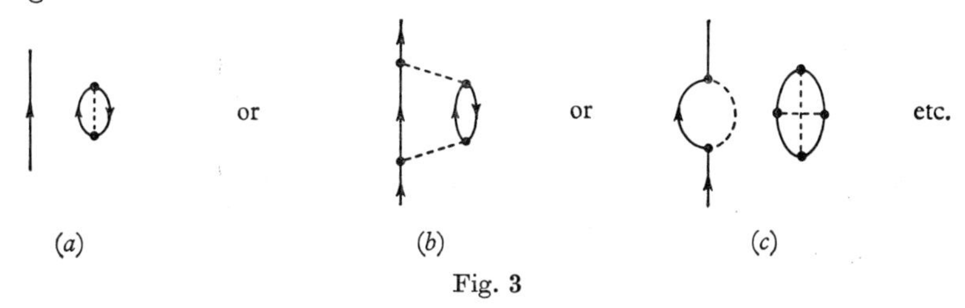

Fig. 3

Each one of these would define a modified propagator that should be inserted into the larger diagram in the appropriate place.

Now think of the sum of all these sub-diagrams, with just two external fermion lines. This is, of course, an infinite sum; but suppose we could evaluate the sum of the corresponding modified propagators, i.e.

$$G''(\mathbf{k}, \mathscr{E}) = \sum_{\text{all self-energy diagrams}} G'(\mathbf{k}, \mathscr{E}). \tag{3.118}$$

What would it mean to use $G''(\mathbf{k}, \mathscr{E})$ for the propagator of the zig-zag line of fig. 2 (c)? It would be equivalent to calculating the sum of all the diagrams that one could make by replacing this line by any of the sub-diagrams of fig. 3—by any sub-diagram corresponding to the possible history of a fermion let loose in the vacuum. Thus, suppose we assign to $G''(\mathbf{k}, \mathscr{E})$ a special type of line $\Longrightarrow$. Then a picture such as is not just a single diagram of the type that we considered in the previous section; it is the sum of an infinite set of such diagrams.

The propagator (3.118) has, however, a straightforward physical significance. It represents every thing that can happen to a free particle, all virtual excitations, etc. as it moves in empty space. Thus, it includes all the effects of such excitations on the energy of the fermion, and is called the *self-energy* part.

We can obviously do the same sort of thing for a boson, and define a new type of line = = = = = =, with propagator D'', corresponding to the sum of all diagrams beginning and ending with just one boson line, and hence capable of replacing the free boson propagator D_0. Again, we can think of all diagrams equivalent to a single vertex, as in

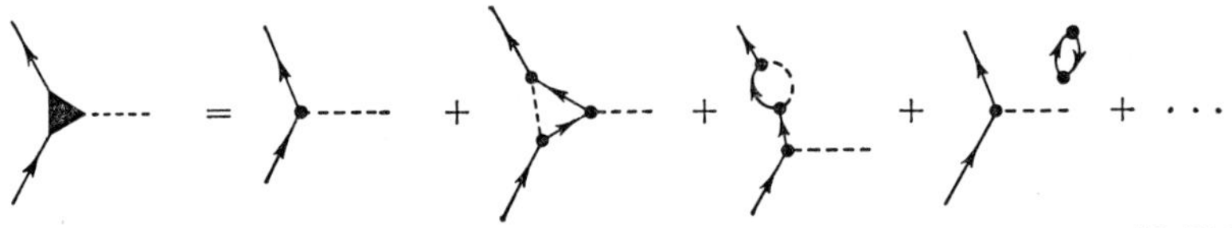

(3.119)

and use a special diagrammatic symbol to indicate that the whole of this sum has been put in place of the usual single vertex. One can easily see that at this modified vertex the conservation of momentum and energy must still hold, but the integral would acquire a special factor, the *vertex part*, $F''(\mathbf{k}, \mathbf{q})$ say, like the form factor in (2.47).

Finally, there will be a *vacuum part*, defined as the sum of all closed graphs, without particles entering or leaving. This would give us S_{vacuum},

the S-matrix for the effect of all polarization fluctuations, excitations of virtual pairs, etc., in empty space.

But topologically there are relations between these various types of sum. Consider, for example, the whole set of self-energy graphs. This will include terms like 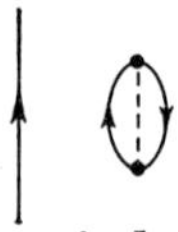 , etc., in which there are fluctuations of the vacuum going on independently of the propagation of our fermion. Now if two parts of a diagram are not connected by any propagator lines, they contribute as quite separate factors, with quite independent variables of integration, in the matrix element of the S-matrix. Thus

$$S_{(AB)} = S_A S_B \tag{3.120}$$

if (AB) is a diagram obtained by just putting two diagrams A and B on to the same page without connecting them.

Think now of the sum of all *connected* self-energy graphs. Any one of these could be combined with any vacuum diagram to give a term in the full collection of self-energy terms. This is to say, in $G''(\mathbf{k},\mathscr{E})$ of (3.118) we get all possible products of the propagator for a connected graph with the S-matrix for a vacuum graph. In other words, if we define another type of fermion propagator

$$G(\mathbf{k},\mathscr{E}) = \sum_{\text{all } connected \text{ self-energy graphs}} G'(\mathbf{k},\mathscr{E}), \tag{3.121}$$

we find the simple algebraic relation

$$G'' = GS_{\text{vacuum}}. \tag{3.122}$$

In a similar way, we can define new boson propagators and vertex parts, D and F, such that only connected graphs are counted in the sums, satisfying similar equations

$$D'' = DS_{\text{vacuum}} \quad \text{and} \quad F'' = FS_{\text{vacuum}}. \tag{3.123}$$

There is a general relationship for the S-matrix:

$$S_{\text{all graphs}} = S_{\text{connected graphs only}} \cdot S_{\text{vacuum}}. \tag{3.124}$$

Let us return now to our basic definition of the S-matrix, (3.61),

$$\begin{aligned} |\Psi_{\text{final}}\rangle &= S\,|\Psi_{\text{initial}}\rangle \\ &= S_{\text{connected}}\,(S_{\text{vacuum}}|\Psi_{\text{initial}}\rangle) \\ &= S_{\text{connected}}|\Psi'\rangle. \end{aligned} \tag{3.125}$$

What is the meaning of this new type of state, defined by

$$|\Psi'\rangle = S_{\text{vacuum}}\,|\Psi\rangle? \tag{3.126}$$

This is a state of the system *as we should really observe it*, including not only some real prescribed particles but also with all fluctuations, virtual pairs, etc. produced in the vacuum by the interaction terms. In fact, this is the true *physical* vacuum of the system; the hypothetical state $|\Psi\rangle$ in which all these fluctuations and excitations have been suppressed, is only a mathematical construction, not an observable condition of the field. So we should always calculate relative to the state $|\Psi'\rangle$ if we want to describe real processes going on in 'real' empty space. That is easy: (3.125) tells us to count *only connected graphs* in the computation of matrix elements, propagators, etc.

Note, however, that this 'correction' is often infinite, or seems to correspond to an infinite change in the basic energy of the system. What this means is that the hypothetical state $|\Psi\rangle$ is not merely inaccessible physically; we have no mathematical description of it in terms of $|\Psi'\rangle$ states. But since this is a constant feature of the theory, and S_{vacuum} is the same for all states, this need cause no real difficulty.

3.10 Dyson's equation and renormalization

From the arguments of the previous section, it is clear that we can only describe processes involving real 'physical' particles in terms of propagators like G and D, corresponding to sums over infinite sets of connected diagrams. Thus, a proper account of, say, the Compton effect would require a diagram such as (i), where all fermion lines, bosons lines, and vertices are drawn as if with all their self-energy corrections, etc. (but now, of course, excluding all vacuum parts). Such a graph is, of course, the sum of an infinity of simple diagrams, but it does not necessarily include *all* simple diagrams that contribute to the phenomenon. Thus, for example, the diagram (ii) would not be counted, and would have to be made the basis of a further infinite set of simple diagrams by using modified propagators and vertex parts.

(i)

Nevertheless, knowledge of the modified propagators would take us a long way towards a solution of the problem. There are several tricks, exploiting the topology of the diagrams, that help.

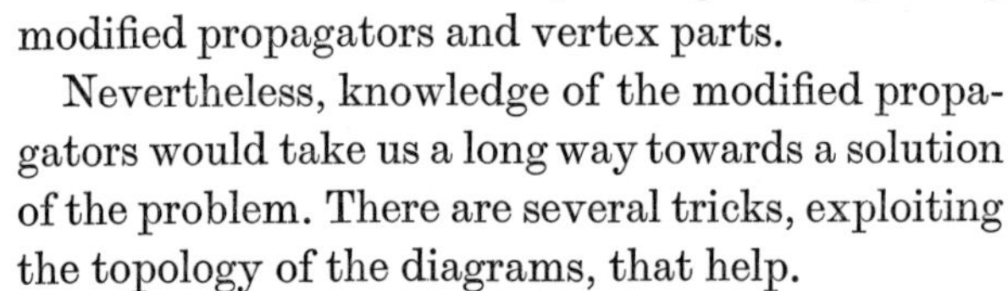

(ii)

For example, suppose we distinguish between *reducible* and *irreducible* self-energy parts. A reducible graph is one which may be severed in two by cutting just one line—for example (iii). Suppose that we write

$$\Sigma = \Sigma_1 + \Sigma_2 + \dots \qquad (3.127)$$

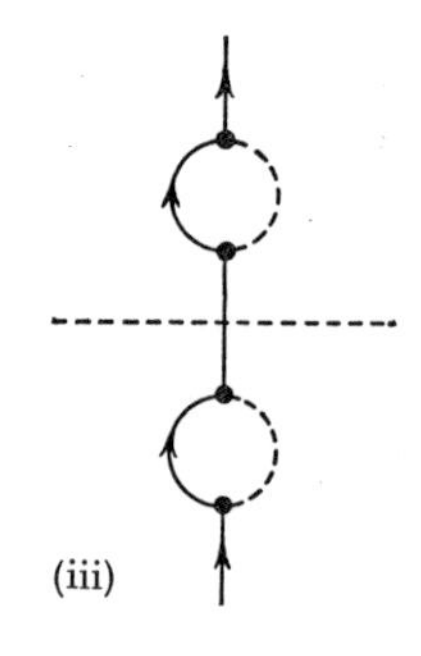

(iii)

for the sum of all different irreducible graphs in the self-energy part G for the propagation of a fermion. Then we have the following algebraic relation

$$G = G_0 + G_0 \Sigma G. \qquad (3.128)$$

This can be proved quite simply. Think of any diagram

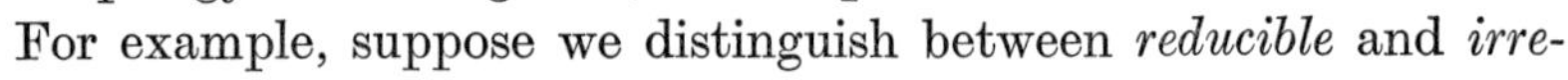

contributing to G. This could be (trivially) just G_0. Otherwise, it is a string of irreducible sub-diagrams Σ_i, Σ_j, etc. linked by free-particle propagators G_0. Suppose now we cut off the first of these, Σ_i. We are left with a diagram beginning and ending with G_0—in fact, just another diagram out of the set G. So we can make up any diagram in G by linking any irreducible diagram out of Σ with any diagram in G. This means that the sum of all diagrams in G can be calculated by taking the product of the sum of all diagrams in Σ with the sum of all diagrams in G. This gives us a *Dyson equation* (3.128).

Notice that this is not merely a symbolic relation concerning sets of diagrams; the symbols G, G_0 and Σ are functions of $(\mathbf{k}, \mathscr{E})$ the momentum-'energy' parameters of the particle, and the multiplication is ordinary algebraic multiplication of these factors as they might occur in the integrand of an element of the S-matrix. Thus, we can solve (3.128), and get

$$G = (G_0^{-1} - \Sigma)^{-1}. \tag{3.129}$$

Now using (3.114) we find

$$G(\mathbf{k}, \mathscr{E}) = \frac{1}{\mathscr{E} - \mathscr{E}(\mathbf{k}) - \Sigma(\mathbf{k}, \mathscr{E}) + \mathrm{i}\delta}. \tag{3.130}$$

This result is very elegant because it shows that the propagator of a 'physical' fermion is mathematically similar to that of an abstract 'free' fermion, except for a change in the energy. That is to say, a physical fermion behaves as if it had the energy

$$\mathscr{E}'(\mathbf{k}) = \mathscr{E}(\mathbf{k}) + \Sigma(\mathbf{k}, \mathscr{E}). \tag{3.131}$$

In other words, we may write

$$G(\mathbf{k}, \mathscr{E}) = \frac{1}{\mathscr{E} - \mathscr{E}'(\mathbf{k}) + \mathrm{i}\delta}, \tag{3.132}$$

just as if it were an elementary free-particle propagator G_0 for a particle with the 'observed' energy $\mathscr{E}'(\mathbf{k})$.

In relativistic calculations, $\mathscr{E}(\mathbf{k})$ becomes a measure of the mass of the particle in the state $\mathbf{k}$. We therefore make the transformation equivalent to (3.131),

$$m' = m + \mathrm{d}m, \tag{3.133}$$

and treat m' as the physical mass of the particle. We then say that the mass has been *renormalized*. There are similar procedures for other parameters of the system, such as the charge.

Here again, we run into the difficulty that $\Sigma(\mathbf{k}, \mathscr{E})$ or $\mathrm{d}m$, may turn out to be infinite, so that we have no mathematical description of the properties of the 'bare' particle, even though we can use the renormalized energy or mass to describe the physical object that we actually observe.

The real problem is to make such arguments fully self-consistent. We need to find quantities that can be treated as the renormalized mass and charge (say) in *all* terms of the S-matrix, in *all* orders of perturbation theory. The investigation of these possibilities, the identification of infinite terms and their systematic elimination, goes beyond our present discussion. It turns out that some systems—especially quantum electrodynamics—can be renormalized whilst others cannot. Perhaps the various examples of fields and their interactions given in § 2.7 may suggest reasons for the differences—the infinities arise, eventually, from the attempt to evaluate non-convergent integrals over all momentum space, and hence depend on form factors and other basic properties of the interacting fields.

Of course, when we have removed the infinities, we still have to work quite hard to evaluate the terms that describe various physical processes, such as the Lamb shift. These may correspond to diagrams of fairly high order, whose enumeration and computation is not a trivial task.

As a final example of the power of the topological approach, let us construct another Dyson equation. This is best represented graphically as follows:

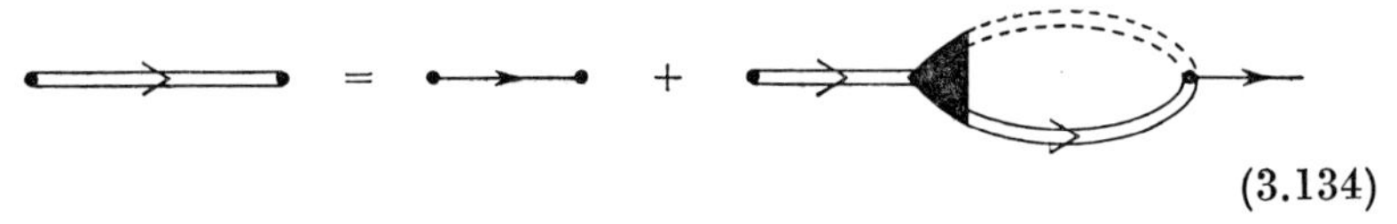

(3.134)

Thus, every diagram in the true fermion propagator is included in the set of diagrams obtained by combining a true fermion propagator with a true boson propagator at a true vertex into a self-energy type of term. This equation is topologically related to the argument for (3.128).

We can express this relation analytically as follows:

$$G(\mathbf{k}) = G_0(\mathbf{k}) + G(\mathbf{k}) \int F(\mathbf{k}, \mathbf{q})\, G(\mathbf{k}-\mathbf{q})\, D(\mathbf{q})\, \mathbf{d}^4\mathbf{q}\, G_0(\mathbf{k}) \quad (3.135)$$

(supposing, for simplicity, that 'energy' is treated as a fourth component of momentum). This is an integral equation, from which $G(\mathbf{k})$ might be calculated if we knew the true boson propagator $D(\mathbf{q})$ and the true vertex part $F(\mathbf{k}, \mathbf{q})$. There is, in fact, a similar integral equation for $D(\mathbf{q})$, corresponding to a similar topological relation for the graphs giving the self-energy of a boson. Unfortunately, the necessary third equation, defining the vertex part in terms of itself and the other propagators, has not been invented; we only have a succession of more and more complicated diagrams to evaluate. But (3.135) might serve as the basis for an approximation scheme, in which particular vertex diagrams are put down for F, and the integral equations solved for $G(\mathbf{k})$ and $D(\mathbf{q})$. Such a procedure may be more powerful than, say, evaluating the successive irreducible diagrams in (3.127).

It is interesting to note the abstract similarity between the various Dyson equations and the Brillouin–Wigner expansion (3.20). Thus, in (3.135) the unknown energy that we want to find to put in the propagator on the left-hand side occurs in a denominator in the integral on the right. The comparison must not be taken too literally, but it suggests that the Brillouin–Wigner series constitutes a partial summation of the terms in the ordinary Rayleigh–Schrödinger perturbation series—and that we can often arrive at many of the results of diagrammatic analysis by direct self-consistency arguments.

CHAPTER 4

GREEN FUNCTIONS

...we have our philosophical persons, to make modern and familiar things supernatural and causeless.

All's Well that Ends Well

4.1 The density matrix

A quantum system is never truly isolated from the remainder of the Universe; we can never entirely ignore the contact that it makes with other systems. But these contacts can never be defined precisely, for then we should have to solve the equations of motion of all the particles in the Universe simultaneously. In other words, the initial and other boundary conditions on the system can only be known imperfectly, in a statistical sense. A theory that always assumes complete information about the wave functions is too exact. How, for example, do we deal with a system whose energy is only approximately constant, because it is in intimate contact with a 'heat bath' or other fluctuating system? We need some analogue of the ensemble-averaging procedure of classical statistical mechanics.

Consider a system with states prescribed in terms of some abstract dynamical variables, which we label x. Let $A(x)$ be the quantum-mechanical operator corresponding to some physical observable. Then the average expectation value of the observable in the state $\psi(x)$ is given by

$$\langle A \rangle = \int \psi^*(x)\, A(x)\, \psi(x)\, \mathrm{d}x. \tag{4.1}$$

But we may not be in a position to assign a state function $\psi(x)$ to the system. For example, the observable may refer to a small region within a large volume of gas, where molecules are always entering and leaving, making collisions, etc. To deal with this case in the conventional framework of quantum theory we should have to embed our system in a much larger assembly which we could assume to be closed —for example, the whole large volume of gas—and which could thus be assigned a state function. Suppose that this assembly requires the specification of an extra set of dynamical variables, q; then we may suppose the existence of a total state function $\Psi(q, x)$ with well-defined equations of motion, sharp boundary conditions, etc. To evaluate the average expectation of the observable $A(x)$, we now need to integrate over both x and q:

$$\langle A \rangle_{\text{ensemble}} = \iint \Psi^*(q, x)\, A(x)\, \Psi(q, x)\, \mathrm{d}x\, \mathrm{d}q. \tag{4.2}$$

We shall call this, in general, an *ensemble average*, upon the supposition that the sort of large assembly that we usually have in mind is a fairly well coupled set of replicas of our original system, as in ordinary statistical mechanics.

Now suppose that we are dealing with a number of quantum mechanical observables that, like $A(x)$, operate only on the variable of our chosen small system. Averages like (4.2) will occur, with always the same quantity to be integrated over the external ensemble variables q. Let us define this standard average as the *density matrix* of the system:

$$\rho(x, x') \equiv \int \Psi^*(q, x')\, \Psi(q, x)\, \mathrm{d}q. \tag{4.3}$$

(It is important, for algebraic consistency, to note the inverted order of the variables in this prescription.)

With this convention, we should write (4.2) in the form

$$\langle A \rangle_{\text{ensemble}} = \int A(x)\, \rho(x, x)\, \mathrm{d}x. \tag{4.4}$$

For an ordinary function $A(x)$, this is clear enough, but when A is an operator, this is not meaningful. Generally speaking, the observable A has a matrix representation $A(x', x)$ in the chosen dynamical variables of our system—it may be in momentum, angular momentum, or even ordinary spatial co-ordinates. We may then calculate as follows:

$$\begin{aligned} \langle A \rangle_{\text{ensemble}} &= \iiint \Psi^*(q, x')\, A(x', x)\, \Psi(q, x)\, \mathrm{d}x\, \mathrm{d}x'\, \mathrm{d}q \\ &= \iint A(x', x)\, \rho(x, x')\, \mathrm{d}x\, \mathrm{d}x' \\ &= \int [A\rho]_{x'x'}\, \mathrm{d}x' \\ &= \operatorname{Tr}[A\rho]. \end{aligned} \tag{4.5}$$

Here we recognize the standard rules for matrix multiplication, and the subsequent definition of the *trace* of the product matrix, i.e. the sum of diagonal elements.

The beauty of this formula is that it is invariant in form. There is a standard theorem of matrix algebra that the trace of a matrix is a canonical invariant, that remains the same under unitary transformation to any other representation. Any such transformation simply changes the matrix elements of A and ρ in such a way that (4.5) is still true; the density matrix can be treated as a 'geometrical object' in the Hilbert space of the variables x whose actual components depend upon the choice of representation that is most convenient for our problem.

Some obvious properties of ρ follow at once. For example $\rho(x, x')$ is a Hermitian matrix, with trace unity, i.e.

$$\begin{aligned} \mathrm{Tr}\,[\rho] &= \int \rho(x, x)\,\mathrm{d}x \\ &= \iint \Psi^*(q, x)\,\Psi(q, x)\,\mathrm{d}q\,\mathrm{d}x \\ &= 1. \end{aligned} \tag{4.6}$$

In an ordinary space representation, where x might refer to the co-ordinate of a single particle, we have

$$\rho(x, x) = \int |\Psi(q, x)|^2\,\mathrm{d}q, \tag{4.7}$$

which is obviously just the probability of finding the particle at x after averaging over all the 'other' contingent variables of the ensemble—in other words just the ordinary particle (probability) density of the Born interpretation of quantum mechanics. But note that this is *not* enough to specify expectation values of observables, such as the momentum operator, which are not diagonal in a space representation. It is essential to know about the correlation and fluctuation effects that are implied by the existence of off-diagonal elements of $\rho(x, x')$. On the other hand, for a member of an ensemble, we cannot learn more than is given by the density matrix; it is all we know, and all we need to known, about the 'state' of the system.

In the Dirac notation, the density matrix becomes a *density operator*

$$\rho = \sum_{mn} |m\rangle \rho_{mn} \langle n|. \tag{4.8}$$

Suppose that the basis states $|m\rangle$ were in fact the eigenstates of this (Hermitian) operator. We should then have, simply

$$\rho = \sum_{m} |m\rangle \rho_m \langle m|, \tag{4.9}$$

showing that the density operator is a generalized projection operator (cf. § 3.1).

In this representation, the ensemble average of any observable takes the form

$$\langle A\rangle_{\text{ensemble}} = \sum_m \rho_m \langle m|\, A\,|m\rangle, \tag{4.10}$$

just as if the system could be thought of as having probability ρ_m of being in one or other of the states $|m\rangle$, where the observable A then appears, on the average, to have its expectation value, without regard to correlation effects involving phase relations with other states. In this special case, therefore, the eigenvalues of the density matrix become equivalent to a classical distribution function.

Since each ρ_m is a probability, it must be positive. From (4.6) we get a useful general rule:

$$\sum_{mn} \rho_m^2 \leqslant 1. \tag{4.11}$$

This is, in fact, just the sum of the diagonal elements of the operator ρ^2, which is algebraically invariant. Thus, in general,

$$\sum_{mn} \rho_{mn}^2 = \operatorname{Tr}[\rho^2] \leqslant 1 \tag{4.12}$$

constrains the elements of the density matrix in any representation.

From its basic definition, the density matrix takes a special form when the system has a definite wave function. This would arise if, for example, the ensemble were to be decoupled into its separate elements, so that the total wave function could be factorized, i.e.

$$\Psi(q, x) = \Phi(q)\,.\,\psi(x). \tag{4.13}$$

We may choose $\psi(x)$ as the basis state $|0\rangle$ of a representation of ρ. It then turns out, by elementary arguments based on (4.3), that the density operator becomes just the projection operator

$$\rho = |0\rangle\,\langle 0|, \tag{4.14}$$

which is diagonal, with eigenvalue $\rho_0 = 1$ and all other elements zero. Symptoms of this special case are that (4.12) attains its upper limit, and that ρ is idempotent:

$$\operatorname{Tr}[\rho^2] = 1; \quad \rho^2 = \rho. \tag{4.15}$$

When the density matrix satisfies these conditions the system is said to be in a *pure state*, and is subject to the ordinary laws of quantum mechanics, with full information supposed about initial conditions, etc. The algebra of expectation values is sometimes simplified and clarified by the use of the density operator formalism in such cases, although this does not really differ from what can be derived from the Schrödinger equation itself. The point is that the trace formula (4.5) expresses the maximum information that can be derived about any physical observable, and hence avoids errors that might arise from the over-definition of unobservable quantities, such as the absolute phase of the wave function.

4.2 Equation of motion of density operator

Consider the general case of a density operator defined for some arbitrary set of basis states, as in (4.8). Now suppose these states vary with time, according to a time evolution operator as in (3.27)

$$|m,t\rangle = U(t)\,|m,0\rangle. \tag{4.16}$$

Then the density operator itself becomes a function of time:

$$\begin{aligned}\rho(t) &= \sum_{mn} |m,t\rangle \rho_{mn} \langle n,t| \\ &= \sum_{mn} U(t)\,|m,0\rangle \rho_{mn} \langle n,0|\, U^*(t) \\ &= U(t)\,\rho(0)\,U^*(t)\end{aligned} \tag{4.17}$$

in the Schrödinger representation.

The Heisenberg representative of a dynamical variable transforms in a somewhat similar fashion (3.34), i.e.

$$A_H(t) = U^*(t)\,A_H(0)\,U(t). \tag{4.18}$$

For an isolated system where the states obey a Schrödinger equation with Hamiltonian H, this is equivalent to the equation of motion (3.36), i.e.

$$i\hbar\frac{\partial A_H(t)}{\partial t} = [A_H(t), H]. \tag{4.19}$$

Applying the same argument, but noting that the order of the time evolution operators is reversed in (4.17), we easily arrive at the equation

$$i\hbar\frac{\partial \rho(t)}{\partial t} = [H, \rho(t)]. \tag{4.20}$$

This is the *equation of motion of the density operator*. The difference between (4.19) and (4.20) in the order of the terms in the commutator is a clear sign that ρ is not a dynamical variable but is a generalized state function, and therefore varies with time in the Schrödinger representation. In the Heisenberg representation the density operator would, of course, remain constant with time.

The above equation can confidently be used for most systems where the general ensemble theory of statistical mechanics is applicable.

Nevertheless, it must be admitted that its precise validation from (4.3), say, where we should only be allowed to discuss the true variation of the state function of the whole ensemble, and where the separation out of a Hamiltonian for each elementary system is not allowed, implies a great deal of careful discussion of assumptions of random phase, ensemble averaging, etc. This equation is, in fact, the analogue of Liouville's equation for the variation of the classical probability density in phase space:

$$\frac{\partial \rho}{\partial t} = -\{\rho, H\}, \tag{4.21}$$

where { } is a Poisson bracket.

4.3 Ensembles in thermal equilibrium

To connect quantum theory with statistical mechanics and thermodynamics, we derive the *canonical density matrix*, corresponding to the canonical distribution of classical theory. We ask for the conditions under which our system will 'remain in equilibrium', i.e. appear unaltered, if isolated from the heat bath and allowed to evolve under its own steam. This obviously requires that the density operator should not change with time; by (4.20) the commutator

$$[H, \rho] = 0 \tag{4.22}$$

must vanish. In other words, the density operator must be a function of the Hamiltonian operator of the system.

The derivation of an appropriate function then follows the same general path as in classical theory. For example, we may argue that two separate systems are at the same temperature if they are in equilibrium with the heat bath, and may be combined into a composite system, whose Hamiltonian is just the sum of the separate Hamiltonians. On the other hand, the density operator of the composite system is a product (strictly speaking, an 'outer product') of the two separate density operators of the component systems. It follows that the relationship between ρ and H must be of the functional form

$$\rho = \alpha\, \mathrm{e}^{-\beta H}, \tag{4.23}$$

where α is a normalization constant, whilst β measures a property shared with the heat bath—i.e. it is a function of the temperature T. More generally, for a system with variable numbers n_i of particles or other elements, one can define a *grand canonical density matrix* of the form

$$\rho = \exp\left(-q + \beta \sum_i n_i \mu_i - \beta H\right), \tag{4.24}$$

where q depends on the normalization and μ_i is the chemical potential of the ith species of particle.

To verify the equivalence with classical theory, let us choose the energy eigenstates

$$H\,|m\rangle = \mathscr{E}_m\,|m\rangle \tag{4.25}$$

as the basis of representation. Then ρ also is diagonal, and takes the form

$$\rho = \alpha \sum_m |m\rangle\, \mathrm{e}^{-\beta\mathscr{E}_m} \langle m| \tag{4.26}$$

just as if the probability of finding the system with energy $\mathscr{E}_m$ were proportional to the familiar Boltzmann factor

$$\begin{aligned} \rho_m &= \alpha\, \mathrm{e}^{-\beta\mathscr{E}_m} \\ &= Z^{-1}\, \mathrm{e}^{-\mathscr{E}_m/\boldsymbol{k}T}. \end{aligned} \tag{4.27}$$

The identification of β with $1/\boldsymbol{k}T$ may be recognized at once, by comparison with thermodynamic definitions of temperature, and turns out to be just the inverse of the partition function Z. Some definitions of thermodynamic quantities may be rephrased in quantum language. It is well known, for example, that the entropy of a classical system distributed with probability p_n into separate classes is given by the general formula

$$\begin{aligned} S &= -\boldsymbol{k} \sum_n p_n \ln p_n \quad \text{over classical states} \\ &= -\boldsymbol{k} \sum_n \rho_n \ln \rho_n \quad \text{over eigenstates of } \rho \\ &= -\boldsymbol{k}\, \mathrm{Tr}\, [\rho \ln \rho] \end{aligned} \tag{4.28}$$

which is now independent of the representation of ρ. The entropy is thus a genuine mathematical invariant of a quantum system. But it must be noted that various conditions and theorems must be satisfied, such as thermal equilibrium, equivalence of time and ensemble averaging, random phasing of state function, etc. before these abstract definitions can be put to use.

In the actual evaluation of thermodynamic averages, the similarity of (4.23) to the expression (3.33) for the time evolution operator

$$U = \mathrm{e}^{-\mathrm{i}(t/\hbar)\, H} \tag{4.29}$$

has often been used. Thus, by putting

$$\beta = \mathrm{i}\tau/\hbar \tag{4.30}$$

one can treat the inverse temperature as an imaginary time variable, and use all the apparatus of perturbation series, diagrams, etc. as sketched in chapter 3. This is at the heart of the technique known as the method of *temperature Green functions* (see §4.7) which has been found very powerful in the theory of the many-body problem. It depends upon the fact that many of the formulae involving propagators, etc. are well-behaved analytically, so that their continuation along the imaginary axes presents no serious difficulties. However, it must be admitted that these methods have not been given a 'physical' interpretation that may be grasped intuitively, and belong to a higher level of mathematical sophistication than can be reached in this book.

4.4 The Kubo formula

The general 'Boltzmann distribution' density matrix (4.23) or (4.24) is appropriate to a system in thermal equilibrium, and is a convenient starting point for the investigation of such quantities as the average energy, or magnetic moment, or quadrupolar field fluctuation in such a system. But we are also interested in the effect of 'external' forces on a system, causing it to flow, conduct heat or electricity, or otherwise respond to the stimulus. The whole question of the formulation of *transport theory* in strict quantum-mechanical language is very difficult and subtle.

Nevertheless, under appropriate conditions, it is possible to make a direct connection with the Onsager theory of irreversible processes, in which it is argued that the response to a general thermodynamic 'force' must be a general thermodynamic 'current' proportional to the force. For example, an oscillating electric field $\mathbf{E}$ of frequency ω must give rise to an electric current $\mathbf{j}$ with cartesian component amplitudes

$$j_{\mu} = \sum_{\nu} \sigma_{\mu\nu}(\omega)\, E_{\nu}, \tag{4.31}$$

where $\sigma_{\mu\nu}(\omega)$ would be the (complex) conductivity tensor of the system. This generalization of Ohm's Law is not *necessarily* true, but should always hold for sufficiently weak applied fields.

To demonstrate this, let us suppose we have a system in equilibrium, with Hamiltonian H_0, so that initially

$$\rho_0 = Z^{-1} \exp\{-\beta(H_0 - \mu N)\}. \tag{4.32}$$

Now apply the perturbation due to the electric field:

$$H'(t) = \mathrm{e}^{\mathrm{i}\omega t}\, \mathbf{E}\cdot\mathbf{X}, \tag{4.33}$$

where we write $\mathbf{X} = \sum_{i} e\mathbf{x}_{(i)}$ as a general symbol for the electrical polarization of the system, which might be, say, a gas of charged particles with co-ordinates $\mathbf{x}_{(i)}$.

The density matrix must now vary with time according to the equation of motion (4.20), i.e. (with $\hbar = 1$, for simplicity)

$$\mathrm{i}\dot{\rho}(t) = [H_0 + H'(t), \rho(t)]. \tag{4.34}$$

This is essentially the same problem as the calculation of the rate of change of a Heisenberg operator in the interaction representation (§ 3.3). We are only interested in terms linear in $H'(t)$, so that we may write

$$\rho(t) = \rho_0 + \Delta\rho(t), \tag{4.35}$$

and drop any terms like $[H'(t), \Delta\rho]$. To this order (i.e. treating the applied field as a first-order perturbation) the equation of motion (4.34) becomes

$$\mathrm{i}\Delta\dot{\rho} \approx [H_0, \Delta\rho(t)] + [H'(t), \rho_0]. \tag{4.36}$$

The second term is now just the inhomogeneous part of a simple time-evolution equation for $\Delta\rho(t)$; elementary calculus gives us the solution

$$\Delta\rho(t) = -\mathrm{i}\int_{-\infty}^{t} \mathrm{e}^{-\mathrm{i}H_0(t-t')}[H'(t'), \rho_0]\,\mathrm{e}^{\mathrm{i}H_0(t-t')}\,\mathrm{d}t'. \tag{4.37}$$

To start this equation off properly in the past, and to keep it convergent, we should assume that ω in (4.33) has a small imaginary part, as in (3.50). We want to calculate the average component of the current density operator j_μ at time t. Using (4.5) we have

$$\begin{aligned}\langle j_\mu(t)\rangle &= \mathrm{Tr}\,[(\rho_0 + \Delta\rho)\,j_\mu] \\ &= \mathrm{Tr}\,[\Delta\rho(t)\,j_\mu],\end{aligned} \tag{4.38}$$

since the average current is presumably zero in the equilibrium condition ρ_0.

It is convenient to introduce the interaction representation of the current operator, relative to the unperturbed Hamiltonian H_0, i.e.

$$j_\mu(t) \equiv \mathrm{e}^{-\mathrm{i}H_0 t}\,j_\mu\,\mathrm{e}^{\mathrm{i}H_0 t}. \tag{4.39}$$

We also use the matrix theorem $\mathrm{Tr}\,(AB) \equiv \mathrm{Tr}\,(BA)$, and make a change of time origin. Thus

$$\begin{aligned}\langle j_\mu(t)\rangle &= -\mathrm{i}\,\mathrm{Tr}\int_{-\infty}^{t} \mathrm{e}^{-\mathrm{i}H_0(t-t')}[\mathbf{E}\cdot\mathbf{X}, \rho_0]\,\mathrm{e}^{\mathrm{i}H_0(t-t')}\,\mathrm{e}^{\mathrm{i}\omega t'}\,j_\mu\,\mathrm{d}t' \\ &= -\mathrm{i}\,\mathrm{Tr}\int_{-\infty}^{t} [\mathbf{X}, \rho_0]\,j_\mu(t-t')\,\mathrm{e}^{\mathrm{i}\omega t'}\,\mathrm{d}t'\cdot\mathbf{E},\end{aligned} \tag{4.40}$$

i.e. the complex conductivity tensor in (4.31) must be of the form

$$\sigma_{\mu\nu}(\omega) = \mathrm{i}\,\mathrm{Tr}\int_0^\infty [X_\nu, \rho_0]\, j_\mu(t')\,\mathrm{e}^{-\mathrm{i}\omega t'}\,\mathrm{d}t'. \tag{4.41}$$

This formula is explicit, but can be transformed into a much more elegant and transparent expression by an analytical device that illustrates the 'imaginary time' interpretation (4.30) of inverse temperature. For any operator A, the following identity may be verified by direct differentiation with respect to an auxiliary variable λ:

$$\begin{aligned}\frac{\mathrm{d}}{\mathrm{d}\lambda}\{\mathrm{e}^{\lambda H_0}[A, \mathrm{e}^{-\lambda H_0}]\} &= H_0\,\mathrm{e}^{\lambda H_0} A\,\mathrm{e}^{-\lambda H_0} - \mathrm{e}^{\lambda H_0} A\,\mathrm{e}^{-\lambda H_0} H_0 \\ &= [H_0, A(-\mathrm{i}\lambda)] \\ &= -\mathrm{i}\dot{A}(-\mathrm{i}\lambda),\end{aligned} \tag{4.42}$$

where $\dot{A}(z)$ means, quite generally, the interaction representation of the rate of change of the operator A, evaluated at the (complex) time z, as defined by (3.36) and (4.39). Now integrate with respect to λ up to the value β, and use the definition (4.32) of the equilibrium density matrix ρ_0: the result is

$$[A, \rho_0] = -\mathrm{i}\rho_0 \int_0^\beta \dot{A}(-\mathrm{i}\lambda)\,\mathrm{d}\lambda. \tag{4.43}$$

But the operator X_ν appearing in (4.41) is just a component of the electric polarization, whose time derivative is the corresponding component of the current density, i.e.

$$\dot{X}_\nu = \sum_i e\dot{x}_{(i)\nu} = j_\nu. \tag{4.44}$$

We may thus replace the commutator in (4.41) by an integral over this component of the current, to give

$$\sigma_{\mu\nu}(\omega) = \mathrm{Tr}\int_0^\infty \mathrm{d}t'\,\mathrm{e}^{-\mathrm{i}\omega t'}\int_0^\beta \rho_0\, j_\nu(-\mathrm{i}\lambda)\, j_\mu(t')\,\mathrm{d}\lambda. \tag{4.45}$$

The trace operation allows us to shift the time origin and then to integrate over λ. With suitable analytical conditions on the functions of the complex variable $t = t' + \mathrm{i}\lambda$, we get

$$\begin{aligned}\sigma_{\mu\nu}(\omega) &= \mathrm{Tr}\int_0^\infty \mathrm{d}t'\int_0^\beta \mathrm{d}\lambda\,\rho_0\, j_\nu(t'+\mathrm{i}\lambda)\, j_\mu(0)\,\mathrm{e}^{-\mathrm{i}\omega(t'+\mathrm{i}\lambda)}\,\mathrm{e}^{-\omega\lambda} \\ &= \tfrac{1}{2}\mathrm{Tr}\int_{-\infty+\mathrm{i}\lambda}^{\infty+\mathrm{i}\lambda} \rho_0\, j_\nu(t)\, j_\mu(0)\,\mathrm{e}^{-\mathrm{i}\omega t}\,\mathrm{d}t \int_0^\beta \mathrm{e}^{-\omega\lambda}\,\mathrm{d}\lambda \\ &= \frac{(1-\mathrm{e}^{-\beta\omega})}{2\omega}\int_{-\infty}^\infty \langle j_\nu(t)\, j_\mu(0)\rangle\,\mathrm{e}^{-\mathrm{i}\omega t}\,\mathrm{d}t,\end{aligned} \tag{4.46}$$

where the product of current operators is averaged over the equilibrium ensemble ρ_0.

This outline derivation of a typical *Kubo formula* illustrates the analytical power of the density matrix in the construction of abstract formulae and in the proof of general theorems. Thus, we see that the conductivity is a property inherent in the quantum-mechanical description of the unperturbed system; the application of a weak electric field merely exposes the time-correlations of the fluctuating components of the electric current in the equilibrium state. Time-reversal (with the reversal of any magnetic fields) allows us to interchange tensor indices, and hence to derive the familiar *Onsager relations* of the thermodynamics of irreversible processes. Kubo formulae have been derived for many different linear response coefficients, transport coefficients, generalized susceptibilities, etc.

It is worth remarking, however, that such a formula may not be the best starting point for the calculation of a transport coefficient in a practical case. When the electrons in a metal, say, are subject to the sort of scattering than can be described by an isotropic relaxation time, then (4.46) becomes simply the quasi-classical *Chambers formula* derived from the Boltzmann equation. But for more complicated scattering mechanisms we have to work quite hard to unravel their effects from the innocent symbol ρ_0 where they lie hidden. It is still essential to find a representation in which the components of the current are nearly diagonal operators, and have reasonably simple equations of motion. The virtue of the density matrix lies in its symbolic mathematical invariance, not in any practical arithmetical simplicity.

4.5 The one-particle Green function

The density matrix is really a rather general operator in Hilbert space: for example, in an N-particle system it would require $6N$ dimensions of representation. In practice we can only manipulate functions of a few space variables, such as the co-ordinates of a single particle. But suppose we write down such a *one-particle density matrix*, e.g.

$$\rho(\mathbf{r},\mathbf{r}') = \iint \dots \Psi^*(\mathbf{r},\mathbf{r}_1,\mathbf{r}_2,\dots)\,\Psi(\mathbf{r}',\mathbf{r}_1,\mathbf{r}_2,\dots)\,\mathbf{d}^3\mathbf{r}_1\,\mathbf{d}^3\mathbf{r}_2\dots \quad (4.47)$$

for a single particle in a large system. This would seem to refer to a *distinguishable* particle, when, in fact, we were interested in average properties involving any one of a large number of *identical* particles.

Thus, we do not, when measuring the density of a gas, follow the motion of a *particular* molecule, but merely ask for the probability that *any* molecule should happen to be in the neighbourhood.

Rather than introducing fully symmetrized or anti-symmetrized wave functions, and then perhaps having to sum over all possible particle co-ordinates to evaluate the trace, we use the language of second quantization. As already shown (§§ 1.11 and 2.3) the expression $\psi^*(\mathbf{r})\,\psi(\mathbf{r})$ considered as a field operator measures the number density of particles (bosons or fermions) of the ψ field at the point $\mathbf{r}$, i.e.

$$\langle|\psi^*(\mathbf{r})\,\psi(\mathbf{r})|\rangle = \rho(\mathbf{r}) \tag{4.48}$$

in the general state $|\,\rangle$ of the system. This immediately suggests a more general expression

$$\rho(\mathbf{r}, \mathbf{r}') = \psi^*(\mathbf{r}')\,\psi(\mathbf{r}), \tag{4.49}$$

which we might call the general density operator on a many-particle wave function and which would give rise to the corresponding *one-particle density matrix*

$$\rho(\mathbf{r}, \mathbf{r}') = \langle|\psi^*(\mathbf{r}')\,\psi(\mathbf{r})|\rangle \tag{4.50}$$

evaluated over some state $|\,\rangle$.

It is easy enough to prove that (4.50) does indeed define a function that could be used in the standard formula (4.5) to evaluate the ensemble average of any single-particle operator—i.e. one that acted on the co-ordinate $\mathbf{r}$ of only one particle at a time, but did not know which one of the numerous particles of the system it was actually looking at.

In principle, it makes no difference whether we calculate (4.50) for a pure state of a very large assembly, or whether we suppose that we have an ensemble of smaller systems, each containing many particles but only to be prescribed statistically as being in a mixed state. For calculations on systems in thermal equilibrium at a finite temperature, the latter description is more convenient. By analogy with the canonical density matrix (4.23), we might write

$$\rho(\mathbf{r}, \mathbf{r}') \propto \mathrm{Tr}\,\{e^{-\beta H}\,\psi^*(\mathbf{r}')\,\psi(\mathbf{r})\}, \tag{4.51}$$

where the trace is to be taken over all states of the system, whose Hamiltonian operator is H. In what follows, however, we shall mostly be concerned with expressions analogous to (4.50), as if we were only interested in the properties of the system in its ground state, at zero temperature, or as if we had some other means of prescribing a pure excited state of the ensemble to correspond to its being at a finite temperature.

But as we have seen in chapter 3, there are many mathematical advantages in not restricting ourselves to Schrödinger field operators, which are independent of the time. To evaluate (4.50), we might want to work in the Heisenberg or interaction representation, as in §§ 3.2 and 3.3, where each of the operators $\psi(\mathbf{r}), \psi^*(\mathbf{r}')$ would acquire a time variation. For reasons that will shortly become apparent, it is best to give each of them a separate time co-ordinate t or t', to go with its separate space co-ordinate, $\mathbf{r}$ or $\mathbf{r}'$. Thus we might consider the properties of

$$\rho(\mathbf{r}, t; \mathbf{r}', t) = \langle | \psi^*(\mathbf{r}', t)\, \psi(\mathbf{r}, t) | \rangle, \tag{4.52}$$

which could, of course, be made to yield the value of $\rho(\mathbf{r}, \mathbf{r}')$ at some special time, t, by simply letting t' tend to t at the end of the calculation.

Now this begins to look like something very familiar. Remember, in the diagrammatic expansion of the S-matrix, where we defined the *free-particle propagator* (§ 3.6)

$$\begin{aligned} G_0(x - x') &\equiv \mathrm{i}\overline{\psi^*(x)\, \psi}(x') \\ &\equiv \mathrm{i}\langle 0 | T\{\psi^*(x)\, \psi(x')\} | 0 \rangle. \end{aligned} \tag{4.53}$$

In this expression, x represents all four space-time co-ordinates, $\mathbf{r}, t$; and $|0\rangle$ means the mathematical ground state for an assembly of non-interacting particles. The symbol T is, of course, *Wick's chronological operator* (3.95), i.e.

$$T\{\psi^*(\mathbf{r}, t)\, \psi(\mathbf{r}', t')\} = \left\{ \begin{array}{ll} \psi^*(\mathbf{r}, t)\, \psi(\mathbf{r}', t') & \text{if} \quad t > t', \\ (-1)^P \psi(\mathbf{r}'t')\, \psi^*(\mathbf{r}, t) & \text{if} \quad t < t', \end{array} \right\} \tag{4.54}$$

where P is the number of exchanges of fermion operators required to get the operators into causal order.

A propagator clearly has the physical effect of destroying a particle at the point $\mathbf{r}'$ at time t', and then creating another particle at $\mathbf{r}$ at the later time t. It also allows one to create a particle—i.e. 'destroy a hole'—at the point $\mathbf{r}$, if t happened to be the earlier time, and then destroy the particle—'create a new hole'—at $\mathbf{r}'$, t'. Thus, it has the effect of carrying one or the other type of excitation forward in time, to some new point in space. The sign factor simply takes care of the fact that fermion field operators always obey anticommutation relations, and therefore always tend to change sign when exchanged like this.

We now generalize the expression (4.53) by asking for the expectation value of a similar combination of operators *in some arbitrary state of the system*, not necessarily the ground state. We define the *one-particle Green function*

$$G(\mathbf{r},t;\mathbf{r}',t') \equiv -\mathrm{i}\langle|T\{\psi(\mathbf{r},t)\,\psi^*(\mathbf{r}',t')\}|\rangle. \tag{4.55}$$

This is the mathematical object with which we shall be concerned for the remainder of this chapter.

For a gas of independent free fermions in the ground state this is just the same as the propagator (4.53), i.e.

$$G(\mathbf{r},t;\mathbf{r}',t') \to G_0(x'-x), \tag{4.56}$$

except that the order of the field operators is reversed in the definition of the Green function.

The connection with the density matrix is easily derived. By (4.50), at the time t,

$$\begin{aligned}\rho(\mathbf{r},\mathbf{r}') &= \langle|\psi^*(\mathbf{r}',t)\,\psi(\mathbf{r},t)|\rangle \\ &= -\langle|T\{\psi(\mathbf{r},t)\,\psi^*(\mathbf{r}',t_+)\}|\rangle \\ &= -\mathrm{i}G(\mathbf{r},t;\,\mathbf{r}',t_+),\end{aligned} \tag{4.57}$$

where $t_+ = t+\delta t$, i.e. a moment infinitesimally later than t. Thus, knowledge of the one-particle Green function brings all available information about the one-particle density matrix, and hence all that can be learnt about ensemble averages of one-particle observables.

The general problem of calculating such averages now seems to require us to find out a great deal about the state $|\rangle$ (which might, for example, be a highly excited 'thermal' state of a gas of interacting electrons) in order that we may evaluate G. But, as we shall see, the diagrammatic expansion of the S-matrix hands us the Green function itself on a plate (albeit in the form of an infinite perturbation series!) so that many uninteresting algebraic steps may be avoided in the derivation of the thermodynamic properties of the system. This is why the Green function method is so important in many-body theory and other statistical problems of quantum mechanics.

4.6 Energy–momentum representation

A Green function, involving two sets of space and time co-ordinates, is obviously a rather complicated mathematical object. In practice it is essential to exploit the symmetry of the physical system so as to simplify the function. Most commonly we deal with systems that are spatially and temporally homogeneous, i.e. which are supposed to be the same everywhere and at all times. Such invariance under translations of $\mathbf{r}$ and t implies that we redefine the Green function in relative co-ordinates, i.e.

$$\begin{aligned} G(\mathbf{r}',t';\mathbf{r}'',t'') &= G(\mathbf{r}'-\mathbf{r}'',t'-t'';0,0) \\ &\equiv G(\mathbf{r}'-\mathbf{r}'',t'-t''). \end{aligned} \tag{4.58}$$

This has a Fourier transform in space and time:

$$G(\mathbf{k},\omega) = \iint G(\mathbf{r},t)\,\mathrm{e}^{-\mathrm{i}\mathbf{k}\cdot\mathbf{r}}\,\mathrm{e}^{\mathrm{i}\omega t}\,\mathbf{d}^3\mathbf{r}\,\mathrm{d}t. \tag{4.59}$$

Now remember that we worked out this transform for a free-particle propagator in the 'bare-vacuum' state $|0\rangle$, i.e. by (3.113)

$$G_0(x'-x) = \frac{1}{(2\pi)^4}\iint \frac{\mathrm{e}^{-\mathrm{i}\{\mathbf{k}\cdot(\mathbf{r}'-\mathbf{r})-\Omega(t'-t)\}}}{\Omega-\mathscr{E}_0(\mathbf{k})+\mathrm{i}\delta}\,\mathbf{d}^3\mathbf{k}\,\mathrm{d}\Omega, \tag{4.60}$$

where δ is a positive small quantity for a fermion state of positive energy $\mathscr{E}_0(\mathbf{k})$ measured from the surface of the Fermi sea. Thus, the Green function for independent free particles has energy–momentum representation

$$G_0(\mathbf{k},\omega) = \frac{1}{\omega-\mathscr{E}_0(\mathbf{k})+\mathrm{i}\delta}. \tag{4.61}$$

This is a characteristic result; the Green function, considered as a function of the energy variable ω, has a pole at the point

$$\omega = \mathscr{E}_0(\mathbf{k})-\mathrm{i}\delta \tag{4.62}$$

which is, of course, in this case known to be the energy of a real excited state of the system with momentum $\mathbf{k}$. By analogy, suppose that the true Green function of a more complex system with strongly interacting particles, etc. turned out to have an energy–momentum representation of the form

$$G(\mathbf{k},\omega) = \frac{1}{\omega-\epsilon(\mathbf{k})+\mathrm{i}\delta}; \tag{4.63}$$

we deduce that the system will look as if it were just an assembly of 'quasi-particles' with energy–momentum relation

$$\mathscr{E}(\mathbf{k}) = \epsilon(\mathbf{k}). \tag{4.64}$$

The reduction of the Green function to a form such as (4.63) with an explicit formula for $\epsilon(\mathbf{k})$, would thus represent major progress in the investigation of the properties of the system.

When the quantity δ is sufficiently small (as assumed, of course, in (4.61)), we may write (4.63) in the form

$$G(\mathbf{k},\omega) = \mathscr{P}\left\{\frac{1}{\omega-\epsilon(\mathbf{k})}\right\} - \mathrm{i}\pi\,\delta\{\omega-\epsilon(\mathbf{k})\}. \tag{4.65}$$

What this means is that whenever we have any occasion to use $G(\mathbf{k},\omega)$, we multiply it by some function of $f(\omega)$ and integrate from 0 to ∞. We then always get an ordinary integral along the real axis whose principal part is taken (as signified by the symbol $\mathscr{P}$) together with an imaginary contribution, which in the limit $\delta \to 0$ turns out to be $\mathrm{i}\pi$ times the value of $f(\omega)$ at the pole, and hence equivalent to a delta function. This can be written

$$\operatorname{Im} G(\mathbf{k},\omega) = -\pi\mathscr{N}(\mathbf{k},\omega); \tag{4.66}$$

in other words, the imaginary part of the Green function gives us the 'density of states of energy ω and momentum $\mathbf{k}$'. This is, in fact, a general formula (the *Lehman spectral representation*) and provides a convenient starting point for some formal theories, but needs more precise discussion when we have 'hole like' excitations as in the degenerate Fermi gas. In a sense, there is nothing very mysterious about it: it is simply a Fourier transform of the wave-function correlation that one would derive from (4.57) and (4.58), i.e.

$$\rho(\mathbf{r},t;0,0) = -\mathrm{i}G(\mathbf{r},t), \tag{4.67}$$

which becomes the mean particle density (4.7) as $\mathbf{r}$ and t tend to zero.

What does it mean when the imaginary part of the denominator of (4.63) is not infinitesimal? Suppose, for example, that the result of a calculation gave

$$G(\mathbf{k},\omega) = \frac{A(\mathbf{k})}{\omega-\epsilon(\mathbf{k})+\mathrm{i}\Gamma(\mathbf{k})}, \tag{4.68}$$

where $\Gamma(\mathbf{k})$ came out to be a well-defined positive quantity, of the dimensions of energy (i.e. inverse time). The momentum–time representation of the Green function will then be given by the Fourier inversion

$$\begin{aligned} G(\mathbf{k},t) &= \frac{1}{2\pi}\int G(\mathbf{k},\omega)\,\mathrm{e}^{-\mathrm{i}\omega t}\,\mathrm{d}\omega \\ &= \frac{1}{2\pi}\int_{-\infty}^{\infty}\frac{A(\mathbf{k})}{\omega-\epsilon(\mathbf{k})+\mathrm{i}\Gamma(\mathbf{k})}\,\mathrm{e}^{-\mathrm{i}\omega t}\,\mathrm{d}\omega. \end{aligned} \tag{4.69}$$

Now when $t < 0$, this integral may be evaluated by completing the contour in the upper half plane. Since no poles are then enclosed the result is zero: this simply expresses the condition that our Green function is 'causal' or 'retarded' (see §4.10). But for $t > 0$, the contour must be completed in the negative half plane, yielding a residue at the pole:

$$G(\mathbf{k}, t) = \mathrm{i}A(\mathbf{k})\,\mathrm{e}^{-\mathrm{i}\epsilon(\mathbf{k})t}\,\mathrm{e}^{-\Gamma(\mathbf{k})t}. \tag{4.70}$$

Thus, finite Γ in (4.68) corresponds to an excitation with finite life-time, decaying at the rate $\Gamma(\mathbf{k})$ per second. It also stands, as we well know, for a state of finite 'energy width', since its time variation will not give sharp selection rules for energy in any transition. We may say that this is a state of 'complex energy' $\epsilon(\mathbf{k}) - \mathrm{i}\Gamma(\mathbf{k})$. Of course, if the imaginary part is too large, we may have difficulty in making sense of the quasi-particle interpretation of the excitation spectrum: it may not then be very profitable to attempt to construct discrete wave packets out of rapidly decaying components such as (4.70).

The theory of the analytical properties of Green functions is quite elaborate, and is written up in many places. For example, the causality condition implies dispersion relations linking the real and imaginary parts. In the present discussion we have confined ourselves, for simplicity, to positive fermion states, but the theory for 'holes' is quite analogous. For bosons, similarly, a Green function theory can easily be constructed. The trouble is that some minor details of the definition and nomenclature of these functions, such as the positions of the poles for positive and negative energies, are not quite standardized and care must be taken to acquaint oneself with the precise conventions adopted by each author.

Although the momentum representation is only fully justified for a homogeneous system such as a gas or liquid, similar theorems can be derived for other cases, provided that suitable basis functions are chosen for the representation of G. Thus, in a perfect crystal lattice one would naturally turn to Bloch functions, or Wannier functions, and in an atom or nucleus to spherical harmonics. In every case, the interpretation of energy denominators such as (4.68) as evidence of quasi-stationary states is still valid.

4.7 Evaluation of Green functions

The special virtue of Green functions is that they can be derived very easily out of the S-matrix perturbation theory. Suppose we have a system in a 'perturbed' state $|\Psi\rangle$, for which we want to evaluate the one-particle density matrix. Suppose that this state is derived from an unperturbed ground state $|\Psi_0\rangle$ by the application of an S-matrix, as in § 3.5, i.e.

$$|\Psi\rangle = S|\Psi_0\rangle. \tag{4.71}$$

By definition of the Green function (4.55), we may write

$$\begin{aligned} G(\mathbf{r},t;\mathbf{r}',t') &\equiv -\mathrm{i}\langle\Psi|\,T\{\psi(x)\,\psi^*(x')\}\,|\Psi\rangle \\ &= -\mathrm{i}\langle\Psi_0|\,S^*T\{\psi(x)\,\psi^*(x')\}S\,|\Psi_0\rangle \\ &= -\mathrm{i}\langle\Psi_0|\,S^*\,|\Psi_0\rangle\langle\Psi_0|\,T\{\psi(x)\,\psi^*(x')\}S\,|\Psi_0\rangle \\ &= -\mathrm{i}\frac{\langle\Psi_0|\,T\{\psi(x)\,\psi^*(x')\}S\,|\Psi_0\rangle}{\langle\Psi_0|\,S\,|\Psi_0\rangle}, \end{aligned} \tag{4.72}$$

where the unitarity of the S-matrix is used, and also some special properties of the ground state.

But as shown in § 3.9, we can reduce the S-matrix by the elimination of all disconnected graphs, and a redefinition of the 'physical vacuum' state by (3.126), i.e.

$$|\Psi'_{\text{initial}}\rangle = S_{\text{vacuum}}|\Psi_0\rangle. \tag{4.73}$$

The result is that the Green function takes the value

$$G(\mathbf{r},t;\mathbf{r}',t') = -\mathrm{i}\langle\Psi'_{\text{initial}}|\,T\{\psi(x)\,\psi^*(x')\}S\,|\Psi_0\rangle, \tag{4.74}$$

where only connected graphs need be counted.

The combination of field operators $T\{\psi(x)\,\psi^*(x')\}$ is, of course, equivalent to a single free-particle propagator G_0. The S-matrix has the power to introduce all possible connected parts into this line, i.e.

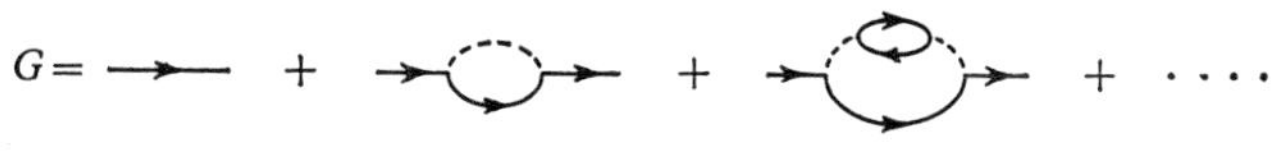

G = —→— + ... + ... + ...

(4.75)

But this is exactly the expansion, in terms of connected graphs, with reducible and irreducible self-energy parts, etc., that was used in § 3.10 to prove Dyson's Theorem, i.e. by (3.128)

$$G = G_0 + G_0\Sigma G. \tag{4.76}$$

In other words, our one-particle Green function turns out to be exactly the modified propagator (3.130), with energy–momentum representation

$$G(\mathbf{k},\omega) = \frac{1}{\omega - \mathscr{E}_0(\mathbf{k}) - \Sigma(\mathbf{k},\omega) + \mathrm{i}\delta}. \tag{4.77}$$

We have already discussed the interpretation of this result as a modification of the apparent energy of the particles by the addition of a correction term $\Sigma(\mathbf{k},\omega)$. From (4.70) we also learn to interpret any imaginary part of Σ as a life-time or energy width. Of course this result is only a formula: in practice we must still sum contributions from all irreducible diagrams, before we can actually evaluate Σ.

Nevertheless, this result is the key to the ubiquitous use of the Green function formalism in the theory of many-body systems. The modified propagator that seems to appear as an auxiliary function in the diagrammatic method turns out to be related to the one-particle density matrix, and hence can tell us all that we can possibly learn about one-particle observations made on the system. Whether or not we use the perturbation expansion itself to evaluate G, this must be the aim of our calculation.

The above discussion is for a system in its ground state; the differences between G and G_0 are due to interactions, modified only by 'zero-point motion' of the particles. But, as we have already remarked, the density matrix of a system in thermal equilibrium at the temperature $T \neq 0$ may be generated by applying the time-evolution operator (4.29) with an imaginary time variable $t = \mathrm{i}\tau$, supposed to vary from, say, 0 to $\mathrm{i}/\boldsymbol{k}T$. This would correspond to the introduction into the theory of a *temperature Green function*, such as

$$G(\mathbf{r},\tau;\mathbf{r}',\tau') \equiv -\mathrm{i}\langle\,|\,T\{\psi(\mathbf{r},\tau)\,\psi^*(\mathbf{r}',\tau')\}\,|\,\rangle, \tag{4.78}$$

where
$$\psi(\mathbf{r},\tau) = \mathrm{e}^{(H-\mu N)\tau}\,\psi(\mathbf{r})\,\mathrm{e}^{-(H-\mu N)\tau}, \tag{4.79}$$

and so on. From the arguments in §§4.3 and 4.4, the reader may perhaps agree that this creates a formal basis for an analytical theory out of which the grand canonical density matrix (4.24) or (4.54) can be derived.

The main point is that this somewhat formal prescription allows one to introduce the whole machinery of the diagrammatic technique, almost exactly as in the case where $T = 0$. Just as in §§3.7 to 3.10, the temperature Green function can be written down as the sum of a series whose terms correspond to topologically distinct diagrams. The difference is that the various 'propagators' cannot be expressed with such analytical simplicity, and the algebraic calculations become much more complicated. Here again, the variety of definitions used by various authors is a major cause of confusion for the beginner.

4.8 Two-particle Green functions

We came upon Green functions from a study of the density matrix. But, of course, the one-particle density matrix (4.50) only answers the very simplest questions about ensemble averages of quantum-mechanical observables. Suppose we had been interested in properties associated with two particles simultaneously, such as their mean energy of interaction. It would be necessary to calculate the *two-particle density matrix*

$$\rho(\mathbf{r}_1, \mathbf{r}_2; \mathbf{r}'_1, \mathbf{r}'_2) = \int \dots \int \Psi^*(\mathbf{r}_1, \mathbf{r}_2, \mathbf{r}_3, \dots, \mathbf{r}_n)\, \Psi(\mathbf{r}'_1, \mathbf{r}'_2, \mathbf{r}_3, \dots, \mathbf{r}_n)\, \mathbf{d}^3\mathbf{r}_3\, \mathbf{d}^3\mathbf{r}_4, \dots, \mathbf{d}^3\mathbf{r}_n. \tag{4.80}$$

This is exactly of the general form (4.3) except that the abstract co-ordinate x now comprises the six vector components of $\mathbf{r}_1$ and $\mathbf{r}_2$, the positions of two distinct particles of the system.

In the case of indistinguishable particles, we naturally use the language of second quantization, and write

$$\rho(\mathbf{r}_1, \mathbf{r}_2; \mathbf{r}'_1, \mathbf{r}'_2) = \langle | \psi^*(\mathbf{r}'_1)\, \psi^*(\mathbf{r}'_2)\, \psi(\mathbf{r}_2)\, \psi(\mathbf{r}_1) | \rangle, \tag{4.81}$$

just as in (4.50). The analogy of (4.52) then suggests the introduction of time co-ordinates and the definition of a *two-particle Green function*:

$$K(1234) \equiv \langle | T\{\psi(\mathbf{r}_1, t_1)\, \psi(\mathbf{r}_2, t_2)\, \psi^*(\mathbf{r}_3, t_3)\, \psi^*(\mathbf{r}_4, t_4)\} | \rangle \tag{4.82}$$

where the product needs to be time-ordered. This is a natural generalization of (4.55). Note that a *two*-particle Green function carries *four* sets of labels, corresponding to the propagation of each particle between two different points in space and time—although in a homogeneous system only relative co-ordinates are eventually of physical significance.

The interpretation of this type of expression in special cases is very simple. Thus when $\mathbf{r}_3 = \mathbf{r}_1$ and $\mathbf{r}_4 = \mathbf{r}_2$ and all the times are nearly the same but correctly ordered, we have

$$\begin{aligned} K(1212) &\to \rho(\mathbf{r}_1, \mathbf{r}_2; \mathbf{r}_1, \mathbf{r}_2) \\ &= \int \dots \int |\Psi(\mathbf{r}_1, \mathbf{r}_2, \mathbf{r}_3, \dots, \mathbf{r}_n)|^2\, \mathbf{d}^3\mathbf{r}_3 \dots \mathbf{d}^3\mathbf{r}_n \\ &= P(\mathbf{r}_1, \mathbf{r}_2), \end{aligned} \tag{4.83}$$

which is just the probability of finding particles at $\mathbf{r}_1$ and $\mathbf{r}_2$, whatever the positions of the other particles. This is what we should call the static *radial distribution function* in the classical statistical theory of liquids and gases.

For a homogeneous system, we may put $\mathbf{r} = \mathbf{r}_1 - \mathbf{r}_2$, and retain a relative time co-ordinate. Thus the *correlation function*

$$\bar{S}(\mathbf{r}, t) = K(\mathbf{r}, t; 0, 0; 0, \delta; \mathbf{r}, t+\delta) \tag{4.84}$$

measures the probability of finding a particle at point $\mathbf{r}$ at time t, given that there was a particle initially at the origin. This is a most important general observable property, since it determines the diffraction of neutrons and X-rays by gases, liquids and solids. Its Fourier transform can be written

$$\bar{S}(\mathbf{q}, t) = \langle | \rho_{-\mathbf{q}}(t) \rho_{\mathbf{q}}(0) | \rangle, \tag{4.85}$$

where $\rho_{\mathbf{q}}$ is the general particle-density at wave-vector $\mathbf{q}$. Thus, the two-particle Green function gives us full information about the time variation and correlation of density fluctuations. Going to an energy representation, we find that the poles of $\bar{S}(\mathbf{q}, \omega)$, like those of the one-particle Green function, correspond to excited states of the system—for example, self-sustaining density fluctuations such as plasma oscillations (see §5.8).

To understand the significance of the two-particle Green function in the language of diagrams and propagators, let us first consider the case of non-interacting particles in the ground state. With a little bit of juggling of selection rules and time ordering, we find that (4.82) can be reduced to a sum of products of one-particle propagators, i.e.

$$\begin{aligned} K_0(1234) &= -\langle 0 | T\{\psi(x_1)\psi^*(x_3)\} | 0 \rangle \langle 0 | T\{\psi(x_2)\psi^*(x_4)\} | 0 \rangle \\ &\quad + \langle 0 | T\{\psi(x_1)\psi^*(x_4)\} | 0 \rangle \langle 0 | T\{\psi(x_2)\psi^*(x_3)\} | 0 \rangle \\ &= G_0(13)\, G_0(24) - G_0(14)\, G_0(23), \end{aligned} \tag{4.86}$$

in an obvious shorthand notation.

But these are merely the first two terms of the diagrammatic series one might write down if one tried to evaluate (4.82) using a prescription like (4.74). Thus

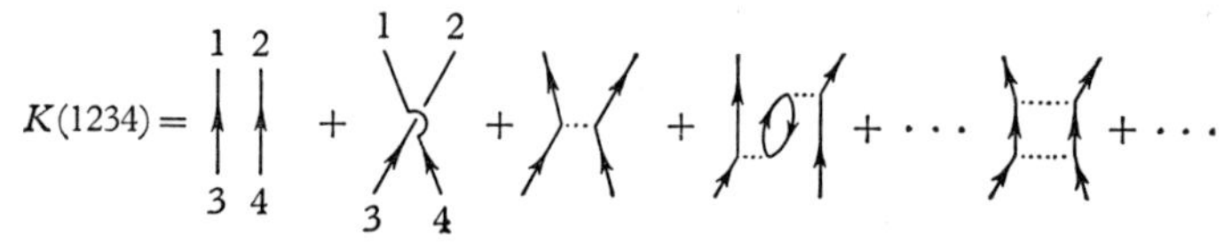

(4.87)

In other words, the two-particle Green function represents *all modes of interaction between pairs of particles*. We might symbolize it as a generalized scattering vertex

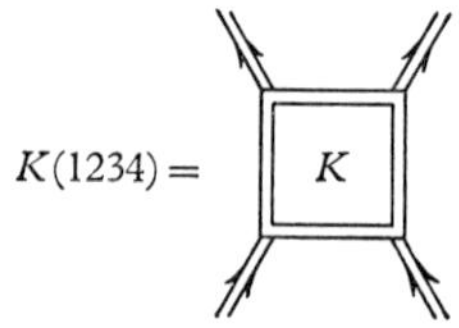

(4.88)

with two renormalized quasi-particles in, and two out, and all connected intermediate processes going on in between.

With (4.86) as our guide, we may guess that in the momentum representation the two-particle Green function ought to be of the algebraic form

$$\begin{aligned} K(1234) = {} & G(13)\,G(24) - G(14)\,G(23) \\ & + G(1)\,G(2)\,G(3)\,G(4)\,\Gamma(1234) \\ & \times \delta(\mathbf{k}_1 + \mathbf{k}_2 - \mathbf{k}_3 - \mathbf{k}_4)\,\delta(\omega_1 + \omega_2 - \omega_3 - \omega_4), \end{aligned} \tag{4.89}$$

where, of course, we must use the modified propagators, as in (4.77),

$$G(13) = G(\mathbf{k}_1 - \mathbf{k}_3, \omega_1 - \omega_3). \tag{4.90}$$

The function $\Gamma(1234)$ then behaves like the Fourier transform of an effective scattering potential between quasi-particle states—a renormalized version of (2.43). The singularities of this function will then appear as singularities in K, and hence as excited states of the system, in addition to those produced by the poles of G. Thus, for example, 'excitons' or 'electron-hole bound pairs' should be produced in this way, as well as the collective modes defined by the poles of the correlation function.

The actual evaluation of such interaction terms is obviously very difficult in general. But, as in § 3.10, one can use topological arguments to prove an analogue of the Dyson equation. Suppose that $J(1234)$ represents the sum of all *irreducible interaction diagrams* in $K(1234)$. Then we may redraw the series (4.87), and replace the later terms by K itself—i.e.

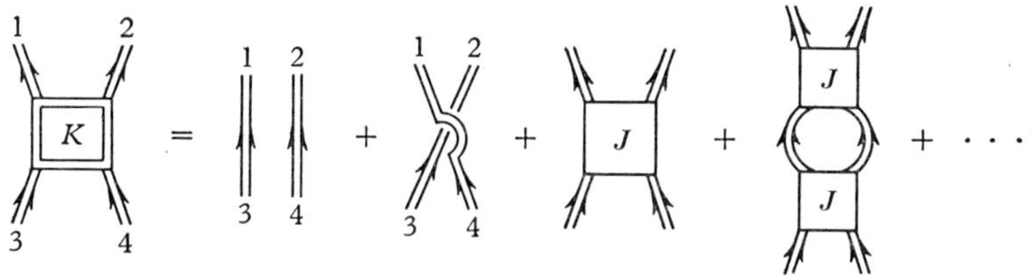

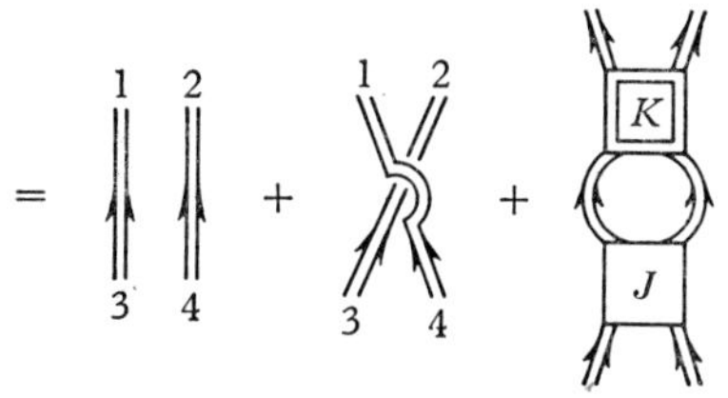

(4.91)

which represents an integral equation of the form

$$K(1234) = G(13)\,G(24) - G(14)\,G(23) + \tfrac{1}{2}\sum_{5,\,6} G(1)\,G(2)\,J(1256)\,K(5634) \quad (4.92)$$

(the factor $\frac{1}{2}$ comes from the equivalence of the two internal lines between J and K).

This rather elegant theorem is called the *Bethe–Salpeter equation*, and gives us some information about the effective interaction function Γ, in terms of irreducible vertex diagrams. Unfortunately, it does not have the simplicity of the Dyson equation as a starting point for analytical or approximate calculations. Indeed, the general mathematical theory of the two-particle Green function is far more complicated than that of the one-particle propagator.

4.9 The hierarchy of Green functions

Consider the case of a system of interacting particles, whose Hamiltonian might be of the form (2.42), i.e.

$$H = H_{\text{ind}} + \iint \psi^*(\mathbf{r})\,\psi^*(\mathbf{r}')\,\mathscr{V}_{\text{int}}(\mathbf{r},\mathbf{r}')\,\psi(\mathbf{r}')\,\psi(\mathbf{r})\,\mathbf{d}^3\mathbf{r}\,\mathbf{d}^3\mathbf{r}', \quad (4.93)$$

H_{ind} being the Hamiltonian for independent particles—including, perhaps, any extraneous fields. Then the field operator $\psi(\mathbf{r},t)$ in the Heisenberg representation satisfies an equation of motion analogous to the time-dependent Schrödinger equation

$$\frac{\hbar}{\mathrm{i}}\frac{\partial\psi}{\partial t}(\mathbf{r},t) + H_{\text{ind}}(\mathbf{r},t)\,\psi(\mathbf{r},t) = -\int \mathscr{V}_{\text{int}}(\mathbf{r},\mathbf{r}'')\,\psi^*(\mathbf{r}'',t)\,\psi(\mathbf{r}'',t)\,\mathbf{d}^3\mathbf{r}''\,\psi(\mathbf{r},t). \quad (4.94)$$

Now let us calculate the time derivative of the one-particle Green function, taking care to allow for the discontinuity associated with the time-ordering factor. Thus, we write, from (4.55),

$$\frac{\hbar}{\mathrm{i}}\frac{\partial G}{\partial t}(\mathbf{r},t;\mathbf{r}',t') = \left\langle \left| \frac{\hbar}{\mathrm{i}}\frac{\partial}{\partial t}(-\mathrm{i})\,\{\psi(\mathbf{r},t)\,\psi^*(\mathbf{r}',t')\,\theta(t-t') - \psi^*(\mathbf{r}',t')\,\psi(\mathbf{r},t)\,\theta(t'-t)\} \right| \right\rangle, \quad (4.95)$$

where $\theta(t-t')$ is a step function, whose derivative must be a delta function, i.e.

$$\frac{\partial}{\partial t}\theta(t-t') = \delta(t-t'). \quad (4.96)$$

When we carry out the differentiation in (4.95), we get, in addition to time derivatives of the field operators themselves, an extra term

$$-\hbar\{\psi(\mathbf{r},t)\,\psi^*(\mathbf{r}',t')\,\delta(t-t')+\psi^*(\mathbf{r}',t')\,\psi(\mathbf{r},t)\,\delta(t'-t)\}$$
$$= -\hbar\,\delta(\mathbf{r}-\mathbf{r}')\,\delta(t-t'). \quad (4.97)$$

This singularity, arising essentially from the 'non-anticommutativity' of the field operators at the same point $(\mathbf{r},t)$ in space-time, is an essential feature of the Green function, and very important in its own right.

The field equation (4.94) is used to evaluate the various time-derivatives of $\psi(\mathbf{r},t)$, etc. that arise in differentiating the Green function. The result may be expressed as follows:

$$\left\{\frac{\hbar}{\mathrm{i}}\frac{\partial}{\partial t}+H_{\mathrm{ind}}(\mathbf{r},t)\right\}G(\mathbf{r},t;\mathbf{r}',t') = -\hbar\,\delta(\mathbf{r}-\mathbf{r}')\,\delta(t-t')$$
$$+\mathrm{i}\int\mathscr{V}(\mathbf{r},\mathbf{r}'')\,K(\mathbf{r}'',t;\mathbf{r},t;\mathbf{r}'',t;\mathbf{r}',t')\,\mathbf{d}^3\mathbf{r}'', \quad (4.98)$$

where the two-particle Green function (4.82) appears in the interaction term.

It is not very difficult to show that the same sort of argument applied to the equation of motion of a *two*-particle Green function yields an integro-differential equation involving the *three*-particle Green function—and so on. Green functions of various orders are thus linked by a hierarchy of such equations, reminiscent of the equations linking distribution functions of successive orders in the statistical theory of liquids and gases. Such equations may be used as the source of approximate calculations of Green functions by making assumptions about the properties of the function of highest order in the hierarchy. Thus, we might replace the function K in (4.98) by some approximate form of (4.89) and hence arrive at a higher approximation to G. The advantage of such methods is that, like the Brillouin–Wigner series (3.19), they are 'self-consistent', and hence may already contain, in compact analytical form, results that would otherwise require the summation of whole classes of perturbation diagrams.

4.10 Time-independent Green functions

The Green function concepts are so ubiquitous in mathematical physics that it has not been easy to decide where to start their discussion. Somewhat perversely, as it may seem to the reader, I have begun on the most elaborate case—*double-time Green functions*—before considering much simpler cases. We might, for example, have set up the diagrammatic expansion of the S-matrix explicitly in this language, and not waited for the interpretation of the propagator in statistical terms. What I have been trying to emphasize, however, is that it is only in problems of a statistical nature, such as occur in many-body theory, that the Green function is an essential mathematical tool. In my opinion, the connection with the density matrix is its deepest and most powerful property.

Nevertheless, the basic idea of a propagator arises already in the much simpler context of the single-particle Schrödinger equation. Take the case of a free particle in empty space, with an ordinary wave function satisfying the ordinary time-dependent equation

$$\left\{\frac{\hbar}{\mathrm{i}}\frac{\partial}{\partial t}+H_0(\mathbf{r})\right\}\psi_0(\mathbf{r},t)=0, \tag{4.99}$$

where $H_0(\mathbf{r})$ is, for example, the kinetic energy operator $(-\hbar^2/2m)\nabla^2$.

The Green function for this system is just a free-particle propagator like (4.53). Because there are no interactions, the equation of motion (4.98) becomes simply

$$\left\{\frac{\hbar}{\mathrm{i}}\frac{\partial}{\partial t}+H_0(\mathbf{r})\right\}G_0(\mathbf{r},t;\mathbf{r}',t')=-\hbar\,\delta(\mathbf{r}-\mathbf{r}')\,\delta(t-t'). \tag{4.100}$$

Think now of some extraneous potential $\mathscr{V}(\mathbf{r},t)$ added to the system—for example, the field of a centre by which the particle might be scattered. The Schrödinger equation now reads

$$\left\{\frac{\hbar}{\mathrm{i}}\frac{\partial}{\partial t}+H_0(\mathbf{r})\right\}\psi(\mathbf{r},t)=-\mathscr{V}(\mathbf{r},t)\,\psi(\mathbf{r},t). \tag{4.101}$$

Conceived as an inhomogeneous equation, this has the following formal solution, as may readily be verified by direct substitution from (4.99) and (4.100):

$$\psi(\mathbf{r},t)=\psi_0(\mathbf{r},t)+\frac{1}{\hbar}\iint G_0(\mathbf{r},t;\mathbf{r}',t')\,\mathscr{V}(\mathbf{r}',t')\,\psi(\mathbf{r}',t')\,\mathbf{d}^3\mathbf{r}'\,\mathrm{d}t'. \tag{4.102}$$

In other words, the Schrödinger differential equation is transformed into an integral equation with the Green function as kernel.

To make this more specific, we must attend to the boundary conditions. In particular, we want to avoid having to integrate over the 'future'; it does not make good physical sense to suppose that values of $\psi(\mathbf{r}',t')$, for $t'>t$, can really influence the value of $\psi(\mathbf{r},t)$, 'now' at the time t. In other words, we eliminate 'advanced' waves by defining the solution of (4.100) to be a *causal Green function* of the form

$$G^{+}(\mathbf{r}, t; \mathbf{r}', t') = 0. \tag{4.103}$$

This condition may easily be achieved by the familiar device used in §3.8. Suppose we introduce an energy representation, by a Fourier transformation

$$G^{+}(\mathbf{r}, t; \mathbf{r}', t') = \int_{-\infty}^{\infty} G^{+}(\mathbf{r}, \mathbf{r}'; \mathscr{E})\, \mathrm{e}^{-\mathrm{i}\mathscr{E}(t-t')/\hbar}\, \mathrm{d}\mathscr{E}. \tag{4.104}$$

Then for $t < t'$ this integral could be completed by a contour in the upper half plane; if $G^{+}(\mathbf{r}, \mathbf{r}'; \mathscr{E})$ has no poles on or above the real axis of $\mathscr{E}$, there is no residue and (4.103) is satisfied. In fact we are only interested in poles that correspond to eigenvalues of the time-independent Schrödinger equation, which are always real. By subtracting a small imaginary part from each of these, as in (4.62), we ensure that the 'retarded' solution of (4.100) is always obtained. In the present context, of course, the complications associated with 'hole' states are expressly avoided.

Put (4.104) into the equation of motion (4.100); this is satisfied for all t and t' if

$$(H_0 - \mathscr{E})\, G^{+}(\mathbf{r}, \mathbf{r}'; \mathscr{E}) = -\delta(\mathbf{r} - \mathbf{r}'). \tag{4.105}$$

Since H_0 is a differential operator in real space, this is just an inhomogeneous partial differential equation with a singular 'source' at the point $\mathbf{r}'$. Here we begin to link up with the original idea of Green. Consider, for example, Poisson's equation for the electrostatic potential $\Phi(\mathbf{r})$ in a region where the charge density is $\rho(\mathbf{r})$:

$$\nabla^2 \Phi(\mathbf{r}) = 4\pi\rho(\mathbf{r}). \tag{4.106}$$

As is well known, this has the solution

$$\Phi(\mathbf{r}) = \int \frac{1}{|\mathbf{r} - \mathbf{r}'|} \rho(\mathbf{r}')\, \mathbf{d}^3\mathbf{r}'. \tag{4.107}$$

In other words, the function

$$G(\mathbf{r}, \mathbf{r}') = \frac{1}{4\pi} \frac{1}{|\mathbf{r} - \mathbf{r}'|}, \tag{4.108}$$

satisfying the differential equation

$$\nabla^2 G(\mathbf{r}, \mathbf{r}') = \delta(\mathbf{r} - \mathbf{r}'), \tag{4.109}$$

is the kernel by which the differential equation (4.106) is transformed into an integral equation. This was the method used systematically by Green to solve various problems in potential theory.

The classical theory of the wave equation requires a slightly more complicated Green function. Thus, for the field distribution at a fixed frequency, ν, we should use the function satisfying

$$(\nabla^2 + \nu^2/c^2)\, G^+(\mathbf{r}, \mathbf{r}'; \omega) = \delta(\mathbf{r}-\mathbf{r}'), \tag{4.110}$$

with a convention at the poles of ν to retain only 'retarded' waves. In fact, this is essentially the same equation as for the free electron; (4.105) usually reads

$$\left(-\frac{\hbar^2}{2m}\nabla^2 - \mathscr{E}\right) G^+(\mathbf{r}, \mathbf{r}'; \mathscr{E}) = -\delta(\mathbf{r}-\mathbf{r}'). \tag{4.111}$$

The main point to notice is that this simple type of propagator is defined at a particular energy; the analogue of (4.102), i.e.

$$\psi(\mathbf{r}) = \psi_0(\mathbf{r}) + \int G^+(\mathbf{r}, \mathbf{r}'; \mathscr{E})\, \mathscr{V}(\mathbf{r}')\, \psi(\mathbf{r}')\, \mathbf{d}^3\mathbf{r}', \tag{4.112}$$

is only satisfied by the solution of the time-independent Schrödinger equation

$$(H_0 + \mathscr{V} - \mathscr{E})\, \psi = 0 \tag{4.113}$$

at the energy $\mathscr{E}$.

4.11 Matrix representation of the Green function

Quite generally, let us suppose that the operator H_0 (not necessarily the Hamiltonian of a free particle in empty space) has eigenvalues and eigenstates given by

$$H_0 \psi_m(\mathbf{r}) = \mathscr{E}_m \psi_m(\mathbf{r}). \tag{4.114}$$

We may represent the Green function with these states as a basis, and write

$$G_0(\mathbf{r}, \mathbf{r}'; \mathscr{E}) = \sum_{m,n} G_{mn} \psi_m(\mathbf{r})\, \psi_n^*(\mathbf{r}'). \tag{4.115}$$

Substituting from (4.115) into (4.105) we get

$$\begin{aligned} -\delta(\mathbf{r}-\mathbf{r}') &= (H_0 - \mathscr{E})\, G_0(\mathbf{r}, \mathbf{r}'; \mathscr{E}) \\ &= \sum_{mn} G_{mn}(\mathscr{E}_m - \mathscr{E})\, \psi_m(\mathbf{r})\, \psi_n^*(\mathbf{r}'), \end{aligned} \tag{4.116}$$

which is easily solved to yield

$$G_{mn} = \frac{\delta_{mn}}{\mathscr{E} - \mathscr{E}_m}; \tag{4.117}$$

i.e., in the language of projection operators,

$$G_0(\mathbf{r},\mathbf{r}';\mathscr{E}) = \sum_m \frac{\psi_m(r)\,\psi_m^*(\mathbf{r}')}{\mathscr{E}-\mathscr{E}_m}$$

$$= \sum_m |m\rangle \frac{1}{\mathscr{E}-\mathscr{E}_m} \langle m|. \tag{4.118}$$

This is the basic theorem from which almost all the properties of the Green function may be derived. In the first place, it tells us that $G_0(\mathscr{E})$ is diagonal in a representation in which the Hamiltonian is diagonal, and has eigenvalues that are just the inverse of the eigenvalues of the operator $\mathscr{E}-H_0$. It is perfectly proper, therefore, to write (4.105) in the form

$$(\mathscr{E}-H_0)\,G_0(\mathscr{E}) = 1 \tag{4.119}$$

with the algebraic solution

$$G_0(\mathscr{E}) = \frac{1}{\mathscr{E}-H_0}. \tag{4.120}$$

In this sense, the Green function is an operator in the Hilbert space of the Hamiltonian.

We also note that $G_0(\mathscr{E})$, considered as a function of $\mathscr{E}$, has poles at the bound states of H_0, just as in the case of the more general propagators discussed in §4.6. To get the proper solution of the full time-dependent Schrödinger equation (or, what is the same thing, to fix the values of quantities that are not fully defined because of the infinite range of time integration in (4.104)) we must, of course, subtract a small imaginary part from each $\mathscr{E}_m$, to get a causal Green function operator

$$G_0^+(\mathscr{E}) = \frac{1}{\mathscr{E}-H_0+i\delta}. \tag{4.121}$$

In the most general way, suppose we are seeking the eigenvalues of any matrix or operator H. We wish to solve the equation

$$(H-z)\,|\,\rangle = 0. \tag{4.122}$$

The solution is obtained by finding the roots of the secular determinant

$$\|(H-z)\| = 0, \tag{4.123}$$

considered as a function of the variable z. But the determinant of the inverse of a matrix is the inverse of the determinant of that matrix. Therefore we are looking for the singularities of the determinant of the matrix

$$R(z) = \frac{1}{H-z}. \tag{4.124}$$

The eigenvalues of H may thus be located by finding the poles of the *resolvent* operator (4.124), considered as a function of the complex variable z. Since $R(z)$ is just a Green function by another name, this method is not essentially different from the techniques already considered here.

The above argument does not depend on the original Hamiltonian being a differential operator in a continuous real space. The symbol H in (4.122) may be any Hermitian operator, of finite or infinite dimensionality. Let us return, for example, to the classical problem of finding the normal modes of vibration in a crystal lattice, as in §1.4. For the displacement $\mathbf{u}_l$ on the lth site, due to interatomic forces with coefficients $\mathbf{F}_{ll'}$, we have to solve the equations of motion

$$M_l \ddot{\mathbf{u}}_l = -\sum_{l'} \mathbf{F}_{ll'} \cdot \mathbf{u}_{l'}, \tag{4.125}$$

i.e. for a mode of fixed frequency ν,

$$\sum_{l'} \{\mathbf{F}_{ll'} - M_l \nu^2 \boldsymbol{\delta}_{ll'}\} \cdot \mathbf{u}_{l'} = 0. \tag{4.126}$$

This set of algebraic equations has a resolvent matrix

$$R(\nu^2) = [\mathbf{F}_{ll'} - M_l \nu^2 \boldsymbol{\delta}_{ll'}]^{-1}, \tag{4.127}$$

the distribution of whose poles yields the frequency spectrum. For a perfect lattice, we should, of course, Fourier transform to a reciprocal lattice representation, but the argument is not restricted to such cases. The study of $R(\nu^2)$ is the basis of the *Green function method* for the calculation of the vibrational properties of imperfect, impure, or disordered crystals. The reader may amuse himself by establishing the connection between (4.127) and (4.110) in the long-wave continuum limit of §1.5.

4.12 Space representation of time-independent Green function

As an example of the use of the basic theorem (4.118), let us calculate the Green function for a free particle, with eigenstates

$$|\mathbf{k}, \mathbf{r}\rangle = e^{i\mathbf{k}\cdot\mathbf{r}}, \tag{4.128}$$

of energy $\mathscr{E}(\mathbf{k}) = k^2$ in appropriate units. In the language of projection operators we may write

$$\begin{aligned} G^+(\mathbf{r}, \mathbf{r}'; \mathscr{E}) &= \sum_{\mathbf{k}} |\mathbf{k}, \mathbf{r}\rangle \frac{1}{\mathscr{E} - \mathscr{E}(\mathbf{k}) + i\delta} \langle \mathbf{k}, \mathbf{r}'| \\ &= \frac{1}{8\pi^3} \int e^{i\mathbf{k}\cdot\mathbf{r}} \frac{1}{\kappa^2 - k^2 + i\delta} e^{-i\mathbf{k}\cdot\mathbf{r}'} \mathbf{d}^3\mathbf{k}, \end{aligned} \tag{4.129}$$

where $\kappa^2 = \mathscr{E}$.

The system is obviously homogeneous so that the Green function depends only on the difference of co-ordinates $\mathbf{R} = \mathbf{r} - \mathbf{r}'$. Thus

$$\begin{aligned} G^+(\mathbf{R};\mathscr{E}) &= \frac{1}{8\pi^3}\int_0^{2\pi} d\phi \int_0^{\pi} \sin\theta\, d\theta \int_0^{\infty} k^2\, dk \frac{e^{ikR\cos\theta}}{\kappa^2 - k^2 + i\delta} \\ &= \frac{1}{4\pi^2}\int_0^{\infty} \frac{1}{ikR}\frac{e^{ikR} - e^{-ikR}}{\kappa^2 - k^2 + i\delta} k^2\, dk \\ &= -\frac{1}{4\pi^2 iR}\frac{1}{2}\int_{-\infty}^{\infty} e^{ikR}\left\{\frac{1}{k - (\kappa + i\delta')} + \frac{1}{k + (\kappa + i\delta')}\right\} dk \end{aligned} \quad (4.130)$$

by various integrations, change of variables, and introduction of a new infinitesimal $\delta' = \delta/2\kappa$ to define the poles.

But now, in closing the contour, the sign of δ' is all-important. Since R is positive definite, we complete with a semicircle in the upper half plane, where only the first pole is included. The result is

$$G^+(\mathbf{R};\mathscr{E}) = -\frac{1}{4\pi}\frac{e^{i\kappa R}}{R}, \quad (4.131)$$

corresponding, as one might have expected, to an outward-going spherical wave. It is obvious that if δ' had been negative we should have constructed the 'advanced' propagator

$$G^-(\mathbf{R};\mathscr{E}) = -\frac{1}{4\pi}\frac{e^{-i\kappa R}}{R} \quad (4.132)$$

corresponding to a wave 'closing in' on the point $\mathbf{R} = 0$. If, on the other hand, we had placed the pole definitely on the real axis, we should have found the 'standing wave' function

$$\begin{aligned} G^0(\mathbf{R};\mathscr{E}) &= \tfrac{1}{2}\{G^+(\mathbf{R};\mathscr{E}) + G^-(\mathbf{R};\mathscr{E})\} \\ &= -\frac{1}{4\pi}\frac{\cos\kappa R}{R}. \end{aligned} \quad (4.133)$$

This analysis shows very clearly the analytical connection between the time-ordering convention in the Green function and the definition of the position of the poles in the complex energy plane. It is also instructive to verify that the definition (4.129), leading to (4.131), (4.132), or (4.133) does indeed produce solutions of the inhomogeneous wave equation (4.110), to which, of course, the free particle Schrödinger equation is reduced by putting $\kappa = \nu/c$.

4.13 The Born series

The standard problem of one-particle quantum theory is the effect produced upon a free particle by a local potential $\mathscr{V}(\mathbf{r})$. We have been taught to tackle this problem by looking for the solutions of the corresponding Schrödinger equation, with various boundary conditions. The Green function method is an alternative procedure which is often more powerful—although not necessarily more physically significant.

For example, the resolvent (4.124) of such a system would be the same as the Green function

$$G(\mathscr{E}) = \frac{1}{\mathscr{E}-H} = \frac{1}{\mathscr{E}-H_0-\mathscr{V}}. \tag{4.134}$$

If $\mathscr{V}$ were in some sense 'small', this could be expanded in an infinite series, i.e.

$$G(\mathscr{E}) = \frac{1}{\mathscr{E}-H_0}+\frac{1}{\mathscr{E}-H_0}\mathscr{V}\frac{1}{\mathscr{E}-H_0}+\ldots = G_0+G_0\mathscr{V}G_0+\ldots, \tag{4.135}$$

where G_0 is an appropriate free-particle propagator. The condition for the singularity of this function of $\mathscr{E}$ then turns out to be the Brillouin–Wigner perturbation series (3.20), which locates the stationary state of the Hamiltonian $H_0+\mathscr{V}$.

Again, by an algebraic identity, we have

$$G = G_0+G_0\mathscr{V}G, \tag{4.136}$$

which is essentially an integral equation from which the Green function G may in principle be derived. In fact, this is the Dyson equation (3.128) for this very simple system. The graphical representation of the perturbation series (4.135) would be, say,

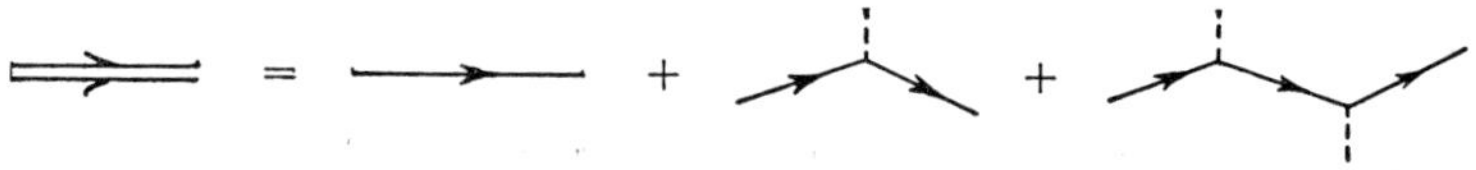

(4.137)

where the broken line represents the perturbation potential $\mathscr{V}$ at each vertex. Since $\mathscr{V}$ is an 'external' field, whose reaction is not treated explicitly, this line does not go to another vertex, and does not, therefore link the various parts of the diagram. In the series, therefore, each term is reducible, so that Σ of (3.127) is just $\mathscr{V}$ itself.

Typically, we are concerned with scattering problems, in which the potential $\mathscr{V}(\mathbf{r})$ is localized. It then becomes important to construct Green functions that refer to natural causal conditions, with, say, a state $|\Psi^+\rangle$ that looks like a simple forward-propagating free-space function $|\Phi\rangle$ at a large distance from the scatterer, and does not have any 'closing in' parts like (4.132). These functions satisfy the integral equation

$$|\Psi^+, \mathbf{r}\rangle = |\Phi, \mathbf{r}\rangle + \int G_0^+(\mathbf{r}, \mathbf{r}'; \mathscr{E})\, \mathscr{V}(\mathbf{r}')\, |\Psi^+, \mathbf{r}'\rangle\, \mathbf{d}^3\mathbf{r}', \qquad (4.138)$$

which is the representation in real space of the *Lippmann–Schwinger equation*

$$|\Psi^+\rangle = |\Phi\rangle + \frac{1}{\mathscr{E} - H_0 + \mathrm{i}\delta} \mathscr{V}\, |\Psi^+\rangle. \qquad (4.139)$$

The similarity of (4.136) to (4.139) is obvious, and can be proved formally. The same type of formula also appeared as (3.15) in the original derivation of the Brillouin–Wigner series.

This language is, of course, very abstract. Each product of operators or 'matrices' in (4.139) may, in fact, involve, an integration over all space, as in (4.138). Nevertheless, for formal manipulations it is very convenient. For example, we can construct an iterative solution of the Lippman–Schwinger equation, analogous to (4.135), i.e.

$$\begin{aligned} |\Psi^+\rangle &= |\Phi\rangle + \frac{1}{\mathscr{E} - H_0 + \mathrm{i}\delta} \mathscr{V}\, |\Phi\rangle + \frac{1}{\mathscr{E} - H_0 + \mathrm{i}\delta} \mathscr{V}\, \frac{1}{\mathscr{E} - H_0 + \mathrm{i}\delta} \mathscr{V}\, |\Phi\rangle + \dots \\ &= \{1 + G_0^+ \mathscr{V} + G_0^+ \mathscr{V} G_0^+ \mathscr{V} + \dots\}\, |\Phi\rangle. \qquad (4.140) \end{aligned}$$

In the context of the scattering problem, this is called the *Born Series*. The term of nth order in $\mathscr{V}$ corresponds to n successive virtual scatterings of the particle by the scatterer, as if the whole scattering effect could be written

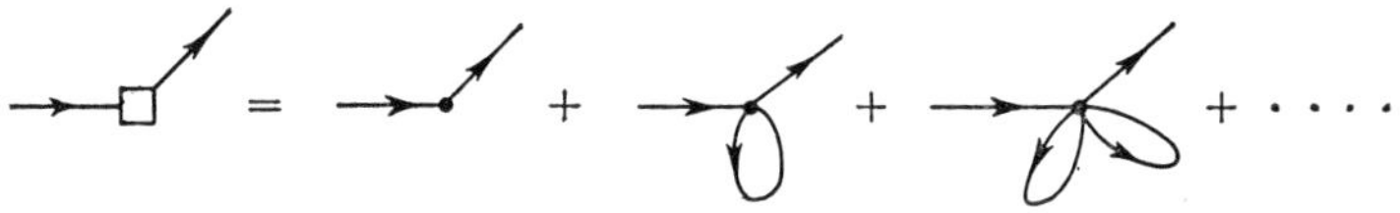

(4.141)

This picture is a form of (4.137), in which we have, so to speak, tried to emphasize that the scattering all takes place in the same region of real space, but have forgotten that there may be any amount of momentum transfer in each virtual process.

To bring this sort of calculation down to earth, let us evaluate the first term. For a plane wave, of momentum $\mathbf{k}$, incident on a localized scatterer, (4.138) becomes

$$\psi^+(\mathbf{r}) = e^{i\mathbf{k}\cdot\mathbf{r}} - \frac{1}{4\pi}\int \frac{e^{ik|\mathbf{r}-\mathbf{r}'|}}{|\mathbf{r}-\mathbf{r}'|}\mathscr{V}(\mathbf{r}')\,\psi^+(\mathbf{r}')\,\mathbf{d}^3\mathbf{r}', \tag{4.142}$$

where the explicit real-space representation (4.131), at the energy $\mathscr{E} = k^2$, has been used for the propagator.

The iterative solution of this equation requires successive integrations over all space, and is therefore somewhat tedious (if it converges!). But the first Born approximation, where the unperturbed plane wave is put in place of the true scattered wave under the integral sign, is easily evaluated. In scattering problems we are only interested in the outgoing wave at a great distance $\mathbf{r}$ from any part of the region where the scattering potential is appreciable. We may use the expansion

$$\frac{e^{ik|\mathbf{r}-\mathbf{r}'|}}{|\mathbf{r}-\mathbf{r}'|} \to \frac{e^{ikr}}{r}\,e^{-i(k/r)\mathbf{r}\cdot\mathbf{r}'} \quad \text{for} \quad r \gg r'; \tag{4.143}$$

the result is

$$\psi^+(\mathbf{r}) \approx e^{i\mathbf{k}\cdot\mathbf{r}} - \frac{1}{4\pi}\int \frac{e^{i\mathbf{k}|\mathbf{r}-\mathbf{r}'|}}{|\mathbf{r}-\mathbf{r}'|}\mathscr{V}(\mathbf{r}')\,e^{i\mathbf{k}\cdot\mathbf{r}'}\,\mathbf{d}^3\mathbf{r}'$$

$$\approx e^{i\mathbf{k}\cdot\mathbf{r}} - \frac{e^{ikr}}{r}\frac{1}{4\pi}\int \mathscr{V}(\mathbf{r}')\,e^{i\mathbf{K}\cdot\mathbf{r}'}\,\mathbf{d}^3\mathbf{r}', \tag{4.144}$$

where $\mathbf{K} = \mathbf{k}-\mathbf{k}'$, and $\mathbf{k}'$ is a vector in the direction of $\mathbf{r}$, but having the same magnitude as the incident wave-vector $\mathbf{k}$. In other words, the outward scattered wave has the same amplitude in this direction as we should get if we were simply to use time-dependent perturbation theory to calculate the matrix element for the transition caused by the potential $\mathscr{V}$ between the plane-wave free-particle states, $|\Phi_{\mathbf{k}}\rangle$ say, and $|\Phi_{\mathbf{k}'}\rangle$.

The Born series (4.140) can, of course, be formally summed, for it is algebraically only a geometric series. Indeed, if we ignore all the operator properties of the symbols in (4.139), we can juggle the Lippman–Schwinger equation into what seems a closed form:

$$|\Psi^+\rangle = |\Phi\rangle + \frac{\mathscr{V}}{\mathscr{E}-H_0-\mathscr{V}+i\delta}|\Phi\rangle. \tag{4.145}$$

But the final term contains what we immediately recognize as the Green function $G(\mathscr{E})$ of the perturbed system, as defined in (4.134), so that we are not really much further ahead.

4.14 The T-matrix

The term-by-term summation of a Born series is a very inefficient way of calculating transition probabilities, unless the convergence is very rapid. Indeed there are many circumstances under which such convergence is not guaranteed at all. Almost the only procedure that then works—and this only for spherically symmetrical potentials—is to separate the Schrödinger equation in spherical harmonics, and then integrate the radial equation outwards to evaluate the *phase shift* $\eta_l(\mathscr{E})$ for the partial wave of angular momentum l at the energy $\mathscr{E}$. The reader will be familiar with the *Faxén–Holtsmark formula*, given in most books on quantum mechanics, which says that the exact solution of (4.142) at large distances is of the form

$$\psi^+(\mathbf{r}) = \mathrm{e}^{\mathrm{i}\mathbf{k}\cdot\mathbf{r}} + \frac{\mathrm{e}^{\mathrm{i}kr}}{r} f(\theta), \tag{4.146}$$

where θ is the angle between the incoming and outgoing wave-vectors, $\mathbf{k}$, and $\mathbf{k}'$, and the scattering amplitude is the complex quantity

$$f(\theta) = \frac{1}{2\mathrm{i}k} \sum_{l=0}^{\infty} (2l+1)(\mathrm{e}^{2\mathrm{i}\eta_l} - 1) P_l(\cos\theta). \tag{4.147}$$

Comparison with (4.144) shows that the first term in the Born series is equivalent to making the approximation

$$4\pi f(\theta) \approx -\langle \Phi_{\mathbf{k}'} | \mathscr{V} | \Phi_{\mathbf{k}} \rangle. \tag{4.148}$$

We now deliberately invert this relation, to ask the following question: what operator, $\mathscr{T}$ say, gives the *exact* scattering amplitude (4.147) for its matrix elements between the free-particle states $|\Phi_{\mathbf{k}}\rangle$ and $|\Phi_{\mathbf{k}'}\rangle$? We study the properties of the *transition matrix*, or T-*matrix*, which is defined so that

$$\langle \Phi_{\mathbf{k}'} | \mathscr{T} | \Phi_{\mathbf{k}} \rangle = -4\pi f(\theta). \tag{4.149}$$

In other words, $\mathscr{T}$ is what $\mathscr{V}$ would have to become if the Born formula (4.144) were to yield the correct transition probability.

Stated thus, this concept is apparently of no greater significance than whatever might have been achieved in summing the Born series. However, the T-matrix does satisfy certain algebraic relations, which provide valuable formal connections between elementary partial-wave scattering theory, Green functions, and the general theory of the S-matrix. These relationships can be written down in many equivalent forms, of which we shall only give a few examples.

To give substance to the definition (4.149), let us start from the Lippman–Schwinger equation in its closed form (4.145). Let $|\Phi_{\mathbf{k}}\rangle$ represent a free-particle wave function, which acquires an outgoing scattered wave when modified into the function $|\Psi_{\mathbf{k}}^{+}\rangle$. Under the influence of the potential $\mathscr{V}$ we ask for the probability that at a very large distance from the scattering centre this should look like another free-particle state $|\Phi_{\mathbf{k}'}\rangle$. We cannot calculate this by simply taking the scalar product $\langle\Phi_{\mathbf{k}'}|\Psi_{\mathbf{k}}^{+}\rangle$, for these two functions do not satisfy the same Schrödinger equation in the neighbourhood of the scatterer. But we may treat $|\Phi_{\mathbf{k}'}\rangle$ as the major part of a function $|\Psi_{\mathbf{k}'}^{-}\rangle$ that is an exact eigenstate of the perturbed system—a state analogous to a scattering solution but with only *incoming* spherical waves. That is to say, $|\Psi_{\mathbf{k}'}^{-}\rangle$ is the solution of a Lippman–Schwinger equation involving $|\Phi_{\mathbf{k}'}\rangle$, but with $-\mathrm{i}\delta$ in the energy denominator, just as in (4.132); it is an 'advanced' solution corresponding to the time-reversal of our ordinary causal solutions. This choice ensures that the scattering process is itself microscopically reversible—that we get the same matrix element in going from $|\Phi_{\mathbf{k}}\rangle$ to $|\Phi_{\mathbf{k}'}\rangle$ as we would in the opposite direction.

The transition probability must therefore be proportional to the square of the inner product of these two state vectors. Some rather finicky algebraic manipulations, using (4.139) and the rules of Hermitian conjugation, can be made to lead to the following:

$$\begin{aligned}\langle\Psi_{\mathbf{k}'}^{-}|\Psi_{\mathbf{k}}^{+}\rangle &= \langle\Phi_{\mathbf{k}'}|\Phi_{\mathbf{k}}\rangle + \left\{\frac{1}{\mathscr{E}-\mathscr{E}'+\mathrm{i}\delta}+\frac{1}{\mathscr{E}'-\mathscr{E}+\mathrm{i}\delta}\right\}\langle\Phi_{\mathbf{k}'}|\mathscr{V}|\Psi_{\mathbf{k}}^{+}\rangle \\ &= \delta_{\mathbf{k}\mathbf{k}'} - 2\pi\mathrm{i}\delta(\mathscr{E}-\mathscr{E}')\langle\Phi_{\mathbf{k}'}|\mathscr{V}|\Psi_{\mathbf{k}}^{+}\rangle, \end{aligned}\tag{4.150}$$

where the analytic representation of a δ-function occurs as in (4.65).

This formula shows that the transition rate is zero unless energy is conserved (more precisely: energy is conserved with uncertainty δ, where $1/\delta$ is the time available for the transition to go to completion). The first Born approximation would arise if we were to replace $|\Psi_{\mathbf{k}}^{+}\rangle$ by $|\Phi_{\mathbf{k}}\rangle$ on the right-hand side. It follows, therefore, that the exact matrix element in (4.150) is equivalent to the matrix element of the T-matrix demanded in (4.149)—i.e.

$$\left.\begin{aligned}\langle\Phi_{\mathbf{k}'}|\mathscr{T}|\Phi_{\mathbf{k}}\rangle &= \langle\Phi_{\mathbf{k}'}|\mathscr{V}|\Psi_{\mathbf{k}}^{+}\rangle,\\ \text{or}\qquad \mathscr{T}|\Phi_{\mathbf{k}}\rangle &= \mathscr{V}|\Psi_{\mathbf{k}}^{+}\rangle.\end{aligned}\right\}\tag{4.151}$$

Now put this symbolism into the Lippman–Schwinger equation (4.139); we get

$$\mathscr{T}\,|\Phi_{\mathbf{k}}\rangle = \left\{\mathscr{V} + \mathscr{V}\,\frac{1}{\mathscr{E} - H_0 + \mathrm{i}\delta}\,\mathscr{T}\right\}|\Phi_{\mathbf{k}}\rangle, \tag{4.152}$$

which seems to hold whatever the state $|\Phi_{\mathbf{k}}\rangle$. We may thus write down various operator equations, analogous to the Born series (4.140):

$$\begin{aligned}\mathscr{T} &= \mathscr{V} + \mathscr{V} G_0^+\,\mathscr{T} \\ &= \mathscr{V} + \mathscr{V} G_0^+\,\mathscr{V} + \mathscr{V} G_0^+\,\mathscr{V} G_0^+\,\mathscr{V} + \ldots\end{aligned} \tag{4.153}$$

or in closed form, by analogy with (4.145) and the Dyson equation (4.136),

$$\mathscr{T} = \mathscr{V} + \mathscr{V} G^+ \mathscr{V}, \tag{4.154}$$

where G^+ is the causal Green function for the total, perturbed system.

These formulae look very powerful and abstract, and are sometimes used as basic definitions from which the T-matrix of a system is to be derived. It must be remembered, however, that the step of dropping the state vector $|\Phi_{\mathbf{k}}\rangle$ from (4.151) is not really warranted. The whole theory of the present sections depends upon the energy $\mathscr{E}$ being the same for all states under consideration; the relationship (4.150) is of no physical significance in (4.149) unless $\mathscr{E}(\mathbf{k}) = \mathscr{E}(\mathbf{k}')$. We may say that the T-matrix is *not defined off the energy shell*, or we use a new symbol, and write, say

$$\langle\Phi_{\mathbf{k}'}|\,T\,|\Phi_{\mathbf{k}}\rangle = 2\pi\mathrm{i}\delta(\mathscr{E} - \mathscr{E}')\langle\Phi_{\mathbf{k}'}|\,\mathscr{V}\,|\Psi_{\mathbf{k}}^+\rangle, \tag{4.155}$$

so as to fix all such arbitrary elements at zero. Unfortunately, there is no settled notation in this field.

For this reason, the actual analytical representation of the T-matrix in real space or momentum space is not usually of much interest. But for potential scattering by a spherically symmetrical field, the angular momentum representation is especially favourable, for the T-matrix is then diagonal. In other words, (4.146) and (4.148) are equivalent to writing, for $\mathbf{r}, \mathbf{r}'$ large,

$$T(\mathbf{r}, \mathbf{r}'; \mathscr{E}) = \sum_l T_l\, j_l(\kappa r)\, j_l(\kappa r')\, Y_{l0}(\hat{\mathbf{r}})\, Y_{l0}(\hat{\mathbf{r}}'), \tag{4.156}$$

where j_l is a spherical Bessel function and $Y_{l0}(\hat{\mathbf{r}})$ a spherical harmonic for the direction of the vector $\mathbf{r}$. The diagonal matrix elements of T in this representation are given by

$$T_l = -\frac{1}{2\mathrm{i}\kappa}\,(\mathrm{e}^{2\mathrm{i}\eta_l(\mathscr{E})} - 1). \tag{4.157}$$

Now the phase shifts $\eta_l(\mathscr{E})$ are always real (for $\mathscr{E} > 0$), so these eigenvalues are complex numbers; *the T-matrix is not Hermitian.* This can be deduced in quite a different way. In the language of § 3.5, the perturbation, turning $|\Phi\rangle$ into $|\Psi^+\rangle$, is equivalent to the action of an S-matrix, i.e.

$$|\Psi^+\rangle = S|\Phi\rangle. \tag{4.158}$$

The definition of the T-matrix given in (3.62) is in fact equivalent to (4.155). The Lippman–Schwinger equation (4.150) allows us to express this in the form

$$S = 1 - 2\pi \mathrm{i}\delta(\mathscr{E} - H_0)T. \tag{4.159}$$

But the S-matrix is unitary; from the relation

$$1 = S^*S = \{1 + 2\pi \mathrm{i}T^*\delta(\mathscr{E} - H_0)\}\{1 - 2\pi \mathrm{i}\delta(\mathscr{E} - H_0)T\} \tag{4.160}$$

we get

$$T^* - T = 2\pi \mathrm{i}T^*\,\delta(\mathscr{E} - H_0)\,T \neq 0 \tag{4.161}$$

which proves that the T-matrix is not Hermitian. Perhaps this connection with the S-matrix should be the basic defining property of T.

It is sometimes convenient, therefore, to define another similar mathematical object, called the *K-matrix*, or *reaction matrix*, which is in fact Hermitian. A suitable modification of (4.159) is

$$\frac{1-S}{1+S} = \pi \mathrm{i}\delta(\mathscr{E} - H_0)\,K. \tag{4.162}$$

It is easy to show that K is Hermitian, and therefore has the real eigenvalues, in the angular momentum representation (cf. (4.157)),

$$K_l = -\frac{1}{\kappa}\tan\eta_l(\mathscr{E}). \tag{4.163}$$

The significance of the K-matrix is that it satisfies an integral equation (4.153) with the 'non-causal Green function' (4.133) in place of G^+. We may write this in the form

$$K = \mathscr{V} + \mathscr{V}\frac{P}{\mathscr{E} - H_0}K, \tag{4.164}$$

where we are to take only the principal part in any integration over $\mathscr{E}$; this is obviously equivalent to (3.15) used in the derivation of the Brillouin–Wigner expansion.

Physically, the K-matrix represents the total effect of the scattering potential when at the heart of a standing wave system—i.e. with both incoming and outgoing spherical waves as described by the propagator (4.133). It is not surprising, therefore, that a singularity of the

K-matrix can occur when one of the phase shifts $\eta_l(\mathscr{E})$ passes through $\pi/2$, for then this spherical standing wave outside the scattering sphere decays most rapidly when matched against the solution within the potential $\mathscr{V}(\mathbf{r})$. We then get a quasi-stationary state, or *resonance* of the system.

This discussion of the T-matrix and its confrères is somewhat abstract; the main point is that these operators provide the correct invariant language for the discussion of scattering problems, even in complicated situations where elastic and inelastic scattering may take place, as of nucleons by nuclei, and where there may be many alternative and competing end products of the reactions that occur in the target region. When we cannot exactly solve the equations—when, indeed the scattering is not describable by a 'potential'—it is very useful to have parameters, such as the elements of these matrices, whose analytical properties may be guessed with some facility (see § 5.6).

4.15 Example: impurity states in a metal

A number of the properties of the elementary time-independent Green function are exemplified rather nicely by the following simplified model calculation, due to A. M. Clogston (*Phys. Rev.* **125**, 439, 1962). Consider a homogeneous system, such as a crystalline metal, where the electron states (Bloch functions) are approximately plane waves each characterized by a wave-vector $\mathbf{k}$ and an energy $\mathscr{E}(\mathbf{k})$. Suppose, however, that the allowed energies fall into a band, so that the density of states function $\mathscr{N}(\mathscr{E})$ vanishes unless $\mathscr{E}_m < \mathscr{E} < \mathscr{E}_M$. What happens if a very sharply localized potential—in practice the field of a single impurity atom, but idealized into a delta function

$$\mathscr{V}(\mathbf{r}) = \mathscr{V}_0 \delta(\mathbf{r}-\mathbf{r}_0) \tag{4.165}$$

of strength $\mathscr{V}_0$—is put at some point $\mathbf{r}_0$ in the metal?

Let us apply the most abstract of our formal equations, the 'Dyson equation', (4.136). In a real space representation, this tells us that the Green function of the perturbed system is derived from the corresponding Green function G_0 in the unperturbed medium through the integral equation

$$\begin{aligned} G(\mathbf{r},\mathbf{r}') &= G_0(\mathbf{r},\mathbf{r}') + \int G_0(\mathbf{r},\mathbf{r}'')\,\mathscr{V}(\mathbf{r}'')\,G(\mathbf{r}'',\mathbf{r}')\,\mathbf{d}^3\mathbf{r}' \\ &= G_0(\mathbf{r},\mathbf{r}') + G_0(\mathbf{r},\mathbf{r}_0)\mathscr{V}_0 G(\mathbf{r}_0,\mathbf{r}') \end{aligned} \tag{4.166}$$

because of the localization (4.165) of the potential. This happens to be a case where the operator $(1-G_0\mathscr{V})^{-1}$, which would generate a formal solution of (4.136), is in fact diagonal in a space-representation. Thus, we can put $\mathbf{r} = \mathbf{r}_0$ in (4.166), and get

$$G(\mathbf{r}_0, \mathbf{r}') = G_0(\mathbf{r}_0, \mathbf{r}') + G_0(\mathbf{r}_0, \mathbf{r}_0)\mathscr{V}_0 G(\mathbf{r}_0, \mathbf{r}')$$

$$= \frac{1}{1-G_0(\mathbf{r}_0, \mathbf{r}_0)\mathscr{V}_0} G_0(\mathbf{r}_0, \mathbf{r}'). \tag{4.167}$$

Putting this back into (4.166), we get the complete solution to the problem:

$$G(\mathbf{r}, \mathbf{r}') = G_0(\mathbf{r}, \mathbf{r}') + G_0(\mathbf{r}, \mathbf{r}_0)\frac{\mathscr{V}_0}{1-G_0(\mathbf{r}_0, \mathbf{r}_0)\mathscr{V}_0} G_0(\mathbf{r}_0, \mathbf{r}'). \tag{4.168}$$

To use this solution, we need an explicit formula for the Green function of the unperturbed system. From the information supplied to us, this may be written as follows, using the spectral representation (4.118):

$$G_0^+(\mathbf{r}, \mathbf{r}'; \mathscr{E}) = \sum_{\mathbf{k}} \frac{\mathrm{e}^{\mathrm{i}\mathbf{k}\cdot(\mathbf{r}-\mathbf{r}')}}{\mathscr{E}-\mathscr{E}(\mathbf{k})+\mathrm{i}\delta}. \tag{4.169}$$

For $\mathbf{r} \neq \mathbf{r}'$, this sum may be transformed into an energy integral involving $\mathscr{N}(\mathscr{E})$, and evaluated by the same procedure as in §4.12; we get the analogue of (4.131), i.e.

$$G_0^+(\mathbf{R}; \mathscr{E}) = -\pi\mathscr{N}(\mathscr{E})\frac{\mathrm{e}^{\mathrm{i}kR}}{kR}, \tag{4.170}$$

where k is the wave-number for which $\mathscr{E}(\mathbf{k}) = \mathscr{E}$. But when $\mathbf{r} = \mathbf{r}' = \mathbf{r}_0$, say, we do not enjoy the convergence factor associated with the spatial wave function, and must use the delta-function representation (e.g. (4.65)) to give

$$G_0(\mathbf{r}_0, \mathbf{r}_0) = G_0^+(0; \mathscr{E}) = \int \frac{\mathscr{N}(\mathscr{E}')\,\mathrm{d}\mathscr{E}'}{\mathscr{E}-\mathscr{E}'+\mathrm{i}\delta}$$

$$= I(\mathscr{E}) - \mathrm{i}\pi\mathscr{N}(\mathscr{E}), \tag{4.171}$$

where $I(\mathscr{E})$ is the principal value of the integral.

Now put (4.170) and (4.171) into (4.168). The Green function for the perturbed system is a little more complicated than (4.170), because of the privileged position of the 'impurity' at $\mathbf{r}_0$. But we notice that G always contains a term with a factor

$$\frac{1}{1-G_0^+(0;\mathscr{E})\mathscr{V}_0} = \frac{1}{\{1-I(\mathscr{E})\mathscr{V}_0\}+\mathrm{i}\pi\mathscr{N}(\mathscr{E})\mathscr{V}_0}, \tag{4.172}$$

which is capable, under some circumstances, of becoming infinite. The theory of the resolvent (4.124) tells us to discover the eigenstates of the system by seeking for the singularities of G as a function of $\mathscr{E}$. From (4.172) these singularities occur when simultaneously

$$1 - I(\mathscr{E})\mathscr{V}_0 = 0, \tag{4.173}$$

$$\mathscr{N}(\mathscr{E}) = 0, \tag{4.174}$$

which must therefore be the conditions that the 'impurity' have a bound state of energy $\mathscr{E}$.

The first of these conditions can always be satisfied by choosing the strength of the perturbation $\mathscr{V}_0$, appropriately; but (4.174) requires that the energy lie outside the limits of the band, i.e. $\mathscr{E} < \mathscr{E}_m$ or $\mathscr{E}_M < \mathscr{E}$. But suppose that we had forgotten that the true definition of the resolvent requires us to use a causal Green function, and that we had carelessly used the 'standing wave propagator' G_0^0, as in (4.133); we should then have lost the imaginary part of (4.172), and hence the condition (4.174) would not have seemed necessary. What significance can be attached to an energy that then happens to satisfy (4.173) *inside* the band of 'Bloch' states?

Let us calculate the effect of the potential $\mathscr{V}(\mathbf{r})$ on a wave function. The spherical symmetry of the system about the point $\mathbf{r}_0$ suggests that we should use an angular momentum representation, and the localization of the perturbation ensures that only the s-wave is affected. For our 'free-particle' state in the Lippmann–Schwinger equation, we therefore use

$$|\Phi\rangle = \frac{\sin k|\mathbf{r}-\mathbf{r}_0|}{k|\mathbf{r}-\mathbf{r}_0|}. \tag{4.175}$$

The solution of (4.145) for the 'scattered' wave follows from (4.167), (4.170), (4.172), i.e.

$$\begin{aligned}
|\Psi^+(\mathbf{r})\rangle &= |\Phi(\mathbf{r})\rangle + \int G(\mathbf{r},\mathbf{r}'')\mathscr{V}(\mathbf{r}'')\,|\Phi(\mathbf{r}'')\rangle\,\mathbf{d}^3\mathbf{r}'' \\
&= |\Phi(\mathbf{r})\rangle + G(\mathbf{r},\mathbf{r}_0)\mathscr{V}_0|\Phi(\mathbf{r}_0)\rangle \\
&= \frac{\sin k|\mathbf{r}-\mathbf{r}_0|}{k|\mathbf{r}-\mathbf{r}_0|} - \frac{\mathscr{V}_0}{1-G_0^+(0;\mathscr{E})\mathscr{V}_0}\pi\mathscr{N}(\mathscr{E})\frac{e^{ik|\mathbf{r}-\mathbf{r}_0|}}{k|\mathbf{r}-\mathbf{r}_0|} \\
&= e^{i\eta(\mathscr{E})}\frac{\sin\{k|\mathbf{r}-\mathbf{r}_0|+\eta(\mathscr{E})\}}{k|\mathbf{r}-\mathbf{r}_0|},
\end{aligned} \tag{4.176}$$

where the phase shift is given by

$$\eta(\mathscr{E}) = \tan^{-1}\left\{\frac{\pi\mathscr{N}(\mathscr{E})\mathscr{V}_0^2}{I(\mathscr{E})\mathscr{V}_0 - 1}\right\}. \tag{4.177}$$

Now suppose that, for a given value of $\mathscr{V}_0$, the denominator of (4.177) vanishes when $\mathscr{E} = \mathscr{E}_r$. We then have, in this region, a formula like

$$\tan \eta(\mathscr{E}) = \frac{W}{\mathscr{E} - \mathscr{E}_r}; \tag{4.178}$$

this is the familiar expression for the phase shift in the neighbourhood of a *resonance* or *virtual state* at $\mathscr{E}_r$. The *width*, W, of the resonance is obviously proportional to $\mathscr{N}(\mathscr{E}_r)$. If the strength of the perturbation $\mathscr{V}_0$ were such as to put $\mathscr{E}_r$ well inside the band, then the resonance would be broad. As $\mathscr{E}_r$ moves towards either edge of the band, $\mathscr{N}(\mathscr{E})$ decreases, and the resonance becomes sharper; as we go outside the band, $\mathscr{N}(\mathscr{E}_r)$ vanishes, and $\mathscr{E}_r$ has become the energy of a true bound state or localized level, as prescribed by (4.173) and (4.174).

The justification of this elementary calculation as a model for an impurity in a metal, and the connection between the behaviour of the model and some actual physical phenomena need not concern us here. The main point is to show that the Green function method can sometimes demonstrate very general properties of systems—note here that only the density of states comes into the final answer, through (4.177) and the integral (4.171)—which might otherwise seem artefacts of the model. It is not necessarily the easiest way of calculating, and sometimes it leads one into making somewhat cavalier assumptions in order to make the equations soluble; but through the formal connections sketched out in this chapter it allows one to look at most problems in quantum mechanics from a single analytical viewpoint, and to unify many apparently diverse aspects—scattering, ensemble averages, bound states, excitation spectra, etc. That is why this technique is so greatly favoured by many theoretical physicists with a taste for mathematical rigour, generality and elegance. On the other hand, there is a price to pay. The propagator, being essentially a function of two vector variables, is inevitably more complicated mathematically than a one-particle wave function, which may be pictured in real space as if it were an optical or acoustic field. Except in its diagrammatic applications, there is a loss of direct 'physical' interpretability if Green functions are used indiscriminately in all quantum-mechanical calculations; if we are not careful, we may find ourselves groping blindly with mechanical algebra, when geometrical vision would lead us straight to the answer.

CHAPTER 5

SOME ASPECTS OF THE MANY-BODY PROBLEM

'Many mickles make a muckle.'

5.1 Quantum properties of macroscopic systems

Most of the 'Laws' of quantum theory have been discovered by watching the behaviour of single particles being scattered, making optical transitions, passing through electric and magnetic fields, etc. Nevertheless, we must learn to apply these laws to the calculation of the properties of matter in bulk. The elucidation of the general mathematical features of such calculations in circumstances where quantum effects are important has been one of the major achievements of theoretical physics in the past decade.

Of course, all macroscopic objects are many-body quantum systems; it often turns out, however, that a hierarchy of single-particle and quasi-classical approximations may be made, which avoids the essential difficulty of the problem in general. Thus, we are quite used to treating each electron in an atom as if nearly independent of the others, and each atom as if interacting through ordinary dynamical forces with a few well-defined neighbours. The genuinely subtle problems concern 'quantum liquids'—systems of identical particles, not closely localized in space and in strong interaction with one another. For electrons in metals, atoms and molecules, for nucleons in nuclear matter, and for the atoms in liquid helium, it is difficult to choose between those particles which are sinned against, and those which are sinning.

The actual physical phenomena to be observed and interpreted in such systems—superconductivity, collective modes of excitation, superfluidity, etc.—are of exceptional interest in themselves, but it would take us much too far afield to set them all out here. In conformity with the general plan of this book, we shall take it for granted that the reader is able to find his way into the descriptive literature of this subject, and we shall concentrate on the basic mathematical techniques that are used to justify and quantify such descriptions. Most of the apparatus of second quantization, Feynman diagrams and

Green functions was originally developed in the context of elementary-particle field theory; most of the recent rapid progress in the study of many-body systems has come from the adaptation and application of these methods. Yet one ought to remember that the many-body problem is a topic in its own right, with its own characteristic methods and results. This chapter, therefore, is meant to be an account of some aspects of many-body theory in its own terms, and not merely as an exemplification of the techniques sketched out in earlier chapters. Thus, we begin with the elementary and semi-intuitive Thomas–Fermi and Hartree methods, which are much more comprehensible as 'physical approximations' than as being formally derivable from general perturbation theory.

5.2 Statistical methods: the Thomas–Fermi approximation

Many-particle theories are necessarily statistical, since they involve enormous numbers of particles, whose co-ordinates cannot be given individual attention. From the discussion in §4.1, it is clear that the simplest possible parameter is the diagonal element of the one-particle density matrix in a co-ordinate representation—the local particle density $n(\mathbf{r})$.

In classical statistical mechanics, the first task is to establish an equation of state, or some equivalent relation, such as the equilibrium value of the chemical potential as a function of the local temperature and density. It is characteristic of fermion systems that the interesting quantum phenomena occur at very low temperatures at energies that are small compared with the large zero-point energy imposed by the Pauli principle. The gross features of the system, such as the density distribution of particles in space, are thus nearly independent of actual temperature, which may be treated as effectively zero throughout the gas or liquid.

But kinetic energy may still be the dominant contribution to the free energy. It is well known, for example, that the Fermi energy of a non-interacting electron gas, of uniform density n particles per unit volume, is given by (cf. §2.7)

$$\mathscr{E}_F = \tfrac{1}{2}(3\pi^2)^{\frac{2}{3}} n^{\frac{2}{3}} \tag{5.1}$$

in atomic units. This is, in fact, the chemical potential for electrons, measured from the bottom of the energy distribution, supposedly at zero electrical potential.

What is the effect of including the interactions between the fermions—coulomb forces between electrons or 'nuclear forces' between nucleons? Can we then write down a more general formula for the energy, free energy, or other equivalent thermodynamic variables, in terms of the particle density? This major goal of many-body theory has not been completely achieved, although we know some results in limiting cases, and perhaps understand most of the principles at stake (see, e.g. §5.11).

It is not difficult, however, to eliminate the major part of the long-range coulomb interaction between electrons. Suppose that the electron density is not uniform, but that $n(\mathbf{r})$ varies with position in space. This gives rise to space-charge fields, whose effect on the electrons may be represented by a potential energy $\Phi(\mathbf{r})$ satisfying Poisson's equation, i.e.

$$\nabla^2\Phi(\mathbf{r}) = 4\pi n(\mathbf{r}) \tag{5.2}$$

(taking $|e| = 1$ for simplicity).

But in computing the chemical potential, μ, this electrostatic potential must be added to the Fermi energy, (5.1). Since μ must be the same everywhere, we obtain the condition

$$\mu = \tfrac{1}{2}(3\pi^2)^{\frac{2}{3}}\{n(\mathbf{r})\}^{\frac{2}{3}} + \Phi(\mathbf{r}). \tag{5.3}$$

Combining (5.3) and (5.2), gives the self-consistent differential equation

$$\nabla^2\Phi(\mathbf{r}) = \frac{8\sqrt{2}}{3\pi}\{\mu - \Phi(\mathbf{r})\}^{\frac{3}{2}} \tag{5.4}$$

derived originally by Thomas and Fermi.

This equation is non-linear, but for simple cases it can be integrated numerically with suitable boundary conditions; for example, in a spherically symmetrical model of an atom with nuclear charge Z we should try to find solutions such that

$$\Phi(\mathbf{r})\begin{cases} \to -\dfrac{Z}{r} & \text{as} \quad r \to 0, \\ \to 0 & \text{as} \quad r \to \infty. \end{cases} \tag{5.5}$$

It is obvious from the above argument that the Thomas–Fermi method is best at discussing deviations from uniformity of charge density, and is quite unsuitable as a basis for the calculation of the total energy of a system of interacting electrons. Nevertheless, the picture that it gives of the general distribution of the electron charge in atoms, molecules and solids is not absurd; such systems cannot deviate too far from the Thomas–Fermi situation without good cause.

The approximation is invalid, however, under two circumstances: the assumption of 'local' density and homogeneity cannot be justified when the fields are supposed to vary very rapidly in space; and the discreteness of electron charge must be allowed for in regions where the electron density ought, in fact, to be very low. The solutions of (5.4) and (5.5) are thus very poor near the nucleus of the atom and in regions well beyond the outermost occupied shell.

Attempts have been made to improve the method by allowing for more subtle effects in the electron gas. Thus, in addition to the pure coulomb energy one might include an 'exchange energy' of the form

$$\mathscr{E}_{\text{ex}} = -\alpha n^{\frac{4}{3}}, \tag{5.6}$$

where α is a constant. Adding this to the chemical potential (5.3), we get a somewhat more complicated differential equation to be solved for $n(\mathbf{r})$ and $\Phi(\mathbf{r})$. This is called the *Thomas–Fermi–Dirac method,* but does not turn out to be very much better. In general, any functional relation by which the energy of the electron gas may be expressed in terms of its density—including generalized non-local relations similar to the integral form of (5.2),

$$\Phi(\mathbf{r}) = \int \frac{n(\mathbf{r}')}{|\mathbf{r}-\mathbf{r}'|}\,\mathbf{d}^3\mathbf{r}' \tag{5.7}$$

—could be used to construct the chemical potential, and hence generate a self-consistent differential or integral equation for $n(\mathbf{r})$. All such procedures suffer from the same weaknesses as the simple Thomas–Fermi approximation; they do not take into account the phase relations between the various electron wave functions which produce correlation and exchange effects in the overall distribution of charge.

5.3 Hartree self-consistent field

The next step is due to Hartree. The average field $\Phi(\mathbf{r})$ produced by the electron distribution is treated as part of the potential energy in the Schrödinger equation for each separate electron. For the ith electron, say, the wave function $\psi_i(\mathbf{r})$ must satisfy the equation

$$-\frac{\hbar^2}{2m}\nabla^2\psi_i(\mathbf{r}) + \mathscr{V}_{\text{s.c.}}(\mathbf{r})\,\psi_i(\mathbf{r}) = \mathscr{E}_i\psi_i(\mathbf{r}). \tag{5.8}$$

The local electron density is then given by

$$n(\mathbf{r}) = \sum_{i=1}^{N} |\psi_i(\mathbf{r})|^2. \tag{5.9}$$

This, in its turn, gives rise to an electrostatic potential, $\Phi(\mathbf{r})$, which must satisfy the Poisson equation (5.2). The total *self-consistent field* acting on the electron must therefore be

$$\mathscr{V}_{\text{s.c.}}(\mathbf{r}) = \mathscr{V}_{\text{ext}}(\mathbf{r}) + \Phi(\mathbf{r}), \tag{5.10}$$

where $\mathscr{V}_{\text{ext}}(\mathbf{r})$ is the potential due to all 'extraneous' charges, such as the nuclei of atoms.

For a given assignment of electrons to the various eigenfunctions of (5.8), these equations usually have a unique solution, which may be arrived at by iteration. We start from some arbitrarily chosen approximate potential, which is put into (5.8) to yield wave functions for the electron density (5.9), and hence, by (5.2) a potential $\Phi(\mathbf{r})$ to put into (5.10). The new estimate of the self-consistent field goes back into (5.8), and the procedure repeated.

The best examples of the use of the Hartree method are for atoms, where the central field of the nucleus is the dominant term, but where each electron may be assigned to an atomic shell, whose field in its turn, helps screen the nuclear charge for electrons in exterior shells. But a little thought about this familiar type of system soon suggests a correction to the argument: in the sum for $n(\mathbf{r})$, and in the calculation of $\Phi(\mathbf{r})$, we ought to exclude that part of the electron density associated with the electron state currently being considered in the Schrödinger equation—for an electron does not experience any force from its own electric charge. In other words, the equations to be solved should read

$$\left.\begin{aligned} -\frac{\hbar^2}{2m}\nabla^2\psi_i(\mathbf{r}) + \{\mathscr{V}_{\text{ext}}(\mathbf{r}) + \Phi^{(i)}(\mathbf{r})\}\,\psi_i(\mathbf{r}) &= \mathscr{E}_i\psi_i(\mathbf{r}), \\ \nabla^2\Phi^{(i)}(\mathbf{r}) &= 4\pi \sum_{\substack{j=1 \\ j\neq i}}^{N} |\psi_j(\mathbf{r})|^2, \end{aligned}\right\} \tag{5.11}$$

where each electron is liable to see a different average field.

The Hartree method is obviously very general. Any theory which provides an expression for 'the energy of a single electron' as a function or functional of the distribution of total electron density might be used as a basis for the definition of a self-consistent field. In the calculation of atomic wave functions for example, exchange effects are often allowed for by the *Slater exchange hole approximation*: much as in (5.6), each electron is supposed to be subject to a local potential energy proportional to $\{n(\mathbf{r})\}^{\frac{1}{3}}$, which is included in the Schrödinger equation (5.8) or (5.11).

For problems involving inhomogeneous charge distributions, such as the electrons in atoms and molecules, the self-consistent field method is often very satisfactory, being quite accurate and well-adapted to numerical computations. But when it is applied to the standard problem of the uniform fermion gas, it fails lamentably, often giving nonsense answers. Since we must have a good theory of the uniform gas or liquid, both in its own right and as a source of energy expressions for the treatment of inhomogeneous systems, it is essential to dig deeper.

Unfortunately the argument given above is 'intuitive', and does not give any hints as to the nature of the mathematical errors, that have been made. We now proceed, therefore, to a more formal derivation of a set of Hartree-like self-consistent field equations from which (5.11) can be deduced as a simple approximation.

5.4 The Hartree–Fock method

Consider an assembly of fermions interacting with one another by a potential $v(\mathbf{r}-\mathbf{r}')$ (for example, an electrostatic coulomb field), and also subject to an 'external' potential $\mathscr{V}_{\text{ext}}(\mathbf{r})$. In the absence of the interaction, the one-particle Hamiltonian of the system would be, say,

$$\mathscr{H}_0(\mathbf{r}) = -\frac{\hbar^2}{2m}\nabla^2 + \mathscr{V}_{\text{ext}}(\mathbf{r}), \tag{5.12}$$

whose eigenstates would provide a basis for various many-particle determinantal wave functions for the whole system, as in (2.2).

Let $|\Psi\rangle$ symbolize the true ground state of the interacting system. The ground state energy must be equal to the expectation value of the total Hamiltonian in this state; using (4.5) and (4.81) we express this in the form

$$\begin{aligned}\langle\Psi|\,\mathscr{H}\,|\Psi\rangle &= \mathrm{Tr}\,[\rho\{\mathscr{H}_0 + v(\mathbf{r}-\mathbf{r}')\}] \\ &= \int \mathscr{H}_0\rho(\mathbf{r},\mathbf{r})\,\mathbf{d}^3\mathbf{r} + \tfrac{1}{2}\iint \rho^{(2)}(\mathbf{r},\mathbf{r}';\mathbf{r},\mathbf{r}')\,v(\mathbf{r}-\mathbf{r}')\,\mathbf{d}^3\mathbf{r}\,\mathbf{d}^3\mathbf{r}'.\end{aligned} \tag{5.13}$$

But actually to calculate the two-particle density matrix $\rho^{(2)}$ in the state $|\Psi\rangle$ requires a solution of the many-body problem. Remember, however, the formula (4.86) for the two-particle Green function for non-interacting particles:

$$K_0(1234) = G_0(13)\,G_0(24) - G_0(14)\,G_0(23). \tag{5.14}$$

In this simple case, the two-particle density matrix must similarly reduce to a function of one-particle density matrices. Let us assume that this reduction is approximately valid for our more complex interacting system—that we may write

$$\rho^{(2)}(\mathbf{r},\mathbf{r}';\mathbf{r},\mathbf{r}') \approx \rho(\mathbf{r},\mathbf{r})\,\rho(\mathbf{r}',\mathbf{r}') - \rho(\mathbf{r},\mathbf{r}')\,\rho(\mathbf{r},\mathbf{r}'), \tag{5.15}$$

where the one-particle density matrix $\rho(\mathbf{r},\mathbf{r}')$ is not necessarily the same as for independent particles, but is to be determined so as to minimize our approximate estimate of the ground state energy

$$\langle \mathscr{H} \rangle \approx \int \mathscr{H}_0(\mathbf{r})\,\rho(\mathbf{r},\mathbf{r})\,\mathbf{d}^3\mathbf{r} + \tfrac{1}{2}\iint \{\rho(\mathbf{r},\mathbf{r})\,\rho(\mathbf{r}',\mathbf{r}') - \rho(\mathbf{r},\mathbf{r}')\,\rho(\mathbf{r},\mathbf{r}')\}\,v(\mathbf{r}-\mathbf{r}')\,\mathbf{d}^3\mathbf{r}\,\mathbf{d}^3\mathbf{r}'. \tag{5.16}$$

Now we express this 'effective one-particle density matrix' in terms of a limited set of 'occupied one-electron wave functions' as yet to be determined; i.e. we write

$$\rho(\mathbf{r},\mathbf{r}') = \sum_{j=1}^{N} u_j^*(\mathbf{r}')\,u_j(\mathbf{r}), \tag{5.17}$$

as if we knew that the N electrons (or other fermions) could be treated as if they occupied the N quasi-independent orbitals $u_j(\mathbf{r})$. Putting this into (5.16) we get

$$\langle \mathscr{H} \rangle \approx \sum_j \int u_j^*(\mathbf{r})\,\{\mathscr{H}_0(\mathbf{r}) + \tfrac{1}{2}\Phi_H(\mathbf{r})\}\,u_j(\mathbf{r})\,\mathbf{d}^3\mathbf{r} + \tfrac{1}{2}\sum_j \iint u_j^*(\mathbf{r})\,\Phi_{\text{exch}}(\mathbf{r},\mathbf{r}')\,u_j(\mathbf{r}')\,\mathbf{d}^3\mathbf{r}\,\mathbf{d}^3\mathbf{r}', \tag{5.18}$$

where

$$\Phi_H(\mathbf{r}) \equiv \int v(\mathbf{r}-\mathbf{r}')\,\rho(\mathbf{r}',\mathbf{r}')\,\mathbf{d}^3\mathbf{r}' \tag{5.19}$$

would, in the case of a coulomb interaction, be just the Hartree self-consistent field potential defined by (5.7) and (5.9), whilst the *exchange field* is defined through a non-local operator

$$\Phi_{\text{exch}}(\mathbf{r},\mathbf{r}') = -v(\mathbf{r},\mathbf{r}')\,\rho(\mathbf{r},\mathbf{r}'). \tag{5.20}$$

The integrals in (5.18) may be considered as functionals of each $u_j(\mathbf{r})$ which may be varied to minimize the energy. In this operation it is important to note that Φ_H and Φ_{exch} are themselves functionals of the $u_j(\mathbf{r})$ and must be varied accordingly. Introducing Lagrangian variational parameters ϵ_j to preserve the normalization of the 'orbitals' $u_j(\mathbf{r})$, we get the set of simultaneous integro-differential *Hartree–Fock equations*

$$\left\{-\frac{\hbar^2}{2m}\nabla^2+\mathscr{V}_{\text{ext}}(\mathbf{r})+\Phi_H(\mathbf{r})\right\}u_j(\mathbf{r})+\int\Phi_{\text{exch}}(\mathbf{r},\mathbf{r}')\,u_j(\mathbf{r}')\,\mathbf{d}^3\mathbf{r}' = \epsilon_j u_j(\mathbf{r}). \tag{5.21}$$

These equations are obviously very similar to the Hartree equations (5.11). Each equation is of Schrödinger type, to be solved for an eigenfunction $u_j(\mathbf{r})$ in the combined field of the external potential $\mathscr{V}_{\text{ext}}(\mathbf{r})$, the average Hartree potential $\Phi_H(\mathbf{r})$, and the non-local exchange potential Φ_{exch}. But they are, in fact, non-linear, because of (5.17), (5.19) and (5.20), and can only be solved by an iterative self-consistency procedure.

It is not difficult to deduce the Hartree–Fock equations from the slightly more elementary assumption that our many-particle wave function may be expressed as a *single* determinant over a restricted set of unknown functions

$$|\Psi\rangle = \begin{vmatrix} u_1(\mathbf{r}_1) & u_1(\mathbf{r}_2) & \dots & u_1(\mathbf{r}_N) \\ u_2(\mathbf{r}_1) & u_2(\mathbf{r}_2) & \dots & \dots \\ \dots & \dots & \dots & \dots \\ u_N(\mathbf{r}_1) & \dots & \dots & u_N(\mathbf{r}_N) \end{vmatrix}. \tag{5.22}$$

We arrive fairly easily at (5.18) for the expectation value of the energy. The derivation in the language of density matrices is a little more compact, and, in (5.15), illustrates the nature of the approximation that has been made.

What is the meaning of the exchange term? This is a subject upon which much has been written. In an algebraic sense it arises from the anticommutativity of fermion field operators, as expressed by the rules for time-ordering—which is another way of saying that the many-fermion state functions are totally antisymmetric for the exchange of particle co-ordinates. The use of a determinantal wave function, as in (5.22), forces certain phase relations upon the wave functions, which appear in the off-diagonal elements of the density matrix and which ensure that no two particles may seem to be occupying the same orbital state. We notice, for example, that the two-particle density matrix (5.15) vanishes when $\mathbf{r} = \mathbf{r}'$, implying that no two particles can be at the same point in space; each electron is surrounded by an 'exchange hole', where other electrons are not to be found. But this argument only applies to electrons of the same spin, and does not prevent two electrons of opposite spin—essentially *distinguishable* fermions—from approaching one another.

The evaluation of the exchange field is often very laborious, and it is usual to make some approximation, such as replacing the non-local operator by its average value in a free electron gas at the appropriate local particle density; as we have remarked, the equations are then of a modified Hartree type. It is worth noticing, however, that the contributions of the jth orbital to the Hartree and exchange fields (5.19) and (5.20) exactly cancel in the 'Schrödinger equation' (5.21) for this particular function. In evaluating the particle density and density matrix from which these fields are to be derived, we could have excluded this term from the sum over occupied states, (5.17). Thus, the Hartree equations (5.11), by supposedly excluding the field of the chosen electron acting on itself, actually makes some allowance for exchange effects.

It is important to remember that the total energy of the system is not to be calculated by simply adding up the 'energy eigenvalues' ϵ_j of the separate electron states. These are merely variational parameters for the non-linear functional (5.18), so that ϵ_j measures the effect of varying u_j in the field of all the other electrons, and adding all the ϵ_j would count the average interaction between the jth and kth electrons twice over. The true estimate of the total energy is obtained by finding the expectation value of the Hartree and exchange fields in (5.18), and can be written

$$E = \langle \mathscr{H} \rangle = \sum_j \epsilon_j - \tfrac{1}{2}\langle \Phi_H + \Phi_{\text{exch}} \rangle. \tag{5.23}$$

5.5 Diagrammatic interpretation of Hartree–Fock theory

We cannot possibly discuss here the numerous applications of the Hartree–Fock method to atoms, molecules and solids. Generally speaking, we may say that it improves upon the more intuitive Hartree method by putting in some genuine exchange effects, and can therefore account for phenomena such as an apparent *exchange interaction* between two fermions, depending on their relative spin orientations.

As an approach to the basic problem of calculating the ground state energy of a uniform fermion gas it goes some of the way towards allowing for correlation effects, by which particles that repel each other at close quarters tend to arrange their lives so as never actually to meet. As we have seen, the exchange hole does this automatically for two fermions of the same spin, but not for those of opposite spin which may suffer, nevertheless, the same antipathy. It is easy to evaluate (5.23), but the true ground state energy of a homogeneous fermion gas must be somewhat lower than this by an amount dubbed the *correlation energy*. For electrons with long range coulomb interactions, this is a relatively familiar topic, discussed at length in connection with the calculation of the cohesive energy of a metal, and not too difficult to allow for by elementary physical arguments. For nuclear matter, however, where the nucleons interact with a very strong repulsion at short distances, this correction is so serious as to make the Hartree–Fock estimate of the binding energy quite misleading.

Even in the case of the electron gas, attempts to calculate the excitation spectrum by treating the residual particle–particle interaction effects as perturbations on the Hartree–Fock ground state also lead to nonsense. It is well known, for example, that the 'exchange correction' to the density of states at the Fermi level is divergent in this approximation.

To understand the techniques that have been evolved to avoid these errors, it is helpful to interpret the Hartree–Fock theory in the language of diagrams, so that we may get some idea of what has been left out.

In this realm of theory there is considerable choice of dialect. Being concerned mainly with the ground state and excitation spectrum of a static system with fixed interaction forces, we could follow some of the important original papers and introduce a time-independent graphical formalism based upon a series expansion of the resolvent (4.124). On the other hand, as pointed out in § 3.8, the familiar time-dependent theory of chapter 3, with Feynman-type graphs generated by the adiabatic switching-on of the interaction $v(\mathbf{r}-\mathbf{r}')$, yields the same results. We represent the fermion–fermion interaction as a 'horizontal' (i.e. instantaneous in time) line, whose label $\mathbf{K}$ measures the amount of momentum transfer from one particle to another in the collision, and is therefore the argument of the factor $v(\mathbf{K})$ introduced into the matrix element of the diagram by each such line, as in

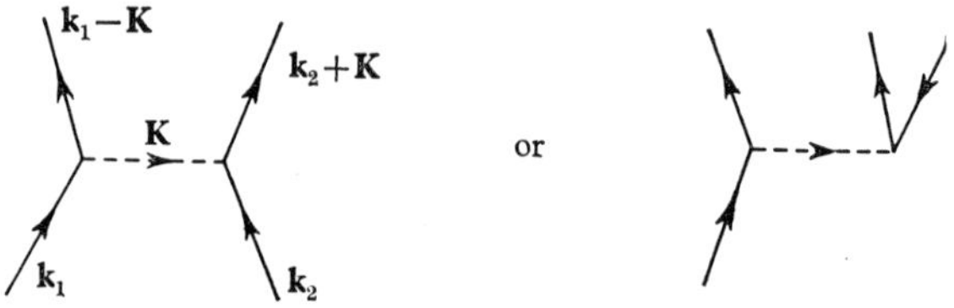

$\mathbf{k}_1-\mathbf{K}$ $\mathbf{k}_2+\mathbf{K}$ $\mathbf{K}$ $\mathbf{k}_1$ $\mathbf{k}_2$ or

It is worth noting that these diagrams are topologically identical with those describing the transfer of virtual bosons between the fermions, although we should then have a somewhat more complicated propagator, such as (3.117), in the algebraic formula corresponding to the diagram. As shown by (3.77), such an exchange of bosons gives rise to a force between the fermions, which might, indeed, be the physical origin of the interaction $v(\mathbf{r}-\mathbf{r}')$. Unless we have sufficient energy to excite a real boson of this field, we do not need to distinguish between these cases (see § 6.11). Electrons in metals at low temperatures, for example, interact by both mechanisms—the direct instantaneous coulomb repulsion and the more subtle force (3.77) due to the electron–phonon interaction.

Using Feynman diagrams, one can prove by a topological argument, essentially the same as was used in § 3.9 to define the 'physical vacuum' of particle theory, that only connected graphs are needed in evaluating genuine physical properties of the system. In the context of the diagrammatic expansion of the resolvent, this is called the *linked cluster theorem*, but does not really need separate proof here.

The ground state energy of the system will be given by the sum of all self-energy diagrams—all linked graphs without 'external lines'. Of these, the leading terms would be

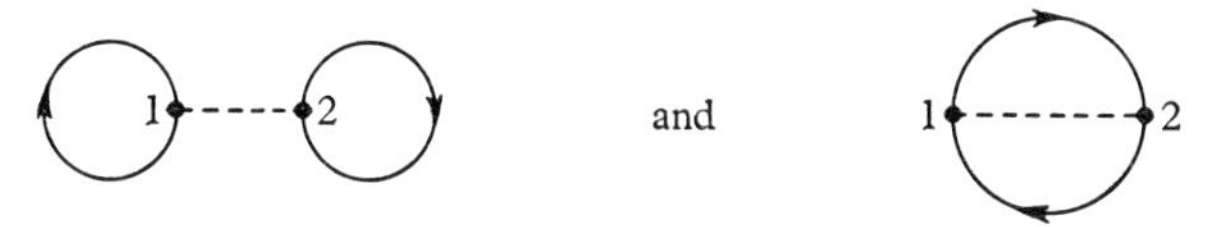

whose algebraic sum is

$$G_0(11)\,v(12)\,G_0(22)-G_0(12)\,v(12)\,G_0(12) = v(12)\,K_0(1212), \quad (5.24)$$

where summation (integration) over variables labelled by repeated indices is implied. We notice that this expression is equivalent to the average value of the interaction potential, upon the assumption that the two-particle density matrix/Green function is the same as for free independent particles, as in (4.86) and (5.14).

But the Hartree–Fock theory does better than this. According to (5.16), we are to evaluate

$$\langle \Phi_H + \Phi_{\text{exch}} \rangle = G(11)\,v(12)\,G(22)-G(12)\;v(12)\,G(21), \quad (5.25)$$

where G is a modified propagator, consistent with this correction to the energy of each state, i.e.

$$G = \frac{1}{\mathscr{E}-\mathscr{E}_0(\mathbf{k})-\langle \Phi_H + \Phi_{\text{exch}} \rangle}. \quad (5.26)$$

These two equations are the basis of the integral equations which are varied and solved self-consistently in (5.17)–(5.21).

The algebraic formula (5.26) for G is in fact equivalent to summing the following Dyson series:

(5.27)

On the left, the symbol for G indicates that the Hartree and exchange fields look like an external potential; the diagrams on the right are obtained by inserting into the free-electron propagator G_0 all possible sub-diagrams corresponding to independent, direct or exchange, virtual excitations of the vacuum—i.e. the diagrams one would obtain from those of (5.24) by opening up one of the fermion lines. In other words, by replacing (5.24) by (5.25) we are replacing an infinite set of self-energy diagrams, such as, for example

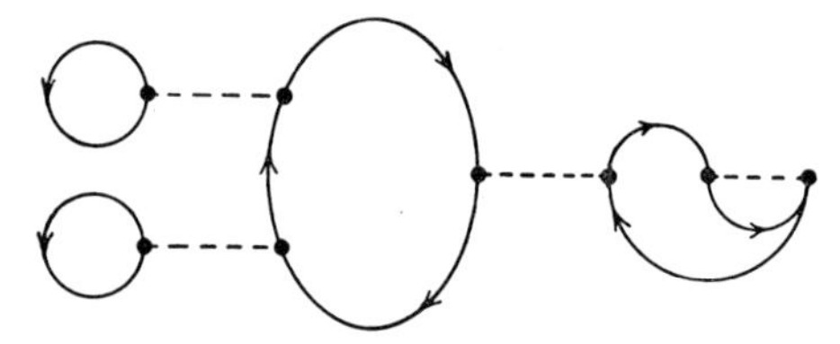

by the two simple 'first order' graphs

and . Indeed, the total energy in the Hartree–Fock approximation is given by just these two diagrams, corrected by a simpler graph which corresponds to the term $-\frac{1}{2}\langle\Phi_H + \Phi_{\text{exch}}\rangle$ in (5.23).

5.6 The Brueckner method

As we have already seen, the Hartree–Fock method fails to give a proper account of correlations between the particles. Indeed, for a uniform gas, the state functions in the determinant (5.22) can only be plane waves (since momentum is a good quantum number) so the modified propagator G only differs from G_0 by a change in the zero of energy. Thus, the two-particle Green function $K(1234)$ is just the same as the free-particle function $K_0(1234)$ of (5.14); the only correlation effect is the 'exchange hole' for particles of the same spin.

The exact two-particle Green function (4.89) cannot, in fact, be represented as a simple difference of products of one-particle propagators, as assumed in (5.15), but contains a special vertex part, labelled $\Gamma(1234)$. This kernel can only be evaluated by summing *all* diagrams representing interactions between a pair of fermions, not merely those included in a series such as (5.27).

For example, the rules by which the Hartree–Fock propagator is constructed do not allow two fermion lines to be joined by more than one interaction line, and thereby exclude *ladder graphs*

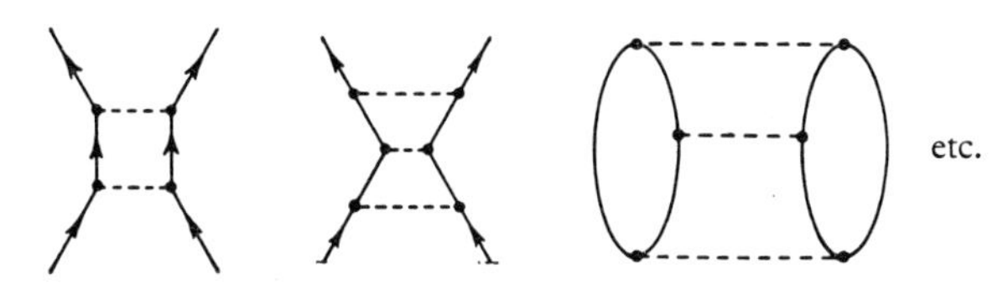

in which the interaction may be repeated several times before the particles go their separate ways. These may occur as sub-diagrams in self-energy diagrams of quite low order, and cannot therefore be safely ignored. Indeed, such diagrams must often be important, for they correspond to the higher-order terms in a straightforward perturbation expansion of the matrix element for the scattering of just two particles, in powers of the interaction potential $v(\mathbf{r}-\mathbf{r}')$. This is, of course, the Born series (4.140), which does not converge at low energies unless the potential is rather weak. It is not surprising, therefore, that the attempt to improve the Hartree–Fock theory by a term-by-term evaluation of these 'corrections' leads to discrepancies and divergences, both for the long-range electrostatic forces between electrons in a metal and the short-range hard-core repulsions in nuclear matter.

But we know that the problem of potential scattering between free particles may be solved directly, without recourse to the Born series. It is an elementary exercise, for example, to calculate the phase shifts, and hence to construct the T-matrix (4.157) for a 'perfectly impenetrable' core, which would of course be grossly singular as a 'perturbation'. The essence of the *Brueckner method* is to replace the sum of all ladder sub-diagrams by the corresponding solution of the two-particle scattering problem, obtained by some other more convergent procedure.

The actual technique of this method is somewhat involved, and by no means cut and dried. We cannot, for example, simply replace the interaction potential v by the corresponding element of the T-matrix, because this is not Hermitian, and would not, therefore, yield real answers for, say, the average 'Hartree–Fock' energy (5.25). In this context, the reaction matrix, or 'K-matrix' (4.162), which is Hermitian, comes into its own.

The chief difficulty is that a 'ladder' occurring as a sub-diagram in a larger closed cluster 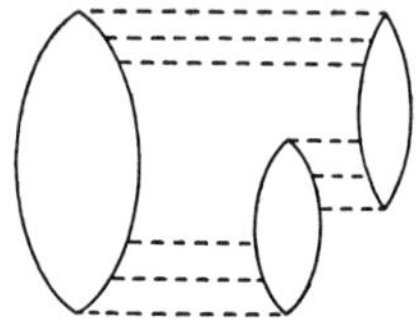 can correspond to a more general 'scattering' process than a real collision between actual fermions. If the process is, so to speak, 'virtual', then it may end or begin at an 'intermediate' state of the perturbation term, and energy need not be conserved in this part of the whole diagram. But neither the T-matrix nor the K-matrix is explicitly defined 'off the energy shell', so we have to make some arbitrary analytical continuation, based perhaps on the integral equation (4.164), to construct a replacement for v in general. Nevertheless, the method has been quite successful in giving estimates of the binding energy for high-density, strongly-interacting fermion systems.

5.7 The dielectric response function

In another light, the Hartree method may be seen as an example of *screening*. In (5.10) we notice that the self-consistent field seen by an electron is not the same as the 'external' applied field; the fermion gas redistributes itself so as to produce new forces on each particle. In other words, the effect of the external field is systematically modified, in magnitude and spatial distribution, by the self-consistent reponse of the many-body system itself. We know, for example, that the field of a static positive charge immersed in an electron gas—say an atom of zinc in metallic copper—draws just sufficient electron charge into its neighbourhood as to be almost neutralized in a very short distance.

The exact solution of the Hartree equations, even in such a simple situation, can only be obtained by laborious numerical computation. Moreover, these equations are not linear; the self-consistent field produced by two external potentials acting together is not merely the superposition of the fields they would each produce separately. The Hartree–Fock theory is thus a *non-linear approximation.*

But suppose we confine ourselves to very weak external fields, and ask for an *exact* theory of the *linear* term in the relationship between $\mathscr{V}_{\text{ext}}(\mathbf{r},t)$, produced by 'external' sources, and the 'total' field $\mathscr{V}_{\text{tot}}(\mathbf{r},t)$ seen by a test particle in the presence of the many-body system. In general, therefore, we hypothesize a *response function* $R(\mathbf{r},t;\mathbf{r}',t')$ such that

$$\mathscr{V}_{\text{tot}}(\mathbf{r},t) = \iint R(\mathbf{r},t;\mathbf{r}',t')\,\mathscr{V}_{\text{ext}}(\mathbf{r}',t')\,\mathbf{d}^3\mathbf{r}'\,\mathrm{d}t', \tag{5.28}$$

and study its analytical properties.

For a homogeneous system, this relation may be Fourier analysed in both space and time; it is natural to write it in the form

$$\mathscr{V}_{\text{tot}}(\mathbf{q},\omega) = \frac{1}{\epsilon(\mathbf{q},\omega)}\mathscr{V}_{\text{ext}}(\mathbf{q},\omega), \tag{5.29}$$

where the analogy of elementary electrostatics (which is really a special case of this same situation) suggests the definition of a generalized *dielectric function* $\epsilon(\mathbf{q},\omega)$ in place of its inverse R assumed in (5.28). In this spirit, we introduce a *generalized polarizability* $\alpha(\mathbf{q},\omega)$, of the many-body system, such that

$$\epsilon(\mathbf{q},\omega) = 1+\alpha(\mathbf{q},\omega). \tag{5.30}$$

These equations imply that the field induced by polarization of the medium is usually opposed to the total field seen by the particles, i.e.

$$\begin{aligned}\mathscr{V}_{\text{ind}}(\mathbf{q},\omega) &= \mathscr{V}_{\text{tot}}(\mathbf{q},\omega)-\mathscr{V}_{\text{ext}}(\mathbf{q},\omega)\\ &= -\alpha(\mathbf{q},\omega)\,\mathscr{V}_{\text{tot}}(\mathbf{q},\omega).\end{aligned} \tag{5.31}$$

In general, $\epsilon(\mathbf{q},\omega)$ is a complex function of ω, and being causal must therefore satisfy dispersion relations such as

$$\operatorname{Re}\left\{\frac{1}{\epsilon(\mathbf{q},\omega)}\right\} = 1+\frac{2}{\pi}\mathscr{P}\int_0^\infty \frac{\omega'}{\omega'^2-\omega^2}\operatorname{Im}\left\{\frac{1}{\epsilon(\mathbf{q},\omega')}\right\}\mathrm{d}\omega'. \tag{5.32}$$

The imaginary part describes phenomena associated with the dissipation of energy in the system, and must therefore be linked with the Kubo theory of irreversible processes discussed in §4.4. But it is important in this context to distinguish between the *transverse* and *longitudinal* dielectric functions. The electrical conductivity calculated in (4.46), say, which would also be applicable to phenomena such as optical absorption, is the response to a transversely polarized electromagnetic field; in the present case we are studying the effect of the longitudinal, 'electrostatic' component of the field, which gives rise to quite different formulae and physical effects.

Quite apart from its use in relating and analysing various physical phenomena, why is the dielectric function important in the general theory of many-body systems? It turns out that $1/\epsilon(\mathbf{q}, \omega)$ is closely connected with the exact two-particle Green function of the system, and has singularities at its exact excited states. Moreover, the simple heuristic device of treating the particle–particle interaction as if it were 'screened'—replacing $v(\mathbf{q})$ by a modified interaction line

$$v_{\text{scr}}(\mathbf{q}, \omega) = \frac{v(\mathbf{q})}{\epsilon(\mathbf{q}, \omega)} \tag{5.33}$$

—is a powerful way of accounting for many correlation effects in the fermion gas, and is equivalent to summing many infinite sets of diagrams in the perturbation expansion.

5.8 Spectral representation of dielectric function

To demonstrate some of these basic properties of the dielectric function, let us calculate the effect of a time-dependent Hamiltonian

$$\mathscr{H}'_{\text{ext}} = \sum_{\mathbf{q}, \omega} \mathscr{V}_{\text{ext}}(\mathbf{q}, \omega) \sum_{j=1}^{N} \exp\{\mathrm{i}(\mathbf{q}\cdot\mathbf{r}_j - \omega t)\} \tag{5.34}$$

acting uniformly on the particles of the fermion gas. Because $\epsilon(\mathbf{q}, \omega)$ is defined as the *linear* response function, we can determine the effect of each Fourier component separately, merely assuming that ω has a small imaginary part $\mathrm{i}\delta$ to prescribe the adiabatic switching-on of the field in the remote past. For the same reason, $\mathscr{H}'_{\text{ext}}$ may be assumed to be infinitesimally weak, so that we need not proceed beyond first-order terms in a perturbation calculation.

For simplicity we consider only the case where the whole interacting fermion gas is initially in its unperturbed (time-dependent, Schrödinger) ground state $|\Psi_0\rangle$, relative to which its *exact* eigenstate $|\Psi_n\rangle$ has energy E_n. The perturbation has the effect of mixing into $|\Psi_0\rangle$ a proportion $a_n(t)$ of each excited state $|\Psi_n\rangle$. By elementary time-dependent perturbation theory, we have

$$a_n(t) = \mathscr{V}_{\text{ext}}(\mathbf{q}, \omega) \frac{\langle \Psi_n | \rho_{\mathbf{q}} | \Psi_0 \rangle}{E_n - \omega - \mathrm{i}\delta} \mathrm{e}^{-\mathrm{i}\omega t}, \tag{5.35}$$

where we introduce the *particle density operator* with Fourier component

$$\rho_{\mathbf{q}} \equiv \sum_{j=1}^{N} \exp(\mathrm{i}\mathbf{q}\cdot\mathbf{r}_j) \tag{5.36}$$

as in (4.85).

The effect of this admixture is to alter the particle density, which now acquires a time-varying expectation value with $\mathbf{q}$ component

$$\begin{aligned}\langle\rho_{\mathbf{q}}(t)\rangle &= \sum_n \{\langle\Psi_0| + a_n^*(t)\langle\Psi_n|\}\rho_{\mathbf{q}}\sum_m\{|\Psi_0\rangle + a_m(t)|\Psi_m\rangle\} \\ &= \mathscr{V}_{\mathrm{ext}}(\mathbf{q},\omega)\sum_n |\langle\Psi_n|\rho_{\mathbf{q}}|\Psi_0\rangle|^2\left\{\frac{1}{E_n-\omega-\mathrm{i}\delta}+\frac{1}{E_n+\omega+\mathrm{i}\delta}\right\} \\ &\qquad + O(\mathscr{V}_{\mathrm{ext}}^2).\end{aligned} \tag{5.37}$$

In deriving this expression we have used the translational invariance of the electron gas which implies that each excited state $|\Psi_n\rangle$ has a unique wave-vector. Since the sum includes only those states for which $\langle\Psi_0|\rho_{\mathbf{q}}|\Psi_n\rangle$ does not vanish, all matrix elements of the form $\langle\Psi_n|\rho_{\mathbf{q}'}|\Psi_0\rangle$ are eliminated except those for which $\mathbf{q}'$ exactly equals our initially prescribed wave-vector $\mathbf{q}$. For the benefit of those who like their algebra in full, it must be admitted that a little subtle juggling of positive and negative values of ω has also been used to collect together expressions that are essentially Hermitian and complex conjugates of one another.

Now this density variation gives rise, in its turn, to a field, for the interaction potential $v(\mathbf{r}-\mathbf{r}')$ acts not only between the particles of the gas but also on our hypothetical test particle. The induced field must have Fourier components

$$\mathscr{V}_{\mathrm{ind}}(\mathbf{q},\omega) = \langle\rho_{\mathbf{q}}(\omega)\rangle\, v(\mathbf{q}). \tag{5.38}$$

But by the definitions (5.29) and (5.31), we have

$$\mathscr{V}_{\mathrm{ind}}(\mathbf{q},\omega) = \left\{\frac{1}{\epsilon(\mathbf{q},\omega)} - 1\right\}\mathscr{V}_{\mathrm{ext}}(\mathbf{q},\omega). \tag{5.39}$$

Thus, from (5.37)–(5.39) we may derive the following very general formula

$$\frac{1}{\epsilon(\mathbf{q},\omega)} = 1 + v(\mathbf{q})\sum_n |\langle\Psi_n|\rho_{\mathbf{q}}|\Psi_0\rangle|^2\left\{\frac{1}{E_n-\omega-\mathrm{i}\delta}+\frac{1}{E_n+\omega+\mathrm{i}\delta}\right\}. \tag{5.40}$$

This formula, by expressing the dielectric function directly in terms of the exact eigenstates of the fully interacting fermion system indicates clearly the importance of this concept in many-body theory.

The derivation that we have sketched here is very simple, and makes no use of a 'self-consistency' argument. The perturbation is the true 'external' field $\mathscr{V}_{\text{ext}}$; not, as might perhaps have been thought, the final 'total' field $\mathscr{V}_{\text{tot}}$. The definition of each excited state $|\Psi_n\rangle$ automatically takes account of any effects on the system due to its own polarization field.

We note, for example, that $1/\epsilon(\mathbf{q}, \omega)$ has a pole in ω at every excited state of the system of wave-vector $\mathbf{q}$. An investigation of the singularities of the inverse dielectric function can lead us to the spectrum of elementary excitations of the system, including both single 'quasi-particle' excitations and collective excitations such as plasma oscillations.

We also observe that the residue at each such pole is directly related to the two-particle Green function of the system. The imaginary part of (5.40), in the limit of small δ, may be written

$$\operatorname{Im}\left\{\frac{1}{\epsilon(\mathbf{q}, \omega)}\right\} = \pi v(\mathbf{q}) \sum_n |\langle\Psi_n| \rho_{\mathbf{q}} |\Psi_0\rangle|^2 \{\delta(E_n + \omega) - \delta(E_n - \omega)\}. \quad (5.41)$$

Now in (4.85) we defined the van Hove correlation function in terms of the density operator (5.36). In the ground state of the system, this has energy-momentum transform

$$\bar{S}(\mathbf{q}, \omega) = \int \langle\Psi_0| \rho_{-\mathbf{q}}(t) \rho_{\mathbf{q}}(0) |\Psi_0\rangle \, \mathrm{e}^{-i\omega t} \, \mathrm{d}t. \quad (5.42)$$

It is an elementary exercise in matrix algebra to insert the unit operator $\Sigma_n |\Psi_n\rangle\langle\Psi_n|$ into this expression, and then to use the time dependence of these (Schrödinger) state vectors to produce a series of terms like (5.41). We find

$$\operatorname{Im}\left\{\frac{1}{\epsilon(\mathbf{q}, \omega)}\right\} = \pi v(\mathbf{q}) \{\bar{S}(\mathbf{q}, -\omega) - \bar{S}(\mathbf{q}, \omega)\}. \quad (5.43)$$

One can give this formula a physical interpretation. The effective interaction of our test particle with the fermions is given by the screened potential $v(\mathbf{q})/\epsilon(\mathbf{q}, \omega)$. The imaginary part of this would correspond to irreversible energy-absorption processes, whose rate would be proportional to

$$\operatorname{Im}\{v(\mathbf{q})/\epsilon(\mathbf{q}, \omega)\} = \pi\{v(\mathbf{q})\}^2 \{\bar{S}(\mathbf{q}, -\omega) - \bar{S}(\mathbf{q}, \omega)\} \quad (5.44)$$

—the energy lost in collisions with the particles of the gas depends upon the square of the matrix element of the interaction with each particle multiplied by the appropriate energy/wave-vector component of the particle distribution function. This sort of theory might, for example, describe neutron diffraction by a liquid or solid many-body system, or the excitation of plasma oscillations by a fast electron passing through a thin film of metal.

Another result that is often derived from (5.43) is as follows. Integrate with respect to ω, to obtain the Fourier space-transform of the static pair-correlation function

$$\begin{aligned} \bar{S}(\mathbf{q}) &= \frac{1}{N} \sum_{i \neq j} \langle e^{i\mathbf{q}\cdot(\mathbf{r}_i - \mathbf{r}_j)} \rangle \\ &= -\frac{1}{\pi v(\mathbf{q}) N} \int_0^\infty \operatorname{Im} \left\{ \frac{1}{\epsilon(\mathbf{q}, \omega)} \right\} d\omega - 1. \end{aligned} \tag{5.45}$$

Now consider the mean 'potential energy' of the interacting particles:

$$\begin{aligned} \langle \mathscr{V} \rangle &= \langle \tfrac{1}{2} \sum_{i \neq j} v(\mathbf{r}_i - \mathbf{r}_j) \rangle \\ &= \tfrac{1}{2} N \sum_{\mathbf{q}} v(\mathbf{q}) \bar{S}(\mathbf{q}) \\ &= -\frac{1}{2\pi} \sum_{\mathbf{q}} \int_0^\infty \operatorname{Im} \left\{ \frac{1}{\epsilon(\mathbf{q}, \omega)} \right\} d\omega - \tfrac{1}{2} N \sum_{\mathbf{q}} v(\mathbf{q}). \end{aligned} \tag{5.46}$$

This does not, of itself, allow us to calculate the total energy of the system, because of unknown corrections to the 'kinematic energy' of the particles in the interaction state. But there is a general theorem (said to have been first used by Pauli) which says: suppose any system, subject to an interaction of strength λv where λ is a parameter between 0 and 1, has ground state vector $|\Psi_0(\lambda)\rangle$; then the ground state energy of the actual interacting system with $\lambda = 1$ is given by

$$E = E_0 + \int_0^1 \frac{d\lambda}{\lambda} \langle \Psi_0(\lambda) | \lambda \mathscr{V} | \Psi_0(\lambda) \rangle, \tag{5.47}$$

where E_0 is the energy without interactions—e.g. $-\frac{3}{5} N \mathscr{E}_F$ for a fermion gas. The proof of this, by differentiation of the expectation value of the Hamiltonian, is not very difficult.

Let us now define $\epsilon^\lambda(\mathbf{q}, \omega)$ as the value of the dielectric function in a system where the interactions have strength λv; from (5.46) and (5.47) we obtain a formula for the ground state energy:

$$E = E_0 - \sum_{\mathbf{q}} \int_0^1 \frac{d\lambda}{\lambda} \left[\frac{1}{2\pi} \int_0^\infty \operatorname{Im} \left\{ \frac{1}{\epsilon^\lambda(\mathbf{q}, \omega)} \right\} d\omega + \tfrac{1}{2} N \lambda v(\mathbf{q}) \right]. \tag{5.48}$$

Of course, merely expressing various measurable physical properties of the system, such as the total energy and particle pair-distribution function, in terms of the dielectric response function does not *solve* the many-body problem; but the use of a reasonable approximation for $\epsilon^{\lambda}(\mathbf{q}, \omega)$ in (5.48), taking serious account of particle correlations, might be expected to yield better answers than the Hartree–Fock theory where these are neglected.

The extension of (5.40) to the case of a system at a finite temperature is relatively simple. The states $|\Psi_n\rangle$ are exact eigenstates of the whole system, so that each should occur in an ensemble with probability given by a Boltzmann factor (4.27). Carrying out the perturbation on this distribution, we get

$$\frac{1}{\epsilon(\mathbf{q}, \omega)} = 1 + \frac{v(\mathbf{q})}{Z} \sum_{mn} e^{-\beta E_m} |\langle \Psi_n | \rho_{\mathbf{q}} | \Psi_m \rangle|^2 \left\{ \frac{1}{E_{nm} - \omega - i\delta} + \frac{1}{E_{nm} + \omega + i\delta} \right\}, \tag{5.49}$$

where $E_{nm} = E_n - E_m$ is the energy change in going from an initial state $|\Psi_m\rangle$ to a virtual excited state $|\Psi_n\rangle$.

This formula, again, may be expressed in terms of the pair-correlation function, just as in (5.43). Moreover, reversing the roles of m and n is equivalent to changing the sign of ω and multiplying by a factor $\exp(\beta\omega)$. Using time-reversal invariance to change $-\mathbf{q}$ into $\mathbf{q}$, we then get

$$\bar{S}(\mathbf{q}, \omega) = e^{\beta\omega} \bar{S}(\mathbf{q}, -\omega), \tag{5.50}$$

and hence, from (5.43),

$$\mathrm{Im}\left\{ \frac{1}{\epsilon(\mathbf{q}, \omega)} \right\} = -\pi v(\mathbf{q}) \{1 - e^{-\beta\omega}\} \bar{S}(\mathbf{q}, \omega). \tag{5.51}$$

This is a case of the *fluctuation–dissipation theorem*, and provides a link with the Kubo formalism (§ 4.4) for irreversible processes. Thus comparing (5.51) with (4.46), we note in each case that the coefficient for a dissipative process has been expressed in terms of the Fourier transform of a two-particle time-correlation function describing fluctuations in the properties of the system.

5.9 Diagrammatic interpretation of dielectric screening

The dielectric function is an *exact* mathematical property of a homogeneous system of interacting particles, but can only be calculated approximately. What does such a calculation imply in the language of Feynman diagrams?

In the Heisenberg or interaction representation, a weak external potential such as (5.34) would be an operator of the general form

$\mathscr{V}_{\text{ext}}\psi^*\psi$. We represent this by a suitably specialized (e.g. zig-zag) horizontal line, leading from an isolated 'external' point to a fermion-scattering vertex. Because $\mathscr{V}_{\text{ext}}$ is infinitesimal, all diagrams with more than two such external lines are to be ignored in evaluating the effect of the perturbation on the energy or other properties of the fermion gas.

For example, to calculate the effect on the ground state energy, we should need to sum all diagrams that are closed except for these interactions—diagrams such as those in fig. 4. If there had been no particle–particle interactions of the sort discussed in § 5.5, only the first diagram, fig. 4(a), would have been needed,† corresponding to an

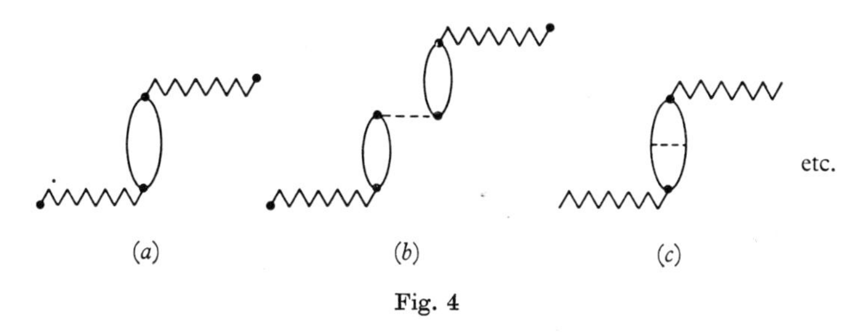

Fig. 4

ordinary Rayleigh–Schrödinger perturbation by $\mathscr{V}_{\text{ext}}$. Our assumption is that the effect of dividing $\mathscr{V}_{\text{ext}}(\mathbf{q}, \omega)$ by $\epsilon(\mathbf{q}, \omega)$ is to include all other possible diagrams in this series describing all possible virtual processes associated with fluctuations of particle density, etc. in the medium.

Now, just as in §§ 3.10, 4.8, 5.5, etc. we undertake a topological analysis of all the graphs in fig. 4. In this case, we look for 'bubble' parts—i.e. sub-diagrams that are closed except for an initial and a final fermion–antifermion vertex. A bubble can, in its turn, be classed as reducible or irreducible, according to whether or not it can be unlinked by cutting a single internal interaction line. Thus, we depict the *irreducible bubble parts* by a 'black' bubble, representing the sum of the series in fig. 5. Algebraically, we suppose that the introduction of a black bubble into a diagram is equivalent to multiplying the corresponding element of the S-matrix by $\pi(\mathbf{q}, \omega)$ which is a function

† The 'self-energy' diagram is removed by a suitable choice of the zero of energy, as in the Hartree–Fock method.

only of the momentum and energy transferred to the system at the initial vertex and eventually recovered at the end of the fluctuation.

In each of the diagrams of fig. 4, the two external potentials may either be joined by an irreducible bubble, as in figs. 4 (*a*) and (*c*), when it will be included in the set of fig. 6 (*a*), or else the intermediate graphs

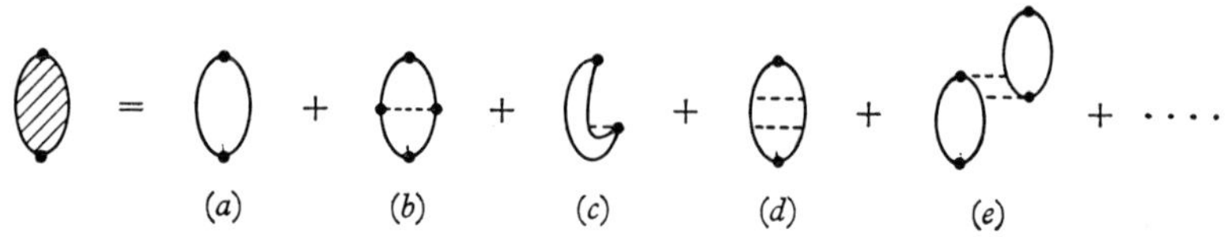

(*a*) (*b*) (*c*) (*d*) (*e*)

Fig. 5

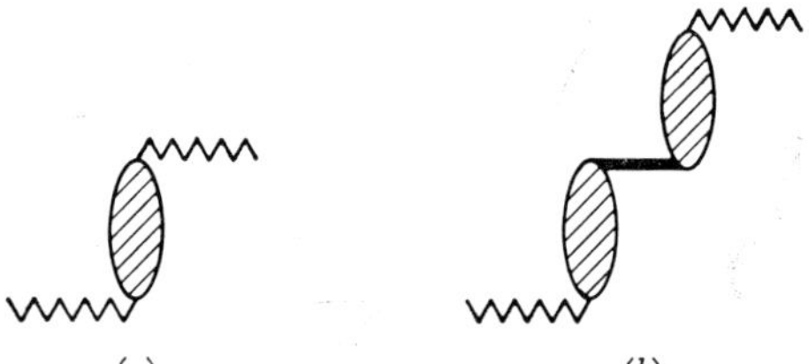

(*a*) (*b*)

Fig. 6

may be analysed into a string of irreducible bubbles, as in fig. 4 (*b*). For the sum of all such strings we introduce a new symbol—the *effective interaction line* defined by

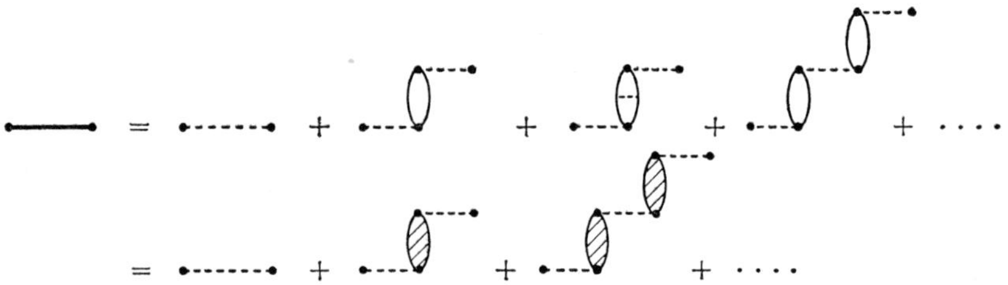

(5.52)

This line represents all possible ways of linking two fermion vertices by internal lines, and therefore behaves like a modified interaction with momentum transform $v'(\mathbf{q})$ in the S-matrix expansion. Putting a black line between two black bubbles, as in fig. 6 (*b*) completes the series of ground state diagrams depicted in fig. 4.

But the series (5.52) is of the familiar geometric type, and may be generated by the topological equation:

(5.53)

In the energy momentum representation, this is a simple algebraic relation

$$v'(\mathbf{k}) = v(\mathbf{k}) + v'(\mathbf{k})\,\pi(\mathbf{k},\omega)\,v(\mathbf{k}), \tag{5.54}$$

by means of which the modified interaction is linked with the magnitude of the irreducible bubble parts. In a similar way, we might define a modified external potential, of strength $\mathscr{V}_{\text{tot}}(\mathbf{k},\omega)$, to include all possible intermediate bubble graphs, by the topological relations:

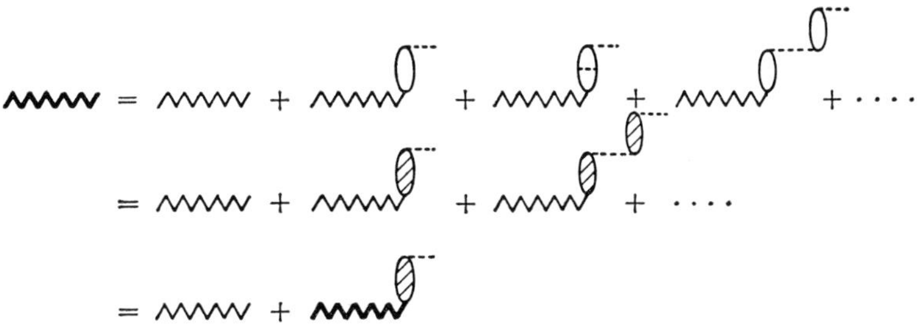

(5.55)

The final summation of this series yields an algebraic equation similar to (5.54):

$$\mathscr{V}_{\text{tot}}(\mathbf{k},\omega) = \mathscr{V}_{\text{ext}}(\mathbf{k},\omega) + \mathscr{V}_{\text{tot}}(\mathbf{k},\omega)\,\pi(\mathbf{k},\omega)\,v(\mathbf{k}). \tag{5.56}$$

These formulae may at once be compared with (5.29); by elementary algebra (5.56) reads

$$\mathscr{V}_{\text{tot}}(\mathbf{k},\omega) = \frac{1}{1 - v(\mathbf{k})\,\pi(\mathbf{k},\omega)}\mathscr{V}_{\text{ext}}(\mathbf{k},\omega), \tag{5.57}$$

i.e.

$$\epsilon(\mathbf{k},\omega) = 1 - v(\mathbf{k})\,\pi(\mathbf{k},\omega). \tag{5.58}$$

In other words, the summation over all irreducible bubble parts gives us the polarizability of the medium: in (5.30),

$$\alpha(\mathbf{k},\omega) = -v(\mathbf{k})\,\pi(\mathbf{k},\omega). \tag{5.59}$$

The meaning of (5.54) is clear: the modified interaction between particles, taking account of polarization of the medium and other particle–particle correlation effects, is what we should call the *screened interaction*,

$$v'(\mathbf{k}) = \frac{1}{\epsilon(\mathbf{k},\omega)}v(\mathbf{k}). \tag{5.60}$$

If we had been dealing with electrostatic forces this simple, heuristically obvious device, which is known to remove all difficulties and divergences due to the long range of the coulomb interaction in the case of electrons in metals, is therefore justifiable as equivalent to a partial summation of the perturbation series.

It is interesting to note that in deriving (5.54) or (5.56) we might have been tempted to stop at the second term in the series (5.52), i.e.

(5.61)

which would read

$$v'(\mathbf{k}) \approx v(\mathbf{k}) + v(\mathbf{k})\,\pi(\mathbf{k},\omega)\,v(\mathbf{k}), \tag{5.62}$$

as if we had approximated to (5.58) by writing

$$\epsilon(\mathbf{k},\omega) \approx \frac{1}{1+v(\mathbf{k})\,\pi(\mathbf{k},\omega)}. \tag{5.63}$$

For weak interactions the difference may not be important, but any investigation of the excitation spectrum of the medium by location of the singularities of $1/\epsilon(\mathbf{k},\omega)$, as in (5.40) will obviously go hopelessly wrong if (5.63) is used instead of (5.58). This is a typical example of the general principle that the analytical properties of a function defined as the sum of an infinite series cannot be discovered by studying the series term by term.

Having given this diagrammatic interpretation of the dielectric response function of a many-body system, we are still not in a position to write down an exact formula for it. The sum over all irreducible bubble parts in fig. 5 to evaluate $\pi(\mathbf{k},\omega)$ is of escalating complexity. But we can see how we might improve on the very first term, and make further systematic partial summations over infinite classes of diagrams. For example, we might use the Hartree–Fock method, as in (5.26), to modify each fermion propagator in this simple bubble, and hence, as shown in (5.27), include all irreducible bubbles containing independent virtual excitations of the vacuum, as in fig. 5(*c*). Again, we might try replacing the interaction potential $v(\mathbf{k})$ by its K-matrix element, as in the Brueckner method of §5.6, to allow for contributions of multiple scattering 'ladders' in more complicated bubbles such as those of figs. 5(*d*) and (*e*).

5.10 The random phase approximation

Let us now consider what appears to be an entirely different approach to the whole problem. In the usual second-quantized momentum representation in terms of fermion field operators, the Hamiltonian of our standard interacting system may be written

$$H = \sum_{\mathbf{p}} (2m)^{-1} b^*_{\mathbf{p}} b_{\mathbf{p}} + \tfrac{1}{2}\sum_{\mathbf{k}} v(\mathbf{k})\,\{\rho^*_{\mathbf{k}}\rho_{\mathbf{k}} - N\}. \tag{5.64}$$

In this expression we have deliberately emphasized the operators generating density fluctuations of various wavelengths, i.e. equivalently to (5.36),

$$\rho_{\mathbf{k}}^* = \sum_{\mathbf{k}'} b_{\mathbf{k}+\mathbf{k}'}^* b_{\mathbf{k}'} \tag{5.65}$$

produces a combination of all 'particle-hole pair' excitations of net momentum $\mathbf{k}$.

We are not here interested in the effects of external fields, but we want to construct an equation of motion for a typical pair excitation operator in (5.65)—for example, the combination $b_{\mathbf{p}+\mathbf{q}}^* b_{\mathbf{q}}$.

To this end, we evaluate the commutator of this operator with the Hamiltonian (5.64), making use of the anticommutation relations of the $b_{\mathbf{k}}^*$, etc. to eliminate as many terms as possible. The result may be expressed as follows

$$[H, b_{\mathbf{p}+\mathbf{q}}^* b_{\mathbf{p}}] = \{\mathscr{E}_0(\mathbf{p}+\mathbf{q}) - \mathscr{E}_0(\mathbf{p})\} b_{\mathbf{p}+\mathbf{q}}^* b_{\mathbf{p}}$$
$$- \sum_{\mathbf{k}} v(\mathbf{k}) \{(b_{\mathbf{p}+\mathbf{q}}^* b_{\mathbf{p}+\mathbf{k}} - b_{\mathbf{p}+\mathbf{q}-\mathbf{k}}^* b_{\mathbf{p}}) \rho_{\mathbf{k}}^* + \rho_{\mathbf{k}}^* (b_{\mathbf{p}+\mathbf{q}}^* b_{\mathbf{p}+\mathbf{k}} - b_{\mathbf{p}+\mathbf{q}-\mathbf{k}}^* b_{\mathbf{p}})\}. \tag{5.66}$$

An equation of motion with this sort of expression in it is obviously insoluble. The behaviour of our single-pair excitation depends upon products of four field-operators—i.e. 'double-pair' excitations—and so on. We are merely on the lowest rung of a hierarchy of equations, as in §4.9. Moreover, our excitation of wave-number $\mathbf{q}$ depends upon the behaviour of fluctuations of all other wave numbers, which makes our equations inseparable.

Both these difficulties are avoided if we make the *random phase approximation* (RPA) of Sawada. Suppose we are studying the properties of the system in the neighbourhood of some eigenstate $|\,\rangle$, or in a mixed state defined by a canonical density matrix at temperature T. Then we replace each miscellaneous product of operators multiplying $\rho_{\mathbf{k}}^*$ on the right by its expectation value or ensemble average. There is a spectacular collapse of innumerable terms (which would correspond to off-diagonal elements randomly out of phase with one another), so that

$$(b_{\mathbf{p}+\mathbf{q}}^* b_{\mathbf{p}+\mathbf{k}} - b_{\mathbf{p}+\mathbf{q}-\mathbf{k}}^* b_{\mathbf{p}}) \approx \langle | b_{\mathbf{p}+\mathbf{q}}^* b_{\mathbf{q}+\mathbf{k}} - b_{\mathbf{p}+\mathbf{q}-\mathbf{k}}^* b_{\mathbf{p}} | \rangle$$
$$= (\bar{n}_{\mathbf{p}+\mathbf{q}} - \bar{n}_{\mathbf{p}})\, \delta_{\mathbf{q},\mathbf{k}}, \tag{5.67}$$

where $\bar{n}_{\mathbf{p}}$ is the mean occupation number of the $\mathbf{p}$th single particle mode in our average basic state. Putting this into (5.66), the *RPA method* gives us

$$[H, b^*_{\mathbf{p}+\mathbf{q}} b_\mathbf{p}] \approx \{\mathscr{E}_0(\mathbf{p}+\mathbf{q}) - \mathscr{E}_0(\mathbf{p})\} b^*_{\mathbf{p}+\mathbf{q}} b_\mathbf{p} - v(\mathbf{q})\,(\bar{n}_{\mathbf{p}+\mathbf{q}} - \bar{n}_\mathbf{p})\rho^*_\mathbf{q}. \quad (5.68)$$

In this approximation, we have neglected all momentum transfers except those of amount $\mathbf{q}$, the wave-vector of the excitation chosen for study. The translational invariance of the system, which singles out the term $\mathbf{k} = \mathbf{q}$ in (5.67), has effectively decoupled excitations of different wavelengths from one another, and hence made the solution of the equations of motion derived from (5.68) relatively simple. We have also neglected fluctuations of particle occupation number about the average $\bar{n}_\mathbf{p}$. The second term in (5.68) may thus be interpreted as the average field $v(\mathbf{q})\rho^*_\mathbf{q}$ produced by all the particles of the system acting on the chosen pair state. The RPA method is, in fact, a generalized self-consistent field procedure, which may be derived by making the Hartree equations (5.8)–(5.10) time-dependent. In addition to calculating the effect of the average *static* field on each single-particle state, we take account of the average oscillatory field that the other particles may produce by slightly correlating their movements.

The solution of the equations of motion of the system in this approximation proceeds as follows. Consider an operator representing some general excitation of wave-vector $\mathbf{q}$, made up of all possible particle-hole pairs:

$$\xi^*_\mathbf{q} = \sum_\mathbf{p} A(\mathbf{p}, \mathbf{q}, \omega_\mathbf{q}) b^*_{\mathbf{p}+\mathbf{q}} b_\mathbf{p}. \quad (5.69)$$

The unknown function $A(\mathbf{p}, \mathbf{q}, \omega_\mathbf{q})$ is to be chosen so as to satisfy a simple harmonic equation of motion at frequency $\omega_\mathbf{q}$, i.e.

$$[H, \xi^*_\mathbf{q}] = \omega_\mathbf{q} \xi^*_\mathbf{q}. \quad (5.70)$$

Substituting from (5.68) and (5.69) in (5.70), we get

$$\sum_\mathbf{p} A(\mathbf{p}, \mathbf{q}, \omega_\mathbf{q}) \{\omega_\mathbf{q} - \mathscr{E}_0(\mathbf{p}+\mathbf{q}) + \mathscr{E}_0(\mathbf{p})\} b^*_{\mathbf{p}+\mathbf{q}} b_\mathbf{p}$$
$$= \sum_\mathbf{p} A(\mathbf{p}, \mathbf{q}, \omega_\mathbf{q}) \{\bar{n}_\mathbf{q} - \bar{n}_{\mathbf{p}+\mathbf{q}}\} v(\mathbf{q}) \rho^*_\mathbf{q}. \quad (5.71)$$

But the left-hand side can also be made a multiple of $\rho^*_\mathbf{q}$ by writing

$$A(\mathbf{p}, \mathbf{q}, \omega_\mathbf{q}) = \frac{1}{\omega_\mathbf{q} - \mathscr{E}_0(\mathbf{p}+\mathbf{q}) + \mathscr{E}_0(\mathbf{p})}. \quad (5.72)$$

These two equations can only be made consistent if

$$1 - v(\mathbf{q}) \sum_\mathbf{q} \frac{\bar{n}_\mathbf{p} - \bar{n}_{\mathbf{p}+\mathbf{q}}}{\omega_\mathbf{q} - \{\mathscr{E}_0(\mathbf{p}+\mathbf{q}) - \mathscr{E}_0(\mathbf{p})\}} = 0. \quad (5.73)$$

This is therefore a dispersion formula for the frequency, $\omega_{\mathbf{q}}$, of a self-consistent quasi-independent excited mode of the system, of momentum $\mathbf{q}$.

But this result can be obtained quite simply from the theory of the dielectric function. Let us evaluate the polarization $\pi(\mathbf{q}, \omega)$ corresponding to the irreducible bubble parts defined in §5.9. In the simplest approximation this is just a single bubble [fig. 5(*a*)]. The rules of §3.8 for evaluating Feynman diagrams must be extended in the case where $T \neq 0$, by including factors such as (2.36) for the relative average occupation numbers of the single-particle modes $\mathbf{p}$ and $\mathbf{p}+\mathbf{q}$ (one of which is to acquire an electron, and the other a hole—or vice-versa), so that

$$\pi(\mathbf{q}, \omega) \approx \sum_{\mathbf{p}} \frac{\bar{n}_{\mathbf{p}}(1-\bar{n}_{\mathbf{p}+\mathbf{q}}) - \bar{n}_{\mathbf{p}+\mathbf{q}}(1-\bar{n}_{\mathbf{p}})}{\omega - \{\mathscr{E}_0(\mathbf{p}+\mathbf{q}) - \mathscr{E}_0(\mathbf{p})\}}$$

$$= \sum_{\mathbf{p}} \frac{\bar{n}_{\mathbf{p}} - \bar{n}_{\mathbf{p}+\mathbf{q}}}{\omega - \{\mathscr{E}_0(\mathbf{p}+\mathbf{q}) - \mathscr{E}_0(\mathbf{p})\}}. \tag{5.74}$$

From (5.58) we have, therefore, an approximation for the dielectric function

$$\epsilon(\mathbf{q}, \omega) = 1 - v(\mathbf{q}) \sum_{\mathbf{p}} \frac{\bar{n}_{\mathbf{p}} - \bar{n}_{\mathbf{p}+\mathbf{q}}}{\omega - \{\mathscr{E}_0(\mathbf{p}+\mathbf{q}) - \mathscr{E}_0(\mathbf{p})\}}. \tag{5.75}$$

The RPA dispersion formula (5.73) is evidently equivalent to finding a pole of this approximation to the inverse dielectric function. This is in accordance with the general theorem (5.40) linking the singularities of $1/\epsilon$ with the spectrum of elementary excitations of the system.

In this context, it is instructive to notice what would have happened if we had tried to use (5.49), say, to evaluate the dielectric function by inserting the single-particle energies $\mathscr{E}_0(\mathbf{p})$ of the *non-interacting gas* in place of the true excitation spectrum E_n of the whole interacting system. The result would have been just the 'first-order perturbation' formula (5.63), with $\pi(\mathbf{q}, \omega)$ given by (5.74). This formula would not, of course, give to $1/\epsilon$ any more singularities than we had put into it in the first place, and therefore would fail to inform us about many collective excitations of the system, although it might not be too bad an approximation to the dielectric function itself under certain circumstances. Perhaps the most confusing aspect of the whole theory of linear response functions is the fact that an expression such as (5.75) is the best *approximate* formula for the dielectric function, whereas equations such as (5.40) and (5.44) are *exact* formulae for its *inverse*.

Having shown that the RPA method is simply equivalent to the most elementary approximation to the polarization bubble in the dielectric function, we ought now to study various applications of these formulae. In practice, this means a discussion of the theory of the electron gas in a metal, where the interaction is the coulomb potential

$$v(\mathbf{q}) = -\frac{4\pi e^2}{\mathbf{q}^2}. \tag{5.76}$$

We discover the elementary theory of dielectric screening, we show the existence of *plasma oscillations* or *plasmons*, and we even evaluate the ground state energy of the system quite accurately by putting (5.75) into (5.48). But this would lead us astray from the rough pastures of mathematical principles into the delightful glades of real physics.

5.11 The Landau theory of Fermi liquids

In the RPA calculation of the dielectric function (5.75), the final result depends upon the mean occupation numbers of the various *free-particle* modes. The next step in improving this formula might be simply to renormalize the one-particle propagator for these lines in the bubble diagram, so as to turn the particle of unperturbed energy $\mathscr{E}_0(\mathbf{k}) = k^2/2m$ into a quasi-particle of energy $\mathscr{E}(\mathbf{k})$ given by, say, a Dyson summation such as (3.131). But strictly speaking, this itself ought to depend on the general state of the system—for example, through the relative degree of occupation of all the other particle modes. Many-body theory is different from elementary-particle field theory in that we often deal with relatively highly excited states of the system, with many quasi-particles present at once. The device of treating the filled states below the Fermi level as a passive 'Dirac sea' from which 'holes' may be excited in small numbers as required (§2.7) cannot deal with, say, the electron distribution in an ordinary metal at room temperature.

As a heuristic principle for dealing systematically with such systems, Landau suggested a model, based upon 'switching-on' an interaction between the particles of a simple Fermi gas. This theory has proved extremely valuable in the discussion of collective excitations and many other observable physical properties of many-fermion systems, and has been justified from first principles, as a canonical approximation procedure, in the language of perturbation expansions and Green functions. Outwardly, it is very simple, seemingly by-passing all the algebraic complexities of advanced quantum mechanics, although this hides many subtleties and dangerous pitfalls for careless thought.

The account that follows is meant merely to indicate the general character of the argument, and is not to be leant on too heavily.

Suppose that we had an assembly of independent fermions, in states labelled by momentum **k** (and, of course, by a spin index which we ignore for simplicity). In the ground state they would occupy a Fermi sphere of radius k_F, and to each one-particle level we could assign an energy, measured relative to the Fermi energy, of the form

$$\begin{aligned} \xi(\mathbf{k}) &= \mathscr{E}(\mathbf{k}) - \mathscr{E}(k_F) \\ &\approx (k - k_F)\, v_F \end{aligned} \tag{5.77}$$

for small $(k - k_F)$. In this expression, v_F is a constant parameter, of the dimensions of velocity. Note that $\xi(\mathbf{k})$ is negative when $k < k_F$; in the Landau theory it is best not to use the 'anti-particle' conventions of §2.7.

Now suppose that a small number of these independent quasi-particles are excited, so that the occupation number of the **k**th level is changed by $\delta n(\mathbf{k})$. Since fermion number is conserved, we have a condition like

$$\sum_{\mathbf{k}} \delta n(\mathbf{k}) = 0. \tag{5.78}$$

But the energy of the whole system would be changed by an amount

$$E \approx \sum_{\mathbf{k}} \delta n(\mathbf{k})\, \xi(\mathbf{k}); \tag{5.79}$$

the excitation spectrum of the whole gas can be expressed as a functional of the occupation numbers $\delta n(\mathbf{k})$.

Landau's basic assumption is that the spectrum of any homogeneous assembly of fermions—a *Fermi liquid*, if the particle–particle interaction is very strong—can be expressed in this form. However, (5.74) can only be the leading term in a Taylor series in powers of δn: to discuss more complicated phenomena we must go to the next order, and write

$$E = \sum_{\mathbf{k}} \delta n(\mathbf{k})\, \xi(\mathbf{k}) + \tfrac{1}{2} \sum_{\mathbf{k}, \mathbf{k}'} f(\mathbf{k}, \mathbf{k}')\, \delta n(\mathbf{k})\, \delta n(\mathbf{k}'), \tag{5.80}$$

where now $f(\mathbf{k}, \mathbf{k}')$ looks like the energy of an additional interaction between our originally independent quasi-particles. We are not saying that $f(\mathbf{k}, \mathbf{k}')$ is the *actual* interaction between the basic fermions (for example, the coulomb interaction between electrons in a metal), but it is what appears as this coefficient in the series when we expand the energy spectrum of the liquid in the form (5.80).

From its very beginning, the Landau theory rests upon the validity of (5.80). Although an expression of this sort is intuitively obvious, a very elaborate diagrammatic analysis in terms of temperature Green functions, etc. is required to show that it is genuinely self-consistent. It turns out that $f(\mathbf{k}, \mathbf{k}')$ is related to the scattering vertex function $\Gamma(1234)$ that occurs in the Bethe–Salpeter equation of §4.8, and therefore correctly represents the residual interaction between quasi-particles when all averaging, correlating and screening effects have been taken into account. But notice that (5.77) is only true for a 'normal' Fermi liquid, and would not be true, for example, in the superconducting state, when an energy gap appears in the excitation spectrum at the Fermi surface.

The chief value of the theory is that it provides a simple canonical description of collective properties of the system. We notice, for example, that the energy required to 'excite one more particle into the level $\mathbf{k}$' is not just (5.77), but must be of the form

$$\mathscr{E}(\mathbf{k}) = \xi(\mathbf{k}) + \int f(\mathbf{k}, \mathbf{k}')\, \delta n(\mathbf{k}')\, \mathrm{d}^3\mathbf{k}'. \tag{5.81}$$

In other words, the effective energy of each single-particle mode depends explicitly on the occupation of the other modes, in a way that we found quite naturally by the Hartree–Fock method, but is here linearized and elevated into a principle. The Landau formalism permits us to derive and describe various collective motions of the liquid—'first sound', 'zero sound', plasmons, spin waves, etc.—and to discuss their variation with temperature and other macroscopic parameters, solely in terms of the 'Fermi velocity' v_F of (5.77) and the quasi-particle interaction function $f(\mathbf{k}, \mathbf{k}')$. It therefore provides a very simple framework for the results of many-body theory, whereby special detailed formulae for particular phenomena may be compared and connected with one another.

It is not the function of this book to give an account of these multifarious phenomena and of the equations governing them. But to demonstrate the power and limitation of the Landau theory, we shall now derive, by elementary means, an almost paradoxical formula governing the calculation of current in a Fermi liquid.

To fix our ideas, let us suppose that each particle of the liquid is of definite mass m, and that we measure the flux of particles by observing the corresponding flow of mass. In other words, we define the current operator as the total momentum divided by m:

$$\mathbf{J} = \frac{1}{m} \sum_i \mathbf{p}_i. \tag{5.82}$$

The same argument will, of course, apply to electric currents carried by charged particles, etc.

A formula for the expectation value of $\mathbf{J}$ in any stationary state of the system can be derived quite generally by an appeal to the principle of translational invariance of the Hamiltonian—true, for example, if all interactions between the particles of the liquid depend only on their relative co-ordinates. Suppose we travel (non-relativistically) with small velocity $\hbar\,\delta\mathbf{q}/m$ relative to the system. For our moving observer this seems to have energy $E(\delta\mathbf{q})$ depending on $\delta\mathbf{q}$. By an elementary transformation of the Hamiltonian to the moving frame, which introduces an apparent term $\delta\mathbf{q}\cdot\mathbf{p}_i/m$ in the kinetic energy operator for each particle, we get the general theorem

$$\langle|\mathbf{J}|\rangle = -\frac{\partial E(\delta\mathbf{q})}{\hbar\,\partial(\delta\mathbf{q})} \tag{5.83}$$

evaluated as $\delta\mathbf{q}\to 0$.

Now apply this theorem to a state of the system containing a quasi-particle in the $\mathbf{k}$th mode, and calculate the contribution of this excitation to the current. This must be of the form

$$\mathbf{j}_{\mathbf{k}} = -\frac{\partial\mathscr{E}(\mathbf{k};\delta\mathbf{q})}{\hbar\,\partial(\delta\mathbf{q})}, \tag{5.84}$$

where $\mathscr{E}(\mathbf{k};\delta\mathbf{q})$ is the energy that would be assigned to this quasi-particle by our hypothetical moving observer.

Although the formula (5.81) is true in any co-ordinate frame, the apparent occupation of the modes labelled by the momentum variable $\mathbf{k}$ may change. In fact, the assumed velocity in real space is equivalent to shifting the apparent origin of $\mathbf{k}$-space by an amount $-\delta\mathbf{q}$. Applying this shift to each term of (5.81), we get two contributions to the current flow. The first, from (5.77), is simply the usual expression for the group velocity

$$-\frac{\partial\xi(\mathbf{k};\delta\mathbf{q})}{\hbar\,\partial(\delta\mathbf{q})} = \frac{\partial\xi(\mathbf{k})}{\partial\mathbf{k}}$$

$$= \mathbf{v}_{\mathbf{k}} \tag{5.85}$$

associated with the $\mathbf{k}$th mode. To evaluate the second term, it is simplest to treat the shift of origin in $\mathbf{k}$-space as creating a thin shell of changed occupation numbers on the Fermi sphere—i.e. for a 'sharp' Fermi distribution ($T = 0$), the change is

$$\delta n(\mathbf{k}) = \delta\mathbf{q}\cdot\mathbf{v}_{\mathbf{k}}\{\mathscr{E}(\mathbf{k})-\mathscr{E}_F\}, \tag{5.86}$$

where (5.77) and (5.85) have again been used. Putting (5.87) and (5.86) into (5.81) and (5.84), we get the theorem

$$j_{\mathbf{k}} = \mathbf{v}_{\mathbf{k}} + \int f(\mathbf{k}, \mathbf{k}')\, v_{\mathbf{k}'}\, \delta\{\mathscr{E}(\mathbf{k}') - \mathscr{E}_F\}\, \mathbf{d}^3\mathbf{k}'. \qquad (5.87)$$

In other words, the current associated with a quasi-particle excitation is *not* simply the momentum derivative of its first-order energy, as it would be in the elementary theory of group velocity, etc.

What this means is that a quasi-particle excitation of a quantum liquid is not as simple as the momentum-representation, wave-like, state function of a single real particle. We cannot, for example, build up a localized wave packet for a separate quasi-particle; the interaction forces 'drag' other particles along with it, so that a complex flow pattern is produced. Since the fluid is infinite in extent, this flow knows no boundary constraints, and can contribute a finite amount to the net particle flux.

We might, on the other hand, have tried to build up a state corresponding to a single quasi-particle being dragged through the system with velocity $\mathbf{v}_{\mathbf{k}}$, without any further net flow of particles through the boundaries: this can only be achieved by producing a 'back-flow' of amount $\mathbf{v}_{\mathbf{k}} - \mathbf{j}_{\mathbf{k}}$ in the surrounding fluid so as to balance the account. The difference between these two situations is quite subtle, and gross errors can occur unless the argument is watched very carefully.

In the context, it is interesting to note that (5.87) seems to contravene a general theorem, derivable from (5.83), whereby the current in an eigenstate is equated with the momentum-derivative of the energy eigenvalue—an exact result. The point is that a quasi-particle excitation level is not an *exact* eigenstate of the many-body Hamiltonian, for it always has a finite lifetime, associated with an imaginary part in the propagator (§ 4.6). The Landau theory, along with all other theories of the many-body problem, implies at some stage an RPA to reduce the equations to computable form. Thus, an equation such as (5.81), where the off-diagonal elements of the density matrix are forgotten, smudges out the delicate phase relations which would have preserved the strict theorem to which I have referred. Although each true eigenstate of the fluid must have a well-defined **k**-vector, it must also be of unimaginable complexity in a single-particle momentum representation, and therefore useless as a basis for calculation. Representation in terms of quasi-particle excitations gives the spectrum correctly, but demands care in the calculation of such other quantities as current, etc.

5.12 The dilute Bose gas

Fermion gases and liquids are familiar in physics; systems of many interacting conserved bosons are rare. But the properties of liquid ^{4}He are of such interest that much effort has been devoted to the theory of this unique substance. Unfortunately, liquid helium has all the theoretical complexity of a dense classical liquid, in addition to its quantal aspects, so that the derivation of the phenomena of superfluidity and phase transitions has been given only for greatly simplified and idealized model systems, or rests upon essentially heuristic arguments. The problems in this field are not those of sheer mathematical technique.

Nevertheless, *Bogoliubov's method* for generating the excitation spectrum of a dilute, weakly interacting boson gas is of interest. Consider, for simplicity, an assembly of N bosons described by the Hamiltonian

$$H = \sum_{\mathbf{k}} (k^2/2m)\, a_{\mathbf{k}}^* a_{\mathbf{k}} + \tfrac{1}{2}\lambda \sum_{\mathbf{k}+\mathbf{k}'=\mathbf{k}''+\mathbf{k}'''} a_{\mathbf{k}'''}^* a_{\mathbf{k}''}^* a_{\mathbf{k}'} a_{\mathbf{k}}. \tag{5.88}$$

The annihilation and creation operators satisfy commutation relations like (1.50). The particle–particle interaction, represented by the energy parameter λ, is assumed to be independent of the relative momenta of the interacting particles, as if only s-wave scattering were important (cf. §1.10).

When the system is at a very low temperature, there will be Bose–Einstein condensation of the particles into the mode $\mathbf{k} = 0$. In other words, we may assume that N_0, the occupation number of this mode, is nearly as large as N. For the energy spectrum near the ground state, we may assume that $(N - N_0)$, the number of 'excited' particles, is much smaller than N_0; we may ignore the interactions of excited particles with one another, and concentrate on their interactions with the condensed particles in the zeroth mode. This argument depends, of course, upon the assumption that the interacting system is similar in this respect to a gas of independent bosons—a point that has to be justified retrospectively when its consequences have been explored.

In algebraic terms, we approximate to the second term in (5.88):

$$H_{\text{int}} \approx \tfrac{1}{2}\lambda[a_0^* a_0^* a_0 a_0 + \sum_{\mathbf{k}\neq 0} \{2a_{\mathbf{k}}^* a_0^* a_{\mathbf{k}} a_0 + 2a_{-\mathbf{k}}^* a_0^* a_{-\mathbf{k}} a_0 + a_{\mathbf{k}}^* a_{-\mathbf{k}}^* a_0 a_0 + a_0^* a_0^* a_{\mathbf{k}} a_{-\mathbf{k}}\}]. \tag{5.89}$$

The separation of all products involving the operators a_0^* and a_0 facilitates a second approximation, which is to treat each of these as a

c-number, equal to $\sqrt{N_0}$. This means, simply, that N_0 is so enormous that the commutator of these two operators (i.e. unity) produces only infinitesimal effects, so that the correspondence principle may be invoked. For (5.89) we now write

$$H_{\text{int}} \approx \tfrac{1}{2}\lambda[N_0^2 + 2N_0 \sum_{\mathbf{k}\neq 0} (a_{\mathbf{k}}^* a_{\mathbf{k}} + a_{-\mathbf{k}}^* a_{-\mathbf{k}}) + N_0 \sum_{\mathbf{k}\neq 0} (a_{\mathbf{k}}^* a_{-\mathbf{k}}^* + a_{\mathbf{k}} a_{-\mathbf{k}})]. \quad (5.90)$$

At this stage, we do not really know the value of N_0; even the ground state of our interacting gas may not be best described by having *all* the bosons in the single zeroth mode. But we are dealing with a gas of real particles, whose total number operator [cf. (1.19)],

$$N = N_0 + \tfrac{1}{2} \sum_{\mathbf{k}\neq 0} (a_{\mathbf{k}}^* a_{\mathbf{k}} + a_{-\mathbf{k}}^* a_{-\mathbf{k}}), \quad (5.91)$$

is conserved. Using this to eliminate N_0 from (5.90), and dropping some terms of higher order, we get the following approximate Hamiltonian for low energy excitations of the system:

$$H \approx \tfrac{1}{2}\lambda N^2 + \tfrac{1}{2} \sum_{\mathbf{k}\neq 0} (k^2/2m + \lambda N)(a_{\mathbf{k}}^* a_{\mathbf{k}} + a_{-\mathbf{k}}^* a_{-\mathbf{k}}) + \tfrac{1}{2} \sum_{\mathbf{k}\neq 0} \lambda N (a_{\mathbf{k}}^* a_{-\mathbf{k}}^* + a_{\mathbf{k}} a_{-\mathbf{k}}). \quad (5.92)$$

The first line of this expression is already diagonal in the number representation of the basic modes, but the second line being of equal order in λ, cannot be ignored. It is evident, however, that now only pairs of modes, $\mathbf{k}$ and $-\mathbf{k}$, are interacting with one another, and it is not difficult to construct a canonical transformation to new operators $\alpha_{\mathbf{k}}$, $\alpha_{\mathbf{k}}^*$, where the off-diagonal terms are eliminated. Thus, try

$$a_{\mathbf{k}} = \frac{1}{\sqrt{(1-B_{\mathbf{k}}^2)}} (\alpha_{\mathbf{k}} + B_{\mathbf{k}} \alpha_{-\mathbf{k}}^*); \quad a_{\mathbf{k}}^* = \frac{1}{\sqrt{(1-B_{\mathbf{k}}^2)}} (\alpha_{\mathbf{k}}^* + B_{\mathbf{k}} \alpha_{-\mathbf{k}}); \quad (5.93)$$

which preserves the usual commutation relations $[\alpha_{\mathbf{k}}, \alpha_{\mathbf{k}'}^*] = \delta_{\mathbf{k}\mathbf{k}'}$, whatever the value of the unknown function $B_{\mathbf{k}}$. Putting these into (5.92) we get the following elaborate expression:

$$\begin{aligned} H = \tfrac{1}{2}\lambda N^2 &+ \sum_{\mathbf{k}\neq 0} \frac{1}{1-B_{\mathbf{k}}^2} \left\{ \left(\frac{k^2}{2m} + \lambda N \right) B_{\mathbf{k}}^2 + \lambda N B_{\mathbf{k}} \right\} \\ &+ \tfrac{1}{2} \sum_{\mathbf{k}\neq 0} \frac{1}{1-B_{\mathbf{k}}^2} \left\{ \left(\frac{k^2}{2m} + \lambda N \right)(1+B_{\mathbf{k}}^2) + 2\lambda N B_{\mathbf{k}} \right\} (\alpha_{\mathbf{k}}^* \alpha_{\mathbf{k}} + \alpha_{-\mathbf{k}}^* \alpha_{-\mathbf{k}}) \\ &+ \tfrac{1}{2} \sum_{\mathbf{k}\neq 0} \frac{1}{1-B_{\mathbf{k}}^2} \left\{ \left(\frac{k^2}{2m} + \lambda N \right) 2B_{\mathbf{k}} + \lambda N(1+B_{\mathbf{k}}^2) \right\} (\alpha_{\mathbf{k}}^* \alpha_{-\mathbf{k}}^* + \alpha_{\mathbf{k}} \alpha_{-\mathbf{k}}). \end{aligned} \quad (5.94)$$

But we are at liberty to choose $B_{\mathbf{k}}$ so that

$$\left(\frac{k^2}{2m}+\lambda N\right)2B_{\mathbf{k}}+\lambda N(1+B_{\mathbf{k}}^2) = 0. \tag{5.95}$$

The awkward terms in the last line of (5.94) are thereby eliminated, so that the Hamiltonian becomes a simple sum of ordinary harmonic oscillator terms.

After solving (5.95) for $B_{\mathbf{k}}$, one gets the final result

$$H = \tfrac{1}{2}\lambda N^2 - \tfrac{1}{2}\sum_{\mathbf{k}\neq 0}\left[\left(\frac{k^2}{2m}+\lambda N\right) - \sqrt{\left\{\left(\frac{k^2}{2m}+\lambda N\right)^2-(\lambda N)^2\right\}}\right]$$
$$+\tfrac{1}{2}\sum_{\mathbf{k}\neq 0}\sqrt{\left\{\left(\frac{k^2}{2m}+\lambda N\right)^2-(\lambda N)^2\right\}}\,(\alpha_{\mathbf{k}}^*\alpha_{\mathbf{k}}+\alpha_{\mathbf{k}}^*\alpha_{-\mathbf{k}}). \tag{5.96}$$

The first term in this expression is what the interaction between the bosons would have given if they had all been strictly condensed into the zeroth mode. But the ground state of our interacting system is more complicated than this; for example, the average occupation numbers of the other single-boson modes do not vanish, although they are always small compared with N_0. The energy of this state is reduced, as shown by the second term in (5.96), by density fluctuations and particle correlations; what we lose in added kinetic energy is more than made up in reduced potential energy of boson–boson interaction.

The energy spectrum near the ground state is generated by the third term in (5.96), which corresponds to independent boson-type excitations of energy

$$\mathscr{E}(\mathbf{k}) = \sqrt{\left\{\left(\frac{k^2}{2m}+\lambda N\right)^2-(\lambda N)^2\right\}}. \tag{5.97}$$

For large values of k, these are not essentially different from the separate boson modes of our non-interacting system, with energy $k^2/2m$. But for small k, we have

$$\mathscr{E}(\mathbf{k}) \to k\sqrt{\frac{\lambda N}{m}}, \tag{5.98}$$

which is the dispersion formula for an ordinary long-wave acoustic vibration of the gas. In fact we can even identify λN^2 as the compressibility of our interacting boson system, so that these waves travel with the usual macroscopic velocity.

This theory is elegant in the mechanism by which the long-wave phonons become ordinary single-particle excitations as the momentum increases, and therefore provides something of a model for the phonon–

roton spectrum of liquid ^{4}He. But, as I have remarked, the proper theory of superfluids is very much more complicated than this.

Actually, the device of simplifying the Hamiltonian to something like (5.92) and then making the transformation (5.93) is older than this particular application. It is used, for example, in the derivation of the spin-wave spectrum of an antiferromagnet, where, again, the zero-point motion in the ground state plays a significant role.

5.13 The superconducting state

The most subtle behaviour exhibited by a many-body system is the transition to the superconducting state. As every schoolboy knows, many metals become superconducting at low temperatures, and display such peculiar properties as the Meissner effect, flux quantization, Josephson tunnelling, etc. The full discussion of these phenomena and their theoretical interpretation demands a treatise to itself, where many excellent examples would be found of applications of the various advanced techniques of modern quantum theory. Indeed, one of the purposes of the present book is to give an account of such techniques, so that such applications may be more readily understood by the lay reader.

But the delights of exploring this field of the intellect must not be anticipated; we must content ourselves here with an outline derivation of the fundamental principle of the whole theory—the demonstration, first given by Bardeen, Cooper and Schrieffer (BCS), that the ground state of an assembly of mutually attracting fermions is separated by an energy gap from the lowest excited levels of the energy spectrum. This not only explains many of the physical phenomena of superconductivity, but is perhaps the most important exemplification of the principle noted in §§5.9 and 5.10, that a many-body system may have states that cannot be reached by a finite sequence of perturbation procedures. The superconducting state of the *BCS theory* is a radically new type of quantum state, which requires special analytical treatment.

The starting point for a full discussion of superconductivity would be the identification of a mechanism for an attractive force between the electrons. However efficiently it may be screened, the direct coulomb interaction is always repulsive. But the positively charged ions provide sources of attraction for the electrons, and may be used as intermediaries: as shown in §3.7, the exchange of phonons between electrons can give rise to an effective electron–electron interaction which is complicated in form but which is negative—attractive—when the energy difference of the two electron states is small. For simplicity it is usual to postulate that this interaction, $v(\mathbf{q})$, is a small negative constant, $\mathscr{V}$ say, for all momentum transfers between electrons whose energies do not differ by more than a small energy w, but is otherwise zero. Since we are dealing with a degenerate Fermi gas near its ground state, we may write the Hamiltonian in the form

$$H = \sum_{\mathbf{k}} \xi(\mathbf{k}) b_{\mathbf{k}}^* b_{\mathbf{k}} + \sum_{\mathbf{k},\mathbf{k}',\mathbf{q}} v(\mathbf{q}) b_{\mathbf{k}+\mathbf{q}}^* b_{\mathbf{k}'-\mathbf{q}}^* b_{\mathbf{k}'} b_{\mathbf{k}}, \tag{5.99}$$

where, as in (5.77), the energy for non-interacting electrons is measured relative to the Fermi level. In this discussion, again, we ignore spin indices for the sake of simplicity.

The next step is to show that a perturbation calculation of the effect of the interaction diverges for any pair of fermions with exactly opposite momenta, $\mathbf{k}$ and $-\mathbf{k}$. This divergence is a sign that this system could form a bound state—a so-called *Cooper pair*—however weak the actual attractive potential $\mathscr{V}$. In the language of diagrams we find that the vertex part of the two-particle Green function (4.89) acquires a singularity in this range of relative momenta. This result may be derived in a sophisticated way from the Bethe–Salpeter equation (4.92), by making certain simplifying assumptions in order to solve for the kernel, but the singularity really arises from the behaviour of the energy denominator in integrals appearing in the very simplest diagrams for this vertex. For a pair of electrons of zero net momentum, both very near the Fermi surface, the number of intermediate scattering states is so enhanced that a bound state (relative coherence of phase) must be produced.

The Cooper-pair phenomenon is not, however, the full story of the superconducting transition: it merely indicates that the normal many-electron state is unstable towards such an effect, and must be replaced by a more complicated description. The following argument, due to Bogoliubov and similar in principle to the method of §5.12, demonstrates one of several equivalent mathematical procedures used to derive this new, more stable ground state.

The first point to notice is that the interaction term in (5.99) is innocuous unless the two interacting particles can form a Cooper pair—i.e. unless $\mathbf{k}' = -\mathbf{k}$. Near the ground state of the system, all other terms in the summations are essentially of random phase, and can therefore be dropped. This is, of course, an approximation, which may be subsequently validated.

Our Hamiltonian is thus of the approximate form

$$H \approx \sum_{\mathbf{k}} \xi(\mathbf{k}) b_{\mathbf{k}}^* b_{\mathbf{k}} + \sum_{\mathbf{k},\mathbf{q}} v(\mathbf{q}) b_{\mathbf{k}+\mathbf{q}}^* b_{-\mathbf{k}-\mathbf{q}}^* b_{-\mathbf{k}} b_{\mathbf{k}}. \tag{5.100}$$

With our experience in diagonalizing the Hamiltonian (5.92) for the *boson* gas, let us make a canonical transformation similar to (5.93), but now preserving the *anticommutation* properties of the *fermion* operators. This we write as follows:

$$\left.\begin{aligned} b_{\mathbf{k}} &= A_{\mathbf{k}} \beta_{\mathbf{k}} + B_{\mathbf{k}} \beta_{-\mathbf{k}}^*; \quad & b_{\mathbf{k}}^* &= A_{\mathbf{k}} \beta_{\mathbf{k}}^* + B_{\mathbf{k}} \beta_{-\mathbf{k}}; \\ b_{-\mathbf{k}} &= A_{\mathbf{k}} \beta_{-\mathbf{k}} - B_{\mathbf{k}} \beta_{\mathbf{k}}^*; \quad & b_{-\mathbf{k}}^* &= A_{\mathbf{k}} \beta_{-\mathbf{k}}^* - B_{\mathbf{k}} \beta_{\mathbf{k}}, \end{aligned}\right\} \tag{5.101}$$

where the proprieties are preserved for the new fermion operators $\beta_{\mathbf{k}}$, etc. provided only that the functions $A_{\mathbf{k}}$ and $B_{\mathbf{k}}$ satisfy

$$A_{\mathbf{k}}^2 + B_{\mathbf{k}}^2 = 1. \tag{5.102}$$

It is worth remarking that (5.101) is equivalent to a real rotation through the angle

$$\theta_{\mathbf{k}} = \cos^{-1}(A_{\mathbf{k}}) \tag{5.103}$$

in the subspace spanned by the operators for $\mathbf{k}$ and $-\mathbf{k}$. The corresponding boson transformation (5.93) represents a rotation by an imaginary angle $i\theta_{\mathbf{k}}$ such that

$$\cosh \theta_{\mathbf{k}} = (1 - B_{\mathbf{k}}^2)^{-\frac{1}{2}}. \tag{5.104}$$

This provides a link with the algebra of spin operators (§ 6.5), which has sometimes been used as a representation for these transformations.

Putting (5.101) into (5.100), we get numerous terms, which we reduce to normal form by pushing all annihilation operators to the right with the aid of the anticommutation relations. The result is as follows:

$$\begin{aligned} H = {} & 2\sum_{\mathbf{k}} \xi(\mathbf{k}) B_{\mathbf{k}}^2 + \sum_{\mathbf{k},\mathbf{q}} v(\mathbf{q}) A_{\mathbf{k}} B_{\mathbf{k}} A_{\mathbf{k}+\mathbf{q}} B_{\mathbf{k}+\mathbf{q}} \\ & + \sum_{\mathbf{k}} \{\xi(\mathbf{k})(A_{\mathbf{k}}^2 - B_{\mathbf{k}}^2) - 2A_{\mathbf{k}} B_{\mathbf{k}} \sum_{\mathbf{q}} v(\mathbf{q}) A_{\mathbf{k}+\mathbf{q}} B_{\mathbf{k}+\mathbf{q}}\} (\beta_{\mathbf{k}}^* \beta_{\mathbf{k}} + \beta_{-\mathbf{k}}^* \beta_{-\mathbf{k}}) \\ & + \sum_{\mathbf{k}} \{2\xi(\mathbf{k}) A_{\mathbf{k}} B_{\mathbf{k}} + (A_{\mathbf{k}}^2 - B_{\mathbf{k}}^2) \sum_{\mathbf{q}} v(\mathbf{q}) A_{\mathbf{k}+\mathbf{q}} B_{\mathbf{k}+\mathbf{q}}\} (\beta_{\mathbf{k}}^* \beta_{-\mathbf{k}}^* + \beta_{\mathbf{k}} \beta_{-\mathbf{k}}) \\ & + O(\beta_{\mathbf{k}+\mathbf{q}}^* \beta_{-\mathbf{k}-\mathbf{q}}^* \beta_{-\mathbf{k}} \beta_{\mathbf{k}}). \end{aligned} \tag{5.105}$$

This formula is obviously analogous to (5.94). The first two lines would produce a spectrum of independent fermion excitations generated by the operators $\beta_{\mathbf{k}}^*$. Acting on the ground state of this system, all other operators in (5.105) would produce zero except the double creation operators in the third line. As in (5.95), we choose $A_{\mathbf{k}}$ so as to make the coefficient of this term vanish, i.e.

$$2\xi(\mathbf{k})\,A_{\mathbf{k}}B_{\mathbf{k}}+(A_{\mathbf{k}}^2-B_{\mathbf{k}}^2)\sum_{\mathbf{q}}v(\mathbf{q})\,A_{\mathbf{k}+\mathbf{q}}B_{\mathbf{k}+\mathbf{q}}=0. \tag{5.106}$$

Putting the solutions of (5.106) and (5.102) back into (5.105) defines the Hamiltonian of the system near the superconducting ground state.

To see what this looks like, let us put into (5.106) the simplified cut-off interaction mentioned above. The following equations can then be used: try

$$\Delta_0=-\mathscr{V}\sum_{-w}^{w}A_{\mathbf{k}+\mathbf{q}}B_{\mathbf{k}+\mathbf{q}}, \tag{5.107}$$

which gives

$$A_{\mathbf{k}}^2=\frac{1}{2}\left[1+\frac{\xi(\mathbf{k})}{\sqrt{\{\Delta_0^2+\xi^2(\mathbf{k})\}}}\right];\quad B_{\mathbf{k}}^2=\frac{1}{2}\left[1-\frac{\xi(\mathbf{k})}{\sqrt{\{\Delta_0^2+\xi^2(\mathbf{k})\}}}\right], \tag{5.108}$$

with the condition that Δ_0 satisfies

$$1=-\tfrac{1}{2}\mathscr{V}\sum_{-w}^{w}\{\Delta_0^2+\xi^2(\mathbf{k})\}^{-\frac{1}{2}}. \tag{5.109}$$

The first row of (5.105) now reads like a small correction to the ground state energy of our whole system; the second row, as in (5.94) defines fermion excitations, of energy

$$\begin{aligned}\mathscr{E}(\mathbf{k})&=\xi(\mathbf{k})\,(A_{\mathbf{k}}^2-B_{\mathbf{k}}^2)-2A_{\mathbf{k}}B_{\mathbf{k}}\sum_{\mathbf{q}}v(\mathbf{q})\,A_{\mathbf{k}+\mathbf{q}}B_{\mathbf{k}+\mathbf{q}}\\&=\sqrt{\{\Delta_0^2+\xi^2(\mathbf{k})\}}.\end{aligned} \tag{5.110}$$

This is the most important consequence of the calculation. It shows that the spectrum of the system near the ground state may be analysed into fermion-like quasi-particles, whose energy tends to a finite value as they approach the Fermi surface, where $\xi(\mathbf{k})=0$. In other words, Δ_0 is the width of the *energy gap*; unlike a boson gas, a superconducting system does not have excitations of vanishingly small energy like the long-wave acoustic modes (5.98).

The actual nature of these excitations may be discovered by working back from (5.108) to (5.101). For $\mathbf{k}$ near the Fermi surface, each quasi-particle is an electron-hole pair in the modes $\mathbf{k}$ and $-\mathbf{k}$ respectively. As $\mathbf{k}$ moves away from the Fermi surface, $\mathscr{E}(\mathbf{k})$ in (5.110) tends to $\xi(\mathbf{k})$, and the quasi-particle becomes essentially an ordinary electron or hole of the simple non-interacting gas. The proximity of the Fermi surface, by inhibiting transitions that violate the Pauli principle, assists the attractive interaction to establish a definite phase relation between electron waves travelling in exactly opposite directions, causing a modest degree of 'condensation' over a narrow range of energies.

From here the discussion might move to the solution of the equation (5.109) for the energy gap Δ_0; it is easy to show that this condition can usually be satisfied when $\mathscr{V}$ is negative, and the familiar exponential formula in terms of the density of states is readily derived. Then we consider terms dropped from (5.105), corresponding, as one would say, to interactions between quasi-particles. These terms are best treated self-consistently in the spirit of the Hartree–Fock/RPA methods, which yield fairly simple integral equations for Δ_0 as a function of T, showing that the superconducting state is unstable above a certain finite temperature. But we must not embark upon the theory of the phase transition, and all the other fascinating properties of these strange systems.

CHAPTER 6

RELATIVISTIC FORMULATIONS

'The Duke didn't kiss Julietta: she kissed the Duke instead.'

6.1 Lorentz invariance

The special theory of relativity may be validated in a variety of ways, to suit almost all philosophical tastes. For our present purposes it is simplest to assert, without much attention to logical precision, that the laws of physics must seem to be the same for all observers moving with uniform velocity relative to one another. This is much the same as the mathematical condition that all equations representing these laws should be capable of expression in *Lorentz-invariant* or *covariant* form.

The essence of elementary three-dimensional vector analysis is that it allows us to write down relations between physical quantities in *rotationally invariant* form. When I use the symbol **E** for the electric field, I want it to represent the same physical force, whether I choose my x and y axes to run east and north respectively, or whether they run ENE and NNW. We allow the *components* to change from one frame of reference to another, but the vector itself preserves its identity.

The most elementary vector is, of course, a mere translation in space—

$$\boldsymbol{\delta}\mathbf{R} = (\delta R_x, \delta R_y, \delta R_z) \tag{6.1}$$

say, in an obvious rotation. In any rotation of co-ordinate axes, without any stretching or straining of space itself, the scalar product of any two such local translations $\boldsymbol{\delta}\mathbf{R}$ and $\boldsymbol{\delta}\mathbf{S}$ remains constant, i.e.

$$\begin{aligned}\delta R_x\,\delta S_x + \delta R_y\,\delta S_y + \delta R_z\,\delta S_z &= \boldsymbol{\delta}\mathbf{R}\cdot\boldsymbol{\delta}\mathbf{S}\\ &= \delta R'_x\,\delta S'_x + \delta R'_y\,\delta S'_y + \delta R'_z\,\delta S'_z,\end{aligned} \tag{6.2}$$

even though the new components $\delta R'_x$, etc. are quite different from the old ones. Since all true three-dimensional vectors transform in the same way as space translations, the same invariance principle holds for all their scalar products, including the modulus squared of any one vector, e.g.

$$|\boldsymbol{\delta}\mathbf{R}|^2 = (\delta R_x)^2 + (\delta R_y)^2 + (\delta R_z)^2. \tag{6.3}$$

Lorentz invariance merely generalizes this condition. The 'events' that are the substance of Being occur at locations in space and at epochs of time that our instruments can measure and compare. One observer may discover two such events to be separated by a spatial vector $\boldsymbol{\delta}\mathbf{R}$ and a time interval δt; another observer may assign to this separation the space co-ordinate $\boldsymbol{\delta}\mathbf{R}'$ and the time difference δt. These observations are not incompatible provided that the quantity

$$\begin{aligned} -(\delta s)^2 &= (\delta R_x)^2 + (\delta R_y)^2 + (\delta R_z)^2 - c^2(\delta t)^2 \\ &= (\delta R'_x)^2 + (\delta R'_y)^2 + (\delta R'_z)^2 - c^2(\delta t')^2 \end{aligned} \tag{6.4}$$

is the same for either set of components.

The constant c in (6.4) is, of course, the velocity of light. In the special case where the two 'events' are the emission and reception of a light signal between two points (such as mirrors on opposite arms of a Michelson interferometer), the *interval* δs is zero. For 'time-like' intervals, such as between points on the trajectory of a moving particle, $(\delta s)^2$ is positive. Every elementary textbook on special relativity gives a detailed derivation of explicit formulae, the *Lorentz transformations* of space and time co-ordinates, that satisfy (6.24), but these are scarcely needed in much of what follows.

Comparison of (6.3) and (6.4) suggests that the symbolic invariance of three-dimensional vector algebra might be generalized to include the time co-ordinate by introducing this as a fourth 'dimension'. Events in space-time are labelled by *4-vectors* with components

$$x^\mu \equiv (x^1, x^2, x^3, x^4) \equiv (x, y, z, ct). \tag{6.5}$$

The condition (6.4) for Lorentz invariance is then written

$$(\delta s)^2 = -g_{\mu\nu}\, \delta x^\mu\, \delta x^\nu, \tag{6.6}$$

where the Einstein summation convention over repeated indices, $\mu = 1 \dots 4$, $\nu = 1, \dots, 4$, is implied. The only extra complication is that the *metric tensor*† $g_{\mu\nu}$. is not simply a unit matrix as in three dimensions, but is diagonal, with $g_{11} = g_{22} = g_{33} = 1$ and $g_{44} = -1$.

Relativistic physics then becomes an exercise in constructing 4-vectors that transform like the co-ordinates under Lorentz transformation and that represent genuine physical properties. The most important rule embodies the analogue of (6.2): any expression of the form of a 'scalar product' of 4-vectors, i.e.

$$g_{\mu\nu} A^\mu B^\nu, \tag{6.7}$$

† The convention $g_{00} = 1$, $g_{11} = g_{22} = g_{33} = -1$, with $x^0 = ct$, is also commonly used.

is a number which is invariant under Lorentz transformation and therefore eligible to appear in any equation about physical quantities.

The starting point of relativistic kinematics and dynamics is the definition of the *momentum-energy 4-vector*, whose components are

$$p^{\mu} \equiv (p_x, p_y, p_z, \mathscr{E}/c); \tag{6.8}$$

in other words, the energy of the particle is added as a *time-like component* to the ordinary three-dimensional momentum. The laws of Conservation of Mass and Energy then are conflated into the principle that an isolated, non-interacting particle may be assigned rest mass, m_0, such that

$$p^{\mu} = m_0 c \frac{\mathrm{d}x^{\mu}}{\mathrm{d}s}. \tag{6.9}$$

This is an example of a *covariant* formula: the momentum-energy 4-vector is equated to a multiple of the *Minkowski velocity*—the rate of change of the co-ordinates of the particle, with interval, along its trajectory in space-time. Under a Lorentz transformation the components of each 4-vector will change, but they will still remain equal to one another, component by component.

For a perfectly free particle, in the absence of forces, one may derive many of the familiar phenomena of relativistic physics, such as the mass-energy relation, time dilatation, etc. by merely asserting that p^{μ} must be constant. But a complete dynamical theory requires the introduction of forces and accelerations. It is natural to express Newton's Laws in the form

$$c\frac{\mathrm{d}p^{\mu}}{\mathrm{d}s} = F^{\mu}, \tag{6.10}$$

say, which reduces to the simple classical relation between the rate of change of momentum and the force, in the limit of low velocities where

$$\mathrm{d}s \approx c\,\mathrm{d}t. \tag{6.11}$$

6.2 Relativistic electromagnetic theory

To proceed further, we need a general theory of 'forces' between our 'particles' which is inevitably cast into the language and formalism of the theory of continuous fields. In elementary classical field theory this formalism is hinged upon the symbol ∇, the vector derivative operator, which measures gradients of scalars and also transforms like a cartesian vector under a rotation of axes.

The corresponding operator in space-time obviously has the derivative $(1/c)\,\partial/\partial t$ as its fourth component, but there is a little difficulty of signs because the metric tensor (6.6) is not positive definite. The number of different conventions that one might introduce at this stage is some combinatorial factor; we shall write

$$\nabla^{\mu} \equiv \left(\frac{\partial}{\partial x}, \frac{\partial}{\partial y}, \frac{\partial}{\partial z}, -\frac{1}{c}\frac{\partial}{\partial t}\right), \tag{6.12}$$

for the 4-*vector gradient operator*, which produces Lorentz-invariant formulae when substituted for an ordinary 4-vector in any expression such as (6.7). It is not difficult (although somewhat messy algebraically) to prove that ∇^{μ} does in fact transform correctly in a Lorentz transformation; the main point to notice is that a space derivative $\partial/\partial x^{\mu}$, being the inverse of a small displacement such as dx^{μ}, is multiplied by the *inverse* transformation matrix in going, say, to primed co-ordinates. In the language of the general tensor calculus, it is a *contravariant* vector, and has to be multiplied by the metric tensor $g_{\mu\nu}$ if it is to be made to behave like the *covariant* vectors which we are using systematically; this explains the change of sign of the time-like component in (6.12). Another way of achieving the same end is to introduce the imaginary time as the fourth component, but this does not link up satisfactorily with the notation of general relativity. So be it!

Our most familiar field theory is that of classical electricity and magnetism, as enshrined in Maxwell's equations in empty space:

$$\left.\begin{array}{ll} (a)\quad \nabla\wedge\mathbf{E} = -\dfrac{1}{c}\dfrac{\partial\mathbf{H}}{\partial t}; & (b)\quad \nabla\wedge\mathbf{H} = \dfrac{1}{c}\dfrac{\partial\mathbf{E}}{\partial t}; \\ (c)\quad \nabla\cdot\mathbf{E} = 0; & (d)\quad \nabla\cdot\mathbf{H} = 0. \end{array}\right\} \tag{6.13}$$

We are used to thinking of the components of the electric and magnetic fields, **E** and **H**, as those of distinct spatial vectors; but this is only true in a fixed frame of reference. The essence of the Lorentz theory is that these fields must be different aspects of one physical entity—the *electromagnetic field.*

To cast Maxwell's equations into relativistic form, we merely recall that both fields may be derived from *potentials*, i.e.

$$\mathbf{E} = -\frac{1}{c}\frac{\partial\mathbf{A}}{\partial t} - \nabla\phi; \quad \mathbf{H} = \nabla\wedge\mathbf{A}. \tag{6.14}$$

The fact that both the *vector potential* **A**, and the *scalar potential* ϕ are required for the electric field, suggests that these are simply different components of a more general 4-vector potential

$$A^{\mu} = (A_x, A_y, A_z, \phi), \tag{6.15}$$

in which the scalar potential has become just the time-like component.

Going back now to (6.14), and using the general gradient operator (6.12), we find that all six components of the electric and magnetic fields occur amongst the set of numbers

$$F^{\mu\nu} = \nabla^{\mu}A^{\nu} - \nabla^{\nu}A^{\mu}; \tag{6.16}$$

for example

$$F^{12} = \frac{\partial A_y}{\partial x} - \frac{\partial A_x}{\partial y} = H_z \tag{6.17}$$

and

$$F^{14} = \frac{\partial \phi}{\partial x} - \left(-\frac{1}{c}\frac{\partial}{\partial t}\right) A_x = -E_x, \quad \text{etc.} \tag{6.18}$$

Thus, the electric and magnetic field strengths appear as elements in the 4×4 array

$$F^{\mu\nu} = \begin{pmatrix} 0 & -H_z & H_y & E_x \\ H_z & 0 & -H_x & E_y \\ -H_y & H_x & 0 & E_z \\ -E_x & -E_y & -E_x & 0 \end{pmatrix}. \tag{6.19}$$

A mathematical object of this type, transforming like the outer product of two 4-vectors, is a *covariant tensor*. The *electromagnetic field tensor* is obviously *antisymmetric* in its two indices. After a Lorentz transformation, the new electric field components E'_z, say, will evidently be linear combinations of components of both the electric and magnetic fields in the original frame of reference.

But so far this proposed notation begs the question of the Lorentz invariance of Maxwell's equations, which we must now write in the same language. Equations (6.13 (a)) and (6.13 (d)) are, of course, identities consequent on the existence of the potentials in (6.14). The other two equations are generated by the relation

$$g_{\mu\nu}\nabla^{\nu}F^{\mu\delta} = 0; \tag{6.20}$$

(6.13 (b)) is equivalent to $\delta = 1, 2, 3$ and (6.13 (c)) to $\delta = 4$. Since (6.20) is manifestly covariant in form, the same relation is true in any coordinate system. If the electric and magnetic fields are defined according to the prescriptions (6.15)–(6.19) then they will automatically obey Maxwell's equations in any chosen frame of reference: the physical laws of electromagnetism will be the same for any observer.

From now on, we shall scarcely be interested in the electromagnetic field quantities themselves, but will concentrate our attention on the potential. Nevertheless, it is worth remarking that *classical electrodynamics*—the theory of the motion of particles in electromagnetic fields—is governed by an equation that can easily be guessed from (6.10) and from the demands of covariance:

$$c\frac{\mathrm{d}p^{\mu}}{\mathrm{d}s} = e g_{\nu\delta}\frac{\mathrm{d}x^{\delta}}{\mathrm{d}s}F^{\mu\nu}, \tag{6.21}$$

where e is the charge carried by the particle. In ordinary three-dimensional notation, this is equivalent to the familiar formula for the Lorentz force

$$\mathbf{F} = e\left(\mathbf{E}+\frac{1}{c}\mathbf{v}\wedge\mathbf{H}\right), \tag{6.22}$$

together with its energy integral.

To complete the theory, we need to show how the field itself stems from the charges. A 4-vector current

$$j^{\mu} \equiv (j_x, j_y, j_z, \rho) \tag{6.23}$$

can be defined out of the components of the ordinary charge-current density $\mathbf{j}$ and charge density ρ. The reader may verify that the usual modifications to Maxwell's equation (6.20) in the presence of currents and charges are expressible in the form

$$g_{\mu\nu}\nabla^{\nu}F^{\mu\delta} = -j^{\delta}. \tag{6.24}$$

6.3 The wave equation and gauge invariance

Maxwell's equations impose upon the potential 4-vector the condition

$$g_{\mu\nu}\nabla^{\nu}\nabla^{\mu}A^{\delta} - g_{\mu\nu}\nabla^{\nu}\nabla^{\delta}A^{\mu} = -j^{\delta}, \tag{6.25}$$

which follows immediately upon substituting (6.16) in (6.24). This is a rather complicated second-order partial differential equation, which can be greatly simplified by a *gauge transformation.*

Looking back at (6.16) we note that the physical observables—the fields $\mathbf{E}$ and $\mathbf{H}$—would come out with exactly the same values if we were to add to the components of the potential the 4-vector gradient of any scalar function—i.e.

$$A^{\mu} \rightarrow A^{\mu} + \nabla^{\mu}\chi. \tag{6.26}$$

It is obvious that we may choose our function χ in such a way as to make the new potential satisfy the *Lorentz condition*

$$g_{\mu\nu}\nabla^{\nu}A^{\mu} = 0. \tag{6.27}$$

Our generalized Maxwell equation (6.25) then reads, simply

$$g_{\mu\nu}\nabla^{\nu}\nabla^{\mu}A^{\delta} = -j^{\delta}. \tag{6.28}$$

In conventional vector notation this is very familiar:

$$\left.\begin{aligned} \square^2\mathbf{A} &\equiv \nabla^2\mathbf{A} - \frac{1}{c^2}\frac{\partial^2\mathbf{A}}{\partial t^2} = -\mathbf{j}, \\ \square^2\phi &\equiv \nabla^2\phi - \frac{1}{c^2}\frac{\partial^2\phi}{\partial t^2} = -\rho. \end{aligned}\right\} \tag{6.29}$$

In other words each component of the 4-vector potential satisfies a wave equation, with the components of the 4-vector current as sources. It is nice to express the wave equation in terms of the *D'Alembertian operator* $\square^2$, which shows clearly that this is the relativistic generalization of the Laplacian ∇^2, and is itself invariant under a Lorentz transformation.

Consider now the situation in empty space. It is natural for us to make a Fourier transformation, just as in the theory of non-relativistic fields. We introduce the frequency as a time-like component of a generalized wave-vector

$$k^{\mu} = (k_x, k_y, k_z, \omega/c), \tag{6.30}$$

and by analogy with (1.90) try a solution of (1.28) in the form

$$A^{\delta}(x) = A^{\delta}(k)\exp(-ig_{\mu\nu}k^{\mu}x^{\nu}), \tag{6.31}$$

i.e.

$$\mathbf{A}(\mathbf{r},t) = \mathbf{A}(\mathbf{k},\omega)\exp\{-i(\mathbf{k}\cdot\mathbf{r}-\omega t)\}. \tag{6.32}$$

When the current j^{δ} vanishes, this gives us, simply, the elementary dispersion law

$$g_{\mu\nu}k^{\mu}k^{\nu} = 0, \quad \text{i.e.} \quad k^2 - \omega^2/c^2 = 0. \tag{6.33}$$

The free space solutions of (6.28) are, indeed, superpositions of plane waves, travelling with the velocity of light.

But the Lorentz condition (6.27) imposes upon the components of the vector potential the relation

$$g_{\mu\nu}k^{\mu}A^{\nu} = 0. \tag{6.34}$$

In empty space, far from any electric charges, we usually assume that the electrostatic potential is zero, i.e.

$$\phi(\mathbf{r},t) \equiv A^4(x) = 0. \tag{6.35}$$

We then deduce from (6.34) that the vector-potential propagates as a *transversely polarized* excitation:

$$\mathbf{k} \cdot \mathbf{A}(\mathbf{k}, \omega) = 0. \tag{6.36}$$

Unfortunately, the arbitrary choice (6.35), although always possible by means of a gauge transformation, is not, itself, Lorentz invariant: The time-like component A'^4 of the potential 4-vector in some different co-ordinate system does not necessarily vanish as a consequence of (6.35). Thus, a different observer may not agree that our light waves are purely transverse, and may assign to them some longitudinal component.

In the classical theory this difficulty can be avoided by associating with each Lorentz transformation an appropriate change of gauge which preserves (6.35): the proof that this is always possible, by a further specialization of the function χ in (6.26), without violating the Lorentz condition (6.27), is quite simple, and need not be given here. In other words, the electric and magnetic fields in a light wave may always be analysed as if derived solely from transversely polarized potentials, even though the potentials might have to be defined differently for different observers. This is important, because it tells us that there are only two degrees of freedom associated with each wave-vector, corresponding to the two orthogonal directions of transverse polarization.

When there are electric charges in the neighbourhood, this theory becomes somewhat more complicated, for we naturally desire to retain the electrostatic potential with these as sources, in conformity with the elementary theory. It then turns out to be convenient to replace the Lorentz condition by one that is not relativistically invariant, i.e. the *coulomb gauge condition*

$$\nabla \cdot \mathbf{A} = 0. \tag{6.37}$$

The scalar potential ϕ then satisfies the usual equations for an electrostatic potential, generating the coulomb force between charges, etc. The proof that this gives the same physical results as one would obtain by retaining the Lorentz condition and allowing for the effects of longitudinal components in $\mathbf{A}(\mathbf{k}, \omega)$, as well as the scalar waves associated with the component ϕ, is to be found in the more advanced treatises.

These difficulties are further compounded when one sets about quantizing the electromagnetic field. Thus, a gauge condition such as

(6.27) must now be applied to a sum of quantum mechanical operators, and can only be given meaning when allowed to act on a state vector. There is an elaborate analysis of this problem (the *Gupta–Bleuler formalism*) showing that we may proceed as if only the transverse photons need be quantized, but this again is largely of technical interest.

6.4 Quantization of relativistic fields

The wave equation (6.28) for electromagnetism in empty space is a special case of the Klein–Gordon equation (1.85), whose relativistic invariance we now express in the form

$$(\Box^2 - m^2)\,\phi(x) = 0. \tag{6.38}$$

The scalar function $\phi(x)$ governs a field of bosons with rest mass m, i.e. the relativistic energy of a particle of this field is given by

$$\omega_{\mathbf{k}} = \pm\,(m^2 + \mathbf{k}^2)^{\frac{1}{2}}. \tag{6.39}$$

The *photons* of electromagnetism are more complicated than this, because they belong to a transverse vector field with two types of polarization, but many of their properties depend mainly upon their having zero rest mass: $m = 0$ in (6.39).

For simplicity, let us take $\hbar = c = 1$, and introduce a more compact relativistic notation in which†

$$kx \equiv -g_{\mu\nu}\,k^\mu x^\mu \equiv -\mathbf{k}\cdot\mathbf{r} + \omega t. \tag{6.40}$$

The space-time solutions of (6.38) are analogous to (6.31), being simple waves of the form

$$\phi(x) = \Phi_{\mathbf{k}}\,\mathrm{e}^{\mathrm{i}kx}, \tag{6.41}$$

with the condition (6.34) imposed on the components of the momentum 4-vector, i.e.

$$k^4 = \omega_{\mathbf{k}}. \tag{6.42}$$

To quantize this field, let us go back to chapter 1, where we learnt to transform a classical field variable into a local field operator of the Schrödinger type. Thus, in the spirit of (1.90), (1.91) and (1.117) we might write

$$\phi(\mathbf{r}) = \sum_{\mathbf{k}} (2\omega_{\mathbf{k}} V)^{-\frac{1}{2}}\{a_{\mathbf{k}}\,\mathrm{e}^{-\mathrm{i}\mathbf{k}\cdot\mathbf{r}} + a_{\mathbf{k}}^*\,\mathrm{e}^{\mathrm{i}\mathbf{k}\cdot\mathbf{r}}\}, \tag{6.43}$$

† The negative sign here makes expressions like k^2 positive definite for particles travelling with velocities $\leqslant c$, and allows us to go over to the conventional expression for the kinetic energy at non-relativistic velocities. Similarly $(\delta s)^2 = \delta x\,\delta x$ is positive for time-like intervals.

where the annihilation and creation operators $a_{\mathbf{k}}$ and $a^*_{\mathbf{k}}$ obey the canonical commutation relations

$$[a_{\mathbf{k}}, a^*_{\mathbf{k}'}] = \delta_{\mathbf{k}\mathbf{k}'}; \quad [a_{\mathbf{k}}, a_{\mathbf{k}'}] = [a^*_{\mathbf{k}}, a^*_{\mathbf{k}'}] = 0. \tag{6.44}$$

If we are to make a relativistic theory, we must obviously include a time co-ordinate in the definition of our field operator. This follows perfectly naturally from (6.4) and (6.42):

$$\phi(x) = \sum_k (2Vk^4)^{-\frac{1}{2}} \{a_k \,\mathrm{e}^{ikx} + a^*_k \,\mathrm{e}^{-ikx}\} \delta(k^4 - \omega_{\mathbf{k}}). \tag{6.45}$$

But this is none other than the Heisenberg or interaction representation of our Schrödinger operator (6.43), as defined in § 3.6, when we introduced the perturbation expansion of the S-matrix. Indeed the relativistic generalization of the whole theory of Feynman diagrams is quite straightforward: the separate notions of 'momentum' and 'energy' are combined into a Lorentz-covariant, momentum 4-vector, but all topological theorems remain unchanged; the main problem is to ensure that all the propagators and vertex factors that might appear in the algebraic expression for a given diagram [e.g. (3.117)] also transform correctly between different frames of reference.

We may check this by actually evaluating a boson propagator of the form (3.106). In a particular co-ordinate frame, the ordered product of field operators is defined as in (3.95)

$$T\{\phi(x)\,\phi(x')\} = \begin{cases} \phi(x)\,\phi(x') & t' < t, \\ \phi(x')\,\phi(x) & t' > t. \end{cases}$$

The propagator is just the expectation value of this operator in the vacuum state. From the definition (6.45), the only terms surviving in the product will be those involving $a_{\mathbf{k}} a^*_{\mathbf{k}}$ in each case. Introducing a step function as in (4.95) to take care of the time ordering, and a factor $-\mathrm{i}$, as in (4.55), we can construct the free-boson Green function

$$\begin{aligned} \Delta_F(x, x') &\equiv -\mathrm{i}\langle 0|\, T\{\phi(x)\,\phi(x')\}\,|0\rangle \\ &= -\mathrm{i}\int \frac{\mathbf{d^3k}}{(2\pi)^3}\frac{1}{2\omega_{\mathbf{k}}}\{\theta(t-t')\,\mathrm{e}^{\mathrm{i}k(x-x')} + \theta(t'-t)\,\mathrm{e}^{-\mathrm{i}k(x-x')}\}. \end{aligned} \tag{6.46}$$

Considering this simply as a function of $(t-t')$, we can achieve all the necessary properties by introducing k^4 as an extra variable of integration which is then constrained to take the value (6.42) by the introduction of appropriate singularities near the real axis. The argument is exactly the same as in (3.112) and (4.130); the reader may verify for himself, as an instructive exercise, that (6.46) is exactly equivalent to

$$\Delta_F(x,x') = \int \frac{d^4k}{(2\pi)^4} e^{ik(x-x')} \frac{1}{k^2 - m^2 + i\delta}, \tag{6.47}$$

where δ is a positive infinitesimal and the integration is along the whole real axis.

From the point of view of diagrammatic expansions, this is simply a special case of the general boson propagator (3.115) with the particular formula (6.39) for $\omega_{\mathbf{q}}$. In other words, the interactions associated with the exchange of virtual mesons obeying the Klein–Gordon equation introduce the factor

$$\Delta_F(k) = \frac{1}{k^2 - m^2 + i\delta} \tag{6.48}$$

in the energy-momentum representation along every meson line.

We notice, moreover, that (6.47) is Lorentz invariant; the value of this number depends only on the actual positions of the two points x and x' relative to one another in space time, not on the co-ordinate system that may have been used to define these points. Thus, our formulation (6.46), which was made in a *particular* frame of reference, turns out to be the same in *any* frame of reference. It is interesting that the poles of (6.48) correspond to both the positive and negative energy roots of (6.39), which are essential in order to maintain the full relativistic invariance of the theory. On the other hand, we must still arbitrarily exclude all negative energy states for *free* Klein–Gordon particles from our description of physical systems, since these would be dynamically unstable.

Although the momentum representation of the propagator (6.48) contains almost all the information of physical significance, it is interesting to study the properties of invariant propagators a little more closely. It is obvious, for example, that (6.47) is the four-dimensional analogue of the Green function (4.105) of the ordinary three-dimensional Schrödinger equation at a definite energy, and must therefore satisfy the inhomogeneous Klein–Gordon equation

$$\{\Box_x^2 - m^2\} \Delta_F(x,x') = -\delta^4(x-x'), \tag{6.49}$$

where the right-hand side is four-dimensional delta function.

Another function that may be derived by very similar arguments is the commutator of the field operators (6.45) at two distinct points in space-time. This must be a scalar invariant from its very definition; but a calculation similar to (6.46) leads to

$$\Delta(x-x') \equiv -\mathrm{i}[\phi(x), \phi(x')]$$
$$= \frac{1}{(2\pi)^3}\int \frac{\mathrm{d}^3\mathbf{k}}{\omega_\mathbf{k}} \sin\{k(x-x')\}. \tag{6.50}$$

Let us choose points in space-time that happen to 'occur' at the same time t for a particular observer. The integral in (6.50) then vanishes automatically, since the integrand, $\sin(\mathbf{k}\cdot\mathbf{r})/\omega_\mathbf{k}$, is an odd function of the ordinary wave-vector $\mathbf{k}$. Thus, observations that appear to be taken simultaneously at different points in space do not interfere with one another. But because $\Delta(x-x')$ is in fact a scalar invariant, this property must extend to all pairs of points in space-time that can be made 'simultaneous' by a suitable Lorentz transformation. In other words, the commutator vanishes for all points that do not lie within the light cone of one another:

$$\Delta(x-x') = 0 \quad \text{for all} \quad (x-x')^2 < 0. \tag{6.51}$$

We recognize in this property a welcome extension of the principle of causality; the formalism only permits dynamical interactions between events that could be linked by a light signal. The formal algebraic proof of this property of an *invariant delta function* (6.50) can be generated by an explicit representation as a four-dimensional integral, similar to the expression (6.47) for $\Delta_F(x,x')$, to which, indeed, $\Delta(x-x')$ is very closely related.

The formulae for these functions in the co-ordinate representation for time-like intervals involves Bessel functions, and is not of great interest. It is encouraging to note, however, that in the photon limit, as $m \to 0$, we arrive at sensible results:

$$-D(x-x') = \lim_{m\to 0} \Delta(x-x')$$
$$= -\frac{1}{4\pi r}[\delta(r-t) - \delta(r+t)], \tag{6.52}$$

where r and t are the real distance and time intervals between the events x and x'. The two terms in this expression correspond, of course, to the two alternative light waves that would have to be propagated from x to x', or vice-versa if the two operators were to interfere with one another. But again, in the general theory of quantum electrodynamics, the main features of the propagation of photons are summed up by assigning to them the momentum representation analogous to (6.48), i.e.

$$\Delta_F(q) = \frac{1}{q^2 + i\delta}. \tag{6.53}$$

A real scalar field ϕ governed by the Klein–Gordon equation corresponds to a *neutral* boson. As shown in § 1.12, a theory of *charged bosons* can be constructed out of a complex field variable, both of whose components satisfy a Klein–Gordon equation. The relativistic quantization of this field follows the same lines as for the neutral boson and introduces no new mathematical difficulties. Similarly, one can generate a free *vector boson* field with three components for each field variable or field operator, each with the same type of propagator (6.48). The physical properties of these various fields depend upon their mutual interactions, which will be discussed more fully in § 6.9.

6.5 Spinors

The Klein–Gordon equation proved not to be a satisfactory basis for a relativistic theory of electrons and nucleons; this was produced, almost by sheer cerebration, by Dirac. Having arrived upon the scene somewhat earlier than, for example, the diagrammatic analysis of perturbation expansions, the Dirac theory is often expounded in relatively elementary texts on quantum theory. For this reason, it is not necessary to repeat the most direct arguments in its favour, nor the various details of its predictions.

A tightly-knit, closed mathematical theory of this sort may always be entered at any of a number of points on its logical circumference. The conventional approach is via the factorization of the Klein–Gordon equation, with the apparently arbitrary device of the Dirac matrices. But this does not bring out the basic relativistic invariance of the formalism, nor does it link up very directly with the group theoretical analysis by which the logical necessity of such a theory can almost be deduced from first principles.

Let us start, therefore, with a discussion of the properties of the Pauli spin operators, which are so familiar in elementary theory. It is argued that the wave function ψ of an electron must have two components (each, of course, complex), corresponding, in some particular frame of reference, to 'spin-up' and 'spin-down' with respect to a chosen axis. In order to manipulate this wave function, we introduce a 2×2 matrix operator $\boldsymbol{\sigma}$ in the Hilbert space which it spans. We assert, moreover, that $\boldsymbol{\sigma}$ is of the nature of a vector in ordinary 3-space, and may therefore be represented as the sum of three separate vectors along the co-ordinate axes, with appropriate components. In other words, we may write

$$\mathbf{R}\cdot\boldsymbol{\sigma} = X\sigma_x + Y\sigma_y + Z\sigma_z \tag{6.54}$$

for the 'spin in the direction of the unit vector $\mathbf{R} = (X, Y, Z)$'.

In this expression we have introduced component spin operators, σ_x, σ_y, σ_z, along the cartesian axes; in the Hilbert space of ψ, these are, of course, capable of representation by the Pauli spin matrices

$$\sigma_x = \begin{pmatrix} 0 & 1 \\ 1 & 0 \end{pmatrix}; \quad \sigma_y = \begin{pmatrix} 0 & -\mathrm{i} \\ \mathrm{i} & 0 \end{pmatrix}; \quad \sigma_z = \begin{pmatrix} 1 & 0 \\ 0 & -1 \end{pmatrix}, \tag{6.55}$$

so that the total spin operator itself is the matrix

$$\mathbf{R}\cdot\boldsymbol{\sigma} = \begin{pmatrix} Z & X-\mathrm{i}Y \\ X+\mathrm{i}Y & -Z \end{pmatrix}. \tag{6.56}$$

Now let us make a rotation of the axes for our frame of reference. It is natural to refer our direction (X, Y, Z) to these new axes—for example, after a rotation by the angle θ about the x-axis, this vector would have components

$$X' = X; \quad Y' = Y\cos\theta + Z\sin\theta; \quad Z' = -Y\sin\theta + Z\cos\theta. \tag{6.57}$$

If now we substitute for X, Y, Z in (6.56), we find that $\mathbf{R}\cdot\boldsymbol{\sigma}$ is no longer of the form (6.54) and (6.55) as a function of the new components X', Y', Z'.

How are we to preserve the directness and simplicity of the prescription (6.54) without setting up some special frame of reference? Let us recall that we never require the wave function itself in any physically observable quantity, but only matrix elements of the form $\langle\phi|f(\boldsymbol{\sigma})|\psi\rangle$, say, of some function of spin operators between two of our two-component wave functions ψ and ϕ. Any transformation that leaves this matrix element unchanged is physically undetectable. For example, any transformation of the form

$$\text{(i)} \quad \psi' = Q\psi; \qquad \text{(ii)} \quad \boldsymbol{\sigma}' = Q\boldsymbol{\sigma}Q^{-1}, \tag{6.58}$$

where Q is a unitary matrix, is permitted.

Let us look a little more closely at the actual transformation induced by (6.57) on the matrix (6.56). It is a matter of elementary algebra to verify that this may be written

$$\begin{pmatrix} Z' & X'-\mathrm{i}Y' \\ X'+\mathrm{i}Y' & -Z' \end{pmatrix}$$
$$= \begin{pmatrix} \cos\frac{1}{2}\theta & \mathrm{i}\sin\frac{1}{2}\theta \\ \mathrm{i}\sin\frac{1}{2}\theta & \cos\frac{1}{2}\theta \end{pmatrix} \begin{pmatrix} Z & X-\mathrm{i}Y \\ X+\mathrm{i}Y & -Z \end{pmatrix} \begin{pmatrix} \cos\frac{1}{2}\theta & -\mathrm{i}\sin\frac{1}{2}\theta \\ -\mathrm{i}\sin\frac{1}{2}\theta & \cos\frac{1}{2}\theta \end{pmatrix}, \tag{6.59}$$

which is exactly of the form demanded by (6.58 (ii)). In other words, if we now assert that the wave function itself is transformed by this particular rotation of axes according to the rule

$$\psi' = \begin{pmatrix} \psi'_\uparrow \\ \psi'_\downarrow \end{pmatrix} = \begin{pmatrix} \cos\frac{1}{2}\theta & \mathrm{i}\sin\frac{1}{2}\theta \\ \mathrm{i}\sin\frac{1}{2}\theta & \cos\frac{1}{2}\theta \end{pmatrix} \begin{pmatrix} \psi_\uparrow \\ \psi_\downarrow \end{pmatrix} = Q_{\theta x}\psi, \tag{6.60}$$

we shall preserve the algebraic form of all equations definining spin effects without attention to the choice of axes in the yz-plane.

The unitary matrix corresponding to this rotation may be written in terms of the Pauli matrices

$$Q_{\theta x} = \cos\tfrac{1}{2}\theta\,.\,1 + \mathrm{i}\sin\tfrac{1}{2}\theta\,.\,\sigma_x, \tag{6.61}$$

so that similar formulae must hold for rotations about the other cartesian axes. It is not difficult, therefore, to generate an explicit unitary matrix that will transform the components of the wave function into the correct form in any new set of co-ordinates. We thus have a rule governing the *spinor* ψ, analogous to the familiar rules for the transformation of the components of vectors, tensors, etc. Any equation or expression containing pairs of products of spinors of the form $\psi_1^*\psi_2$ will thus be invariant under any ordinary rotation of the axes.

The relativistic generalization of this concept is quite simple. Let us add to the components of the vector $\boldsymbol{\sigma}$ defined by (6.54) a 'time component' with basis matrix 1—the 2×2 unit matrix. Thus, we have a 4-vector matrix operator

$$\sigma^\mu = (\sigma_x, \sigma_y, \sigma_z, 1). \tag{6.62}$$

Now it turns out that under any unitary transformation Q in the spin space, as in (6.58), the *norm* of the product of this vector with any 4-vector (X, Y, Z, T) is preserved.

This can be seen from an elementary algebraic identity. Construct the determinant of the matrix $R\sigma$:

$$\det\|R\sigma\| = \det\begin{vmatrix} T-Z & X+\mathrm{i}Y \\ X-\mathrm{i}Y & T+Z \end{vmatrix} = T^2 - X^2 - Y^2 - Z^2. \tag{6.63}$$

In any unitary transformation Q, as in (6.58), the numbers (X, Y, Z, T) are mixed to form new components (X', Y', Z', T'). But since the determinant of a matrix is invariant under a unitary transformation, the function on the right-hand side of (6.63) is conserved. The transformation Q in the spin space induces a linear transformation L in space-time conserving the norm of the 4-vector (X, Y, Z, T). This is the definition of a Lorentz transformation, as discussed in §6.1.

Our task now would be to find the unitary transformation Q, to act on our spinor ψ, when we make an arbitrary Lorentz transformation L. This we have done, for example, in (6.61), in the special case of a space rotation; the appropriate generalization is not difficult to devise. But we discover at once a most interesting feature of this relation; to a given L there correspond two different values of Q.

This ambiguity is obvious in (6.61). The space rotation might have been through a positive angle $\theta' < 2\pi$—or it might have been thought of as being through a larger angle $\theta = \theta' + 2\pi$. Because the matrix depends on $\frac{1}{2}\theta$, we get both $Q_{\theta'}$ and $-Q_{\theta'}$ as allowed solutions. This result is quite general; the same Lorentz transformation L is induced by the action of either Q or $-Q$ on the spinor components. Fortunately, this ambiguity causes no difficulty, provided we establish a convention of sign and keep to it throughout.

There still remains, however, another formal ambiguity in the definition of a spinor. We have tacitly confined our argument to the case of *proper* Lorentz transformations, in which the spatial axes, for example, are not inverted from right-handed to left-handed, nor is it possible to reverse the sign of a time-like displacement and thus, so to speak, travel backwards in time. Although such transformations are forbidden physically, in that they are not accessible by moving in frames of reference at any relative velocity less than that of light, many of the equations of dynamics are invariant in form after such mathematical relabelling of co-ordinates. It is important, therefore, to decide a convention for spinors under this transformation. Suppose that I represents this inversion transformation; then I^2 is the identity operation, and I itself can only be one of the square roots of unity. In other words, we can prescribe either

$$I\psi = \psi \quad \text{or} \quad I\chi = -\chi. \tag{6.64}$$

The first prescription makes ψ a true spinor, whilst the function χ should be called a *pseudospinor*, by analogy with the distinction between true (polar) vectors, and *pseudovectors*.

6.6 The Dirac equation

Having investigated the Lorentz invariance of spinors, we are now in a position to construct equations of motion for spinor fields. Our natural aim is to write down a relativistic generalization of the time-dependent Schrödinger equation for a fermion, described presumably by a two-component wave function ψ. If this is to be linear in the operator $\mathrm{i}\,\partial/\partial t$, then it must also be linear in the momentum operator $\mathbf{p} = -\mathrm{i}\nabla$. We might, for example, try

$$\mathrm{i}\frac{\partial\psi}{\partial t} = m\psi + \boldsymbol{\sigma}\cdot\mathbf{p}\psi; \tag{6.65}$$

if ψ is a spinor, then this equation is certainly invariant under an ordinary rotation of the axes, as in (6.58).

Unfortunately, this equation cannot be made Lorentz invariant. Moreover, it is not invariant under space inversion, for $\boldsymbol{\sigma}$, being a form of angular momentum, is really a pseudovector. To acquire sufficient flexibility under all these transformations, we have to introduce a pseudospinor field χ, coupled to a spinor field ϕ, by the pair of equations

$$\left.\begin{aligned} \mathrm{i}\frac{\partial\phi}{\partial t} &= m\phi + (\boldsymbol{\sigma}\cdot\mathbf{p})\chi, \\ -\mathrm{i}\frac{\partial\chi}{\partial t} &= m\chi - (\boldsymbol{\sigma}\cdot\mathbf{p})\phi. \end{aligned}\right\} \tag{6.66}$$

These are essentially the Dirac equation for a fermion field of half-integral spin. Their justification is not, of course, that they are in some sense the simplest possible equations with the necessary invariance properties, but that they give results in agreement with physical observation.

We may check, for example, by a simple elimination procedure, that each component of ϕ and χ satisfies a Klein–Gordon equation (6.38). Thus, from (6.66),

$$\begin{aligned} \left(\mathrm{i}\frac{\partial}{\partial t} + m\right)\left(\mathrm{i}\frac{\partial}{\partial t} - m\right)\phi &= (\boldsymbol{\sigma}\cdot\mathbf{p})^2\phi \\ &= -\nabla^2\phi, \end{aligned} \tag{6.67}$$

from the elementary properties of the Pauli matrices (6.55). The solution of (6.66) in free space must therefore be a system of coupled plane waves, of the form

$$\begin{pmatrix}\phi \\ \chi\end{pmatrix} = \begin{pmatrix}\phi_{\mathbf{k}} \\ \chi_{\mathbf{k}}\end{pmatrix} \mathrm{e}^{\mathrm{i}(\mathbf{k}\cdot\mathbf{r}-\omega t)}, \tag{6.68}$$

whose frequency is the positive or negative root of the standard formula (6.39) for the energy of a relativistic particle of rest mass m:

$$\omega = \pm|\mathscr{E}(\mathbf{k})| = \pm\sqrt{(m^2+|\mathbf{k}|^2)}. \tag{6.69}$$

But the Dirac equation is a factorization of the Klein–Gordon equation, imposing further conditions upon the various components of ϕ and χ. Taking the two different roots separately, we find either

$$\chi_{\mathbf{k}}^{(+)} = \frac{\boldsymbol{\sigma}\cdot\mathbf{k}}{|\mathscr{E}(\mathbf{k})|+m}\phi_{\mathbf{k}}^{(+)} \quad \text{or} \quad \phi_{\mathbf{k}}^{(-)} = -\frac{\boldsymbol{\sigma}\cdot\mathbf{k}}{|\mathscr{E}(\mathbf{k})|+m}\chi_{\mathbf{k}}^{(-)}. \tag{6.70}$$

For small values of $\mathbf{k}$ (i.e. for velocities much less than the velocity of light), the positive energy solution has $\phi_{\mathbf{k}}^{(+)}$ much larger than $\chi_{\mathbf{k}}^{(+)}$; in other words, an electron of low energy is almost described by the upper spinor field ϕ. On the other hand, the 'negative energy solution' is mainly in the pseudospinor field $\chi_{\mathbf{k}}^{(-)}$ at low momenta. We recognize here the familiar electron and positron states, the latter, of course, being interpreted as a hole in the negative energy solutions as in §2.7. But when the kinetic energy of the particle is comparable with its rest mass, we cannot ignore the mixing of the ϕ and χ fields implied by (6.70).

The above argument is reminiscent of the elementary proof that each component of the electric and magnetic field obeys the wave equation, by direct elimination of $\mathbf{E}$ or $\mathbf{H}$ from Maxwell's equations (6.13)—which are, indeed, analogous to the coupled linear differential equations (6.66). Of course, by working in terms of the potentials we automatically preserve some of the further coupling conditions on the field vectors of electromagnetism although not enough to avoid all the difficulties associated with gauge transformations.

For a fermion such as an electron or nucleon, the complete four-component field (ϕ, χ) is necessary if we are to describe relativistic phenomena. But in the special case $m = 0$, the two fields ϕ and χ satisfy exactly the same equation of motion. This presents no special mathematical difficulties—but it is then profitable to investigate the simpler *Weyl equation*

$$\mathrm{i}\frac{\partial\psi}{\partial t} = (\boldsymbol{\sigma}\cdot\mathbf{p})\psi \tag{6.71}$$

governing an ordinary two-component spinor field ψ. Although this is Lorentz invariant, it was rejected originally because it was not invariant under the improper Lorentz transformation of space inversion. The discovery that parity was not conserved in weak interactions opened the way for the reinstatement of this formula as the equation of motion of the *neutrino* field. The two components of the plane-wave solution of (6.71) correspond to the two states of a particle of zero rest mass, whose half-integral spin may point either parallel or antiparallel to its direction of propagation.

6.7 The Dirac matrices

Free fermions may be described easily enough by the pair of coupled field equations (6.66), but when we come to study the interactions between, say, electrons and photons, we need to express the theory in manifestly covariant form. The time derivative must, surely, appear merely as a component of the 4-vector gradient operator (6.12), whose space components are, of course, proportional to the momentum operator **p**. But in each of the Dirac equations (6.66), the time and space derivatives do not act on the same spinor field; we are forced to combine the two fields into a single symbol with four components, the *Dirac spinor* or *bi-spinor*

$$\psi \equiv \begin{pmatrix} \phi \\ \chi \end{pmatrix} = \begin{pmatrix} \psi_1 \\ \psi_2 \\ \psi_3 \\ \psi_4 \end{pmatrix}. \tag{6.72}$$

Notice, however, that the number of components of this field does not have anything to do with the dimensionality of space-time; the transformation properties of these symbols under a Lorentz transformation are given by the spinor transformations (6.58), (6.60), (6.64), etc. not by the ordinary 4-vector transformations of the Lorentz group.

Looking now at the equations (6.66), we see that an invariant term $m\psi$ will give the correct multiple of each component in each equation. The time derivatives of the components of χ—i.e. of ψ_3 and ψ_4—are reversed in sign. This is achieved formally by multiplying

$$\mathrm{i}\frac{\partial}{\partial t} = -\mathrm{i}\nabla^4 \tag{6.73}$$

by a diagonal 4×4 matrix

$$\gamma^4 \equiv \begin{pmatrix} 1 & \cdot & \cdot & \cdot \\ \cdot & 1 & \cdot & \cdot \\ \cdot & \cdot & -1 & \cdot \\ \cdot & \cdot & \cdot & -1 \end{pmatrix} \tag{6.74}$$

acting in the 'space' of the components of ψ.

The space components of the 4-vector $-i\nabla^\mu$ occur in combination with the Pauli spin matrices. In the bi-spinor space, these must be represented by 4×4 matrices, which also exchange the ϕ and ψ fields with a change of sign. Following (6.55) we introduce 'space components' of γ^4,

$$\gamma^{1,2,3} \equiv \left(\begin{array}{cc|cc} \cdot & \cdot & \multicolumn{2}{c}{\sigma_{x,y,z}} \\ \cdot & \cdot & & \\ \hline & & \cdot & \cdot \\ \multicolumn{2}{c|}{-\sigma_{x,y,z}} & \cdot & \cdot \end{array}\right); \quad \text{thus} \quad \gamma^1 \equiv \begin{pmatrix} \cdot & \cdot & \cdot & 1 \\ \cdot & \cdot & 1 & \cdot \\ \cdot & -1 & \cdot & \cdot \\ -1 & \cdot & \cdot & \cdot \end{pmatrix}, \text{ etc.} \tag{6.75}$$

The reader may now verify that the derivatives that occur in the coupled Dirac equations (6.66) are various components of the differential/matrix operator

$$\begin{aligned} i(\gamma^1\nabla^1+\gamma^2\nabla^2+\gamma^3\nabla^3-\gamma^4\nabla^4) &\equiv i g_{\mu\nu}\gamma^\mu\nabla^\nu \\ &\equiv -i\gamma\nabla, \end{aligned} \tag{6.76}$$

acting on the field ψ. In the notation of (6.40), the Dirac equation now reads

$$(i\gamma\nabla+m)\,\psi = 0, \tag{6.77}$$

which is manifestly Lorentz covariant. This is the analogue, for fermions, of the Klein–Gordon equation (6.38) for scalar bosons. As is well known, Dirac originally obtained this equation directly as a factorization of the Klein–Gordon equation (6.67), by the introduction of a 4-component field subject to multiplication by matrices similar to γ^μ. The derivation given above seeks to justify and explain this formal theory through the more explicit demands of relativistic invariance.

The *Dirac matrices*, γ^μ, play a most important role in advanced quantum theory. Some of their properties are now more or less obvious. For example, in a Lorentz transformation, we may treat them as the components of a 4-vector, and obtain a new set, γ'^μ in a new coordinate system, whilst leaving the field ψ unaltered. On the other hand, following the argument of §6.5, we may treat the γ^μ as if they were unaltered in the transformation, but then make a unitary transformation such as (6.60) on the components of the bi-spinor ψ. The physical results, as expressed in the values of quantum mechanical observables, will be the same in either case; it is convenient, therefore, to assume that the γ^μ are invariant matrices having always the explicit forms (6.74) and (6.75).

But even these formulae are not necessary. We recall that the Pauli spin matrices may be derived from the condition that they satisfy the commutation relations for the components of angular momentum;

$$[\sigma_x, \sigma_y] = 2\mathrm{i}\sigma_z, \quad \text{etc.} \tag{6.78}$$

The corresponding *anticommutation* relations for the Dirac matrices may be laboriously worked out; the resulting equations may all be summed up in the formula

$$\gamma^\mu\gamma^\nu + \gamma^\nu\gamma^\mu = -2g_{\mu\nu}1, \tag{6.79}$$

where 1 is a unit 4×4 matrix. Although this equation is not quite strictly Lorentz covariant as it stands, the substitution of $g^{\mu\nu}$ for $g_{\mu\nu}$ (to which it is equal) will do the trick. These anticommutation relations alone are sufficient to define all relevant physical properties of the Dirac matrices (i.e. of these generators of the Lorentz group, as defined in §7.8), without explicit reference to a particular representation such as (6.74) and (6.75).

The Dirac matrices, being fourth roots of unity, appear in pure mathematics as generators of a hypercomplex algebra. From the standpoint of physics it is interesting to identify the various independent mathematical expressions that may be constructed by combining their products, and to ask how they transform relativistically. But all physical observables occur eventually as 'matrix elements' involving products of 'bra' and 'ket' vectors. It is obvious that the bi-spinor field ψ must have a Hermitian conjugate $\overline{\psi}$, satisfying the equation conjugate to (6.77), i.e. (allowing ∇ to operate to the left),

$$\overline{\psi}(-\mathrm{i}\gamma\nabla + m) = 0. \tag{6.80}$$

The rules of matrix multiplication require $\overline{\psi}$ to be a 'row matrix' in spinor space, so that the combination

$$\overline{\psi}\psi = \sum_{i=1}^{4} \overline{\psi}_i\psi_i \tag{6.81}$$

is in fact a scalar field. The invariance of this quantity under Lorentz transformation, which can be proved from the unitarity of the transformation matrices in (6.58), etc. allows it to play the role of the probability density in the Dirac theory.

The combination

$$V^\mu = \overline{\psi}\gamma^\mu\psi \tag{6.82}$$

must transform like an ordinary 4-vector, whilst

$$T^{\mu\nu} = \tfrac{1}{2}\overline{\psi}(\gamma^\mu\gamma^\nu - \gamma^\nu\gamma^\mu)\psi \tag{6.83}$$

must be an antisymmetric tensor analogous to (6.19). The most interesting combination is constructed with the aid of the matrix symbolized by

$$\gamma^5 \equiv \gamma^1\gamma^2\gamma^3\gamma^4 = \mathrm{i}\begin{pmatrix} \cdot & \cdot & \cdot & 1 \\ \cdot & \cdot & 1 & \cdot \\ \cdot & 1 & \cdot & \cdot \\ 1 & \cdot & \cdot & \cdot \end{pmatrix}. \tag{6.84}$$

The quantity
$$P = \bar{\psi}\gamma^5\psi \tag{6.85}$$

is a sum of products of a spinor component with a pseudospinor component, and therefore, by (6.64) changes sign with reflection of axes; we thus construct a *pseudoscalar* field.

6.8 Quantization of the Dirac field

The Dirac wave function ψ presumably represents the behaviour of a single electron or nucleon. We naturally look for solutions of the equation in empty space, with given momentum-energy 4-vector p. Substitute

$$\psi = u(p)\,\mathrm{e}^{\mathrm{i}px} \tag{6.86}$$

into (6.77); the symbol $u(p)$ must be a Dirac spinor satisfying the condition

$$(\gamma p - m)\,u(p) = 0. \tag{6.87}$$

Since γ is a Dirac matrix, this is really a set of four linear equations for the components of $u(p)$. The compatibility condition for these equations—the vanishing of their determinant—turns out to be the condition that each component of ψ should satisfy the Klein–Gordon equation (6.67). Thus, solutions of the Dirac equation of the form (6.86) only exist when

$$\{\mathscr{E}(\mathbf{p})\}^2 \equiv (p^4)^2 = m^2 + |\mathbf{p}|^2. \tag{6.88}$$

Let $\mathscr{E}(\mathbf{p})$ be defined as a positive number. For each root, $+\mathscr{E}(\mathbf{p})$ and $-\mathscr{E}(\mathbf{p})$ of this equation, we can solve for the ratios of the various components of $u(p)$. Indeed, in the special case of a state of definite momentum in free space, the solutions found in (6.70) can always be transformed by a unitary transformation into a *Foldy–Wouthuysen representation* where $u(p)$ has only the two upper or two lower spinor components; the fermion is, so to speak, 'purely an electron' or 'purely a positron'. Moreover, for each sign of the energy, there are exactly two independent solutions, corresponding to the two distinct spin states of the particle. In other words, for any given value of the ordinary momentum $\mathbf{p}$, we can define four solutions of (6.86), $u_+^{(1)}$, $u_+^{(2)}$

and $u_-^{(1)}$, $u_-^{(2)}$, of which the first two are of positive energy and the second two of negative energy.

We must not pause here to discuss the physical properties of particles in these states, which have been introduced merely as a basis for the representation of second quantized field operators. The Schrödinger operator for the Dirac field must be analogous to (2.22), i.e.

$$\psi(\mathbf{r}) = \sum_{\mathbf{p}} \sqrt{\left(\frac{m}{V\mathscr{E}(\mathbf{p})}\right)} e^{i\mathbf{p}\cdot\mathbf{r}} \sum_{n=1}^{2} \{b_+^{(n)}(\mathbf{p})\, u_+^{(n)}(\mathbf{p}) + b_-^{(n)}(\mathbf{p})\, u_-^{(n)}(\mathbf{p})\}. \quad (6.89)$$

The factor $\sqrt{\{m/V\mathscr{E}(\mathbf{p})\}}$ is a normalization coefficient, equivalent to the factor $1/\sqrt{\omega_{\mathbf{k}}}$ in the analogous boson formula (6.43); the operator character of (6.89) is established by the annihilation operators $b_+^{(1)}(\mathbf{p})$, etc. governing the occupation of the corresponding modes

$$u_+^{(1)}(\mathbf{p})\, e^{i\mathbf{p}\cdot\mathbf{r}}, \quad \text{etc.}$$

We are faced with the difficulty of the negative energy solutions, which can only be avoided by Dirac's hypothesis that these are all occupied in the basic vacuum state of the whole system. To use this assumption, we employ the device of (2.71); each annihilation operator $b_-^{(n)}(\mathbf{p})$ becomes a 'hole creation operator' $\tilde{b}^{*(n)}(-\mathbf{p})$ in a state labelled with the reverse momentum. In other words, we may write

$$\psi(\mathbf{r}) = \sum_{\mathbf{p}} \sqrt{\left\{\frac{m}{V\mathscr{E}(\mathbf{p})}\right\}} \sum_{n=1}^{2} \{b_+^{(n)}(\mathbf{p})\, u_+^{(n)}(\mathbf{p})\, e^{i\mathbf{p}\cdot\mathbf{r}} + \tilde{b}^{*(n)}(\mathbf{p})\, \tilde{u}^{(n)}(\mathbf{p})\, e^{-i\mathbf{p}\cdot\mathbf{r}}\}, \quad (6.90)$$

where now $\tilde{u}^{(n)}(\mathbf{p}) \equiv u_-^{(n)}(-\mathbf{p})$ is a new name for the negative energy spinor of momentum $-\mathbf{p}$.

When applied to the assumed 'vacuum', the first term of (6.90) 'destroys electrons', whilst the second term 'creates positrons'. In going from the Schrödinger representation to the Heisenberg representation these ought to acquire time-varying factors of positive and negative energy, respectively. But this is automatically taken care of by the reversal of sign of $\mathbf{p}$; that is to say, we may write

$$\psi(x) = \sum_{\substack{p;\\ p^4=|\mathscr{E}(p)|}} \sqrt{\left(\frac{m}{Vp^4}\right)} \sum_{n=1}^{2} \{b^{(n)}(p)\, u^{(n)}(p)\, e^{-ipx} + \tilde{b}^{*(n)}(p)\, \tilde{u}^{(n)}(p)\, e^{ipx}\} \quad (6.91)$$

as the form of the Dirac field operator for fermions in the Heisenberg or interaction representation. Electrons, generated by b^*, and positrons, generated by $\tilde{b}^*$, are now on the same footing and all energies are positive. The similarity to (6.45), for bosons, is obvious.

The way is now open for a great many formal developments of the theory. We assume, naturally, that the annihilation and creation operators satisfy the usual anticommutation relations (2.13) and (2.19); that is to say, all anticommute with one another except the annihilator and creator associated with any one mode. From (6.80) we are impelled to construct the Hermitian adjoint operator field $\overline{\psi}$ which can be obtained from (6.91) by (i) interchanging 'starred' and 'unstarred' operations, (ii) replacing i by $-$i, and (iii) replacing each spinor by its Hermitian adjoint. We can then follow the lines of §6.4 in a discussion of the anticommutation relations of these two fields. The main difference between fermions and bosons resides in the algebraic properties of the spinors in (6.91), giving rise to somewhat more complicated invariant delta functions than those defined in (6.47) and (6.50).

For most practical purposes, however, we only need to know the energy-momentum representation of the propagator, for use in the graphical theory. We could start from the time-ordered product (4.55), and follow the argument used to derive (6.47); but this procedure can be short-circuited by use of the fundamental principle of §4.10 that the Green function is a solution of the basic equation of motion of the wave function with a delta-function source. As we saw in (6.48) and (6.49), the propagator for scalar bosons satisfies the inhomogeneous Klein–Gordon equation

$$(\Box_x^2 - m^2)\,\Delta_F(x, x') = -\,\delta^4(x - x'). \tag{6.92}$$

By a four-dimensional Fourier transformation, with attention to causality, this has the solution, in an energy-momentum-representation,

$$\Delta_F(k) = \frac{1}{k^2 - m^2 + \mathrm{i}\delta}. \tag{6.93}$$

The corresponding inhomogeneous Dirac equation, by analogy between (6.38) and (6.77), must be

$$(\mathrm{i}\gamma\nabla + m)\,G_F(x, x') = -\,\delta^4(x - x'), \tag{6.94}$$

where G_F and the right-hand side must have implied spinor properties. The Fourier transform of this equation is just

$$(\gamma p - m)\,G_F(p) = 1, \tag{6.95}$$

which defines the energy-momentum representation of the fermion propagator.

To solve this equation, we need to construct the inverse of a spinor. But if we multiply right through by the conjugate operator $\gamma p+m$ we produce a scalar quantity like the denominator of (6.93). Thus the Feynman propagator for a fermion may be written

$$G_F(p)=\frac{1}{\not{p}-m+\mathrm{i}\delta}\equiv\frac{\not{p}+m}{p^2-m^2-\mathrm{i}\delta}, \tag{6.96}$$

where the conventional symbolism

$$\not{p}\equiv\gamma p\equiv -g_{\mu\nu}\gamma^\mu p^\nu \tag{6.97}$$

conveys the spinor/Lorentz-invariant characteristics of this variable. When we are dealing with phenomena at relativistic energies, this is the expression that ought to appear in place of G_0 in formulae such as (3.114), (3.128), (4.61), etc. It must be emphasized, however, that although the formalism has carried us smoothly up through many successive levels of abstraction, expressions such as (6.96) are really rather complicated, and a great deal of algebraic detail needs to be attended to when one tries to evaluate diagrams, matrix elements, etc. involving spinors.

6.9 Interactions between relativistic fields

Like a rabbit out of a hat, the Dirac equation has been produced without any discussion of its dynamical aspects. The canonical procedure for the generation of classical field equations was outlined in §1.6: relying upon Hamilton's principle, we vary the space-time integral of a Lagrangian density; from the resulting Euler equations, we derive, in turn, the Hamiltonian of the field, which then acts as a pivot for various representations of quantized field operators, as discussed for example, in §§3.2 and 3.3. We need a similar formalism for our various types of relativistic field.

In §1.8 the Lagrangian and Hamiltonian densities for the scalar boson field were defined, whilst in §1.12, the corresponding theory for charged bosons was sketched out. Since these all obey the Klein–Gordon equation, there is no difficulty in checking that they are Lorentz invariant.

For the photon field, we turn to standard treatises on classical electromagnetism, where it is shown that Maxwell's equations in free space, (6.13), may be derived canonically by varying the components of potential in a Lagrangian. In the notation of §6.2 this has the density

$$\mathscr{L}_{\mathrm{e.m}}=-\tfrac{1}{2}g_{\mu\nu}g_{\nu\delta}\nabla^\mu A^\nu\nabla^\gamma A^\delta, \tag{6.98}$$

which is manifestly relativistically invariant and is less clumsy than it looks.

Finally, for free fermions, the Dirac equation (6.77) must be generated. This is of such simplicity that the possible forms of Lagrangian density are easily investigated; the correct formula may be written, in the compact notation of § 6.7,

$$\mathscr{L}_{\text{ferm}} = \frac{1}{2\mathrm{i}}\{\bar{\psi}\gamma(\nabla\psi) - (\nabla\bar{\psi})\gamma\psi\} - m\bar{\psi}\psi; \tag{6.99}$$

variation with respect to $\bar{\psi}$ yields (6.77), whilst variation with respect to ψ gives the conjugate equation (6.80). The verification of this statement, using (1.68), is a good exercise in the use of the spinor formalism.

We must now make our various fields interact dynamically. As shown in § 1.1, such effects are described by adding to the separate free-field Hamiltonians an interaction term, which is a function of the various field operators. Once we know this term, we can use the methods of chapter 3 to calculate a variety of phenomena associated physically with the coupling of the fields. The main task left to us in this chapter, therefore, is to demonstrate that there are only a few types of *simple* interaction terms consistent with relativistic invariance, etc.

This is best undertaken by the canonical procedure, starting from a Lagrangian or Lagrangian density. Consider, for example, the familiar electrodynamic interaction between a charged particle and the photon field. For a classical point particle this problem was solved long ago. The combined system is described by a total Lagrangian

$$L = L_{\text{e.m}} + L_{\text{particles}} + L_{\text{interact}}, \tag{6.100}$$

where we have added to the separate Lagrangians of the particles and electromagnetic field an interaction term, whose density is simply

$$\mathscr{L}_{\text{interact}} = -g_{\mu\nu}\, j^{\mu} A^{\nu}. \tag{6.101}$$

This was the type of interaction used naïvely in § 2.7.

If (6.100) is varied with respect to the components of potential, A^{ν}, we get Maxwell's equations (6.28) including current terms. If, on the other hand, we write

$$j^{\mu} = e\frac{\mathrm{d}x^{\mu}}{\mathrm{d}s} \tag{6.102}$$

for a classical particle of charge e, and vary with respect to the components of position, we introduce the Lorentz force on our classical particle in the electromagnetic field. Again, following through the further steps of the canonical procedure, we find the well-known result that the effect of the electromagnetic field upon the particle is summed up by writing, in the notation of (6.40),

$$H = \frac{1}{2m}(p - eA)^2 \tag{6.103}$$

for its Hamiltonian, in terms of its canonical momentum variable p.

We now have clues to the corresponding formulae for particles described by quantized fields. We recall that the quantum analogue of momentum is the space gradient of the field; in relativistic language

$$p = -\mathrm{i}\nabla. \tag{6.104}$$

The implication of (6.103) is that eA should be subtracted from this operator wherever it occurs; in other words, *particles of charge* e *are coupled to the electromagnetic field by substituting* $\nabla^\mu - \mathrm{i}eA^\mu$ *for* ∇^μ *in the expression for the Lagrangian density of the free-particle field.*

Applying this principle to the fermion field, with Lagrangian density (6.99), we see that this is equivalent to adding an interaction term

$$\mathscr{L}_{\text{interact}} = -e\bar{\psi}\gamma A\psi \tag{6.105}$$

to the sum of (6.98) and (6.99). From this we can easily construct the Dirac equation for a fermion in an electromagnetic field, and also, by analogy with (6.101), find suitable expressions for the current density operator in a Schrödinger representation.

But having learnt the language and technique of second quantization it is natural to proceed directly, by means of the canonical theory of § 1.6, to the definition of a corresponding interaction Hamiltonian density. Since (6.105) does not contain derivatives of either fermion or photon fields, it is carried over without change, to yield

$$\mathscr{H}_{\text{e.m-fermion}} = e\bar{\psi}A\!\!\!/\psi, \tag{6.106}$$

in the notation of (6.97). From this expression we can build up all the phenomena of quantum electrodynamics using the interaction representation. We notice, for example, that each vertex of a Feynman diagram must have an incoming and outgoing fermion line, with one photon line, as assumed in § 3.6. Along with our general expressions (6.53) and (6.96), for boson and fermion propagators, we have all the makings of the perturbation theory, leading eventually to mass and charge renormalization, etc.

A minor pitfall in this procedure is exemplified by trying to use the elementary formula (1.126) for the current density in a charged boson field, in the classical interaction Lagrangian (6.101). The result is not the same as we get by the more fundamental rule of substituting $\nabla - ieA$ for ∇ in the Lagrangian density (1.113) for the free particles. We find, in effect, that the current—the quantity that is always conserved by virtue of the field equations—acquires an extra term $-2eA\phi^*\phi$. This point is familiar in the elementary theory of electrons in a magnetic field, where the Hamiltonian (6.103) produces an extra term in $e^2A^2\psi$ in the Schrödinger equation. Having avoided this danger, we can easily set up a theory of the electromagnetic interactions of charged bosons, such as pions.

But we must also allow for interactions between other types of field. We discover in nature that nucleons—which are, of course, fermions of spin $\frac{1}{2}$—interact strongly with pions, which might, for example, be supposed to be bosons described by a pseudoscalar field ϕ. Then by analogy with (6.106) a suitable form of interaction term in the Lagrangian or Hamiltonian might be

$$\mathscr{H}_{\text{strong}} = -g\bar{\psi}\gamma^5\psi\phi. \tag{6.107}$$

By (6.85) this would be invariant under proper and improper Lorentz transformations and would describe elementary processes in which a pion is emitted or absorbed by a nucleon. Some consequences of this interaction are discussed in §1.11, but unfortunately, the value of the coupling constant g (dimensionless, in units of $\hbar c$) in this *strong interaction* appears to be much larger than the value of e, which is about $1/\sqrt{137}$. As is well known, the attempt to construct a renormalized theory involving this type of interaction has not been successful, and even the lowest order terms in a perturbation expansion are suspect!

On the other hand, the characteristic phenomena of beta decay always involve four field operators—typically, a neutron (ψ_n) decays into a proton (ψ_p) with emission of an electron (ψ_e) and a neutrino (ψ_ν). The *weak interaction* might therefore be of the form

$$\mathscr{H}_{\text{weak}} = -G(\bar{\psi}_n O_1 \psi_p)(\bar{\psi}_\nu O_2 \psi_e), \tag{6.108}$$

where G is constant, and O_1 and O_2 are suitable operators out of the various combinations of the γ^μ discussed in (6.81)–(6.85). Since G is very small, perturbation theory works well in the lowest order. The main physical problem here is to identify the symmetry types of the various fields, and by an analysis of the observed phenomena to discover what combination of constants, symmetry symbols and field operators yields the best results. This is the sort of situation where the theory of chapter 7 is peculiarly helpful.

6.10 Relativistic kinematics

Almost all observations on elementary particles are scattering experiments: typically, two particles are made to collide with high relative energy, and the products of the interaction are recorded, at a distance, as separate independent entities. The characteristics of any such event are described by an S-matrix, as discussed in §§ 3.5 and 4.14. The whole effort of the diagrammatic approach, as outlined in chapter 3, is to calculate this quantity as the sum of a perturbation series.

Unfortunately, this type of series does not converge for strong interactions such as those between pions and nucleons. Yet the S-matrix itself certainly exists, and can, so to speak be explored as a function of the various parameters of the various collision processes that occur between its initial particles. This function is not entirely arbitrary; considerable effort has therefore been expended on the delineation of the mathematical conditions imposed upon it by general physical principles.

Kinematic restrictions, such as conservation of energy and momentum, are by no means trivial at relativistic velocities. It is well known, for examples, that the electron–photon interaction (6.106) does not actually permit the direct production or absorption of a free photon by a free electron in empty space; the lowest order allowed process is Compton scattering (§ 3.7). On the other hand, direct production of *phonons* in a metal is allowed (§ 2.6) because the electron velocity is much greater than the velocity of sound.

A systematic analysis of these kinematic restrictions on the interactions of massive particles in the relativistic case is therefore important, not merely as a guide to experiment but as a source of canonical variables on which the S-matrix must eventually depend. To illustrate the argument, let us consider a hypothetical phenomenon in which four particles, of rest masses m_1, m_2, m_3, m_4, converge on a point and are mutually annihilated there without trace. Let these have energy-momentum 4-vectors p_1, p_2, p_3, p_4 respectively; the mass energy relation for each particle imposes the condition (6.39) which reads

$$p_i^2 = m_i^2 \quad (i = 1, 2, 3, 4), \tag{6.109}$$

in the notation of (6.40).

Conservation of each component of the total energy and momentum reads

$$p_1+p_2+p_3+p_4 = 0; \tag{6.110}$$

when all these conditions have been satisfied, and when allowance has been made for the arbitrariness of the frame of reference, we are left with only three scalar invariants

$$\left.\begin{aligned} s &= (p_1+p_4)^2 = (p_2+p_3)^2, \\ t &= (p_2+p_4)^2 = (p_1+p_3)^2, \\ u &= (p_3+p_4)^2 = (p_1+p_2)^2, \end{aligned}\right\} \tag{6.111}$$

which are not independent but satisfy the linear relation

$$s+t+u = m_1^2+m_2^2+m_3^2+m_4^2. \tag{6.112}$$

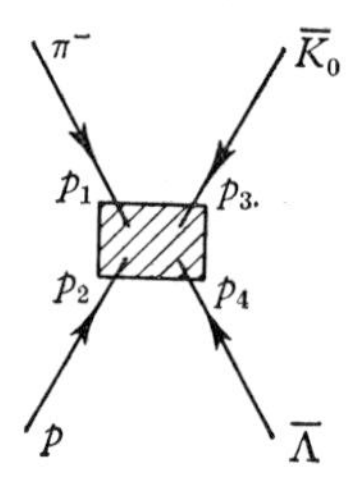

In other words, a given collision process may be characterized by the values of two of these parameters, each of which is the square of the total energy of a chosen pair of particles in a frame in which their centre of mass is at rest.

Now of course we cannot contrive to make four genuine particles converge and annihilate at a point. But the same algebraic relations apply if we use the Feynman convention that a particle is the same as its antiparticle moving backwards in time. Suppose, for example, that we identify the four energy-momenta p_1, p_2, p_3, p_4, with those of four different particles π^-, p, $\bar{K}_0$, $\bar{\Lambda}$, respectively. Then our four-body collision describes the possible physical process

$$\pi^-+p \to K_0+\Lambda. \tag{6.113}$$

All we need to do is to assign the reversed spatial parts of p_3 and p_4 to the product particles K_0 and Λ.

But if this process is really to take place, the total energy of the two 'incident' particles must at least exceed their combined rest masses; according to (6.111), we must have

$$u \geqslant (m_\pi+m_p)^2. \tag{6.114}$$

Similar considerations, of some complexity in the general case, restrict the permitted range of the variables s and t; for example, if all four masses are equal, each of these variables is minus the square of the momentum transfer in the collision, and hence necessarily negative.

These restrictions are most readily visualized by plotting the values of s, t, u as homogeneous co-ordinates of a point in a plane, i.e. as distances from the sides of an equilateral triangle, on such a scale that (6.112) is automatically satisfied. Then our process (6.113) can only

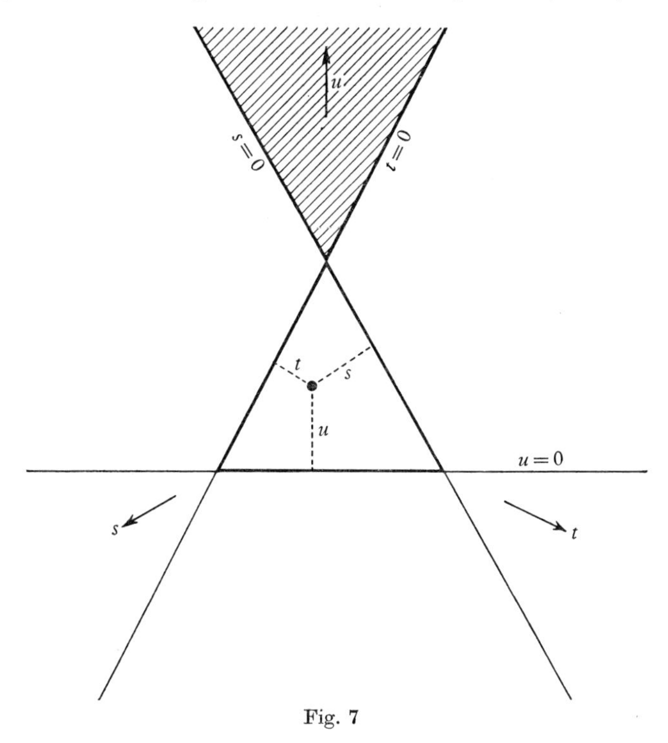

Fig. 7

occur when this point lies within a well-defined region on the plane—for example, in the case of equal masses, within the wedge shaded in fig. 7.

It is obvious, however, that there is a certain symmetry between the roles of the variables s, t and u in this analysis; we might have chosen p_1 and p_4 to be our 'incident' particles, whose barycentric energy variable s must be positive, and so on. In other words, there exist other 'physical' regions of the (s, t, u) plane where the kinematic restrictions can be satisfied. Again in the simple case of equal rest masses, these regions are easily delineated, as in fig. 8. But unless all our particles were in fact identical, these different physical regions would describe quite different scattering processes. Thus, with our present conventions, the region where s exceeds $4m^2$ would describe the reaction

$$\pi^- + \overline{\Lambda} \rightleftharpoons K_0 + \overline{p}, \tag{6.115}$$

whilst the corresponding t region would refer to yet another reaction

$$p + \overline{\Lambda} \rightleftharpoons \pi^+ + K_0. \tag{6.116}$$

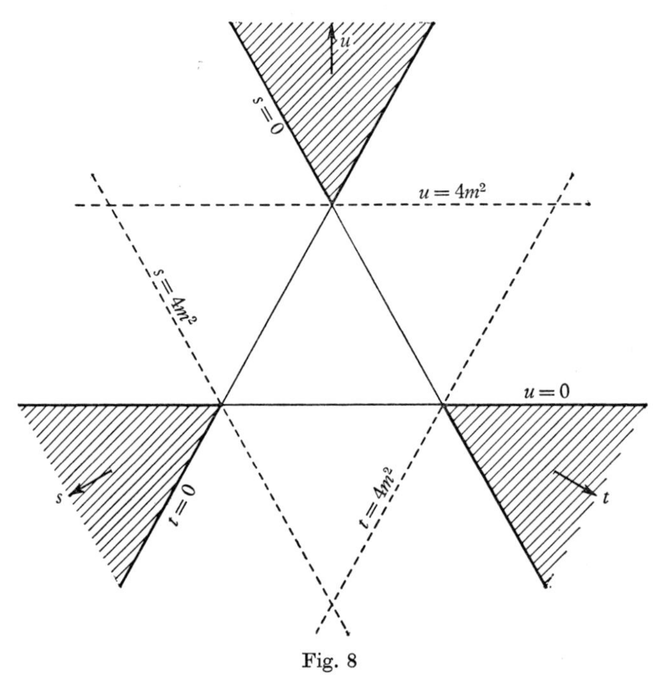

Fig. 8

Because of the actual differences of mass between the various particles in such reactions, the *physical regions* will not be so simple and symmetrical as we have drawn them in the elementary case, but the general principle is still valid; in the *Mandelstam diagram*, well-defined regions exist corresponding to the various channels by which various fundamental reaction may take place. Fig. 9, for example, shows such a diagram, which includes a central physical region corresponding to the decay of one of the particles into the other three.

This geometrical construction suggests very forcibly a most powerful conjecture; the different reactions (6.113), (6.115) and (6.116) say, are all to be described by the same S-matrix, which is essentially only a function of the canonical variables s, t, u (or any two of these, because of (6.112)). Any rule that can be used to connect the values of this function in the various physical regions will then define a *crossing symmetry* relating the observable cross-sections for two apparently dissimilar physical phenomena.

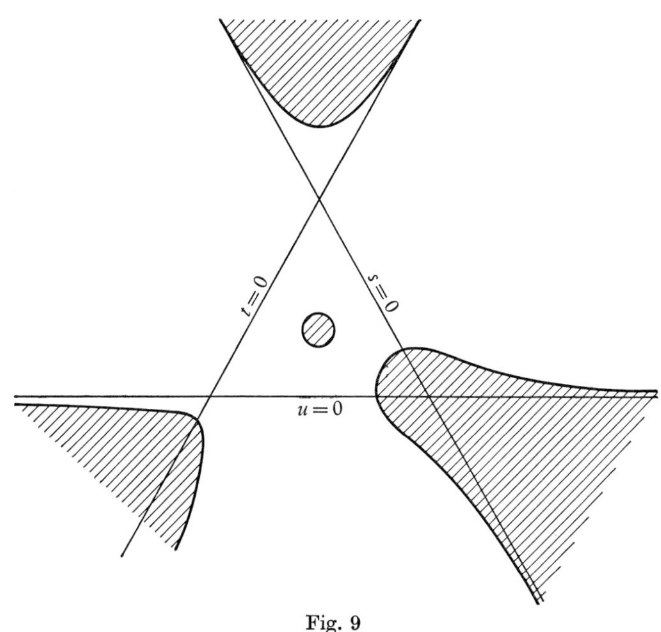

Fig. 9

6.11 The analytic S-matrix

The S-matrix must be consistent with several general physical principles, such as the superposition property of probability amplitudes, the short range character of the forces, and the conservation of probability. These impose mathematical restrictions; for example, conservation of probability requires unitarity (as in (3.29)) which leads to algebraic relationships like (4.161) for the T-matrix or *scattering amplitude* of the process under consideration.

The most subtle physical principle to be obeyed by the S-matrix is that of *causality*; we must be certain that the equations of motion implied by a collision process do not transfer information about 'future' states back into the description of the 'past'. This principle has already been invoked on several occasions (§§ 3.6, 3.8, 4.10, 6.4) and shown to be satisfied by what may be called the '$+i\epsilon$ rule'; in the momentum representation, each causal propagator is an analytic function of a complex energy variable, and is evaluated by taking the limit as the imaginary part of this variable tends to zero from above the real axis. It is argued, by analogy, experience, and proof under special circumstances, that transition amplitudes must have the same mathematical property; it is held to be an axiom of quantum theory that the *S-matrix is analytic*. As we have seen, relativistic invariance requires that the S-matrix be a function of the invariant energy parameters of the process, such as the variables (s, t, u) of (6.111). The causality condition is therefore considered to be satisfied in practice if we treat each of these parameters as a complex variable, and then show that the transition amplitude for a given process is the value of an analytic function of these variables in the neighbourhood of their actual real values.

The mathematical implications of this condition, even for a function of a single variable, cannot be summed up in a single sentence; applied to functions of two or more variables they are subtle and complicated in the extreme. Typically, however, they direct one's attention towards the construction of *dispersion relations*, which are essentially generalizations of the *Kramers–Kronig relations* of elementary macroscopic physics, e.g. (5.32).

The mathematical prototype of all such relations is the *Hilbert transform* of analysis, which is an elementary deduction from Cauchy's theorem. Suppose we have a function $f(w)$ of the complex variable $w = u+iv$, which is known on and near the real axis, beyond some point $w = a$. Suppose, moreover, that $f(w)\to 0$ for $w\to\infty$ in all directions, and that all the singularities of $f(w)$ are known to be concentrated within the stretch along the real axis where the function is known. Then by Cauchy's theorem, if z is any point within the contour C of fig. 10,

$$f(z) = \frac{1}{2\pi \mathrm{i}}\int_C \frac{f(w)}{w-z}\,\mathrm{d}w$$

$$= \frac{1}{2\pi \mathrm{i}}\int_0^\infty \frac{f(u+\mathrm{i}\epsilon)-f(u-\mathrm{i}\epsilon)}{u-z}\,\mathrm{d}u. \tag{6.117}$$

The Hilbert transform

$$f(z) = \frac{1}{\pi}\int_a^\infty \frac{\operatorname{Im} f(u)}{u-z}\,\mathrm{d}u \tag{6.118}$$

then follows, as a special case, if $f(u)$ is real for $u \leqslant a$.

The power of this theorem is evident. Consider, for example, the process (6.113), which takes place, according to (6.114) if the invariant energy u exceeds some threshold value. Suppose we think of $f(w)$ in (6.117) as being the analytic continuation of the scattering amplitude for this process into the 'complex energy' plane $w = u+iv$. The behaviour of $f(w)$ along the real axis beyond this threshold is observable physically. If the other mathematical conditions of the theorem were certainly satisfied (there's the rub!), we could calculate $f(w)$ throughout the complex plane, and especially along the negative

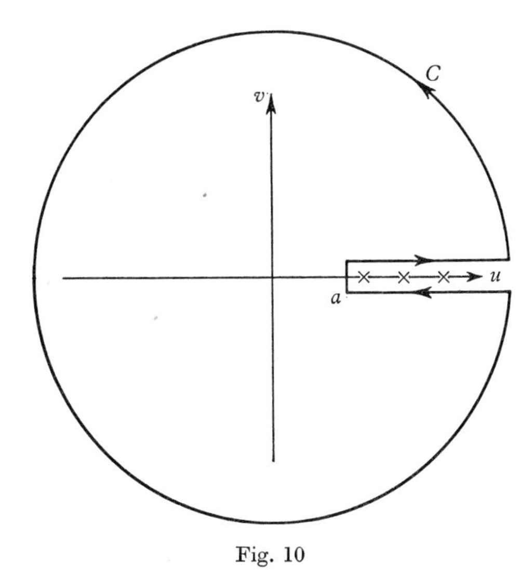

Fig. 10

real axis $u < 0$. But as we see from fig. 9, this is the region of the Mandelstam diagram in which u is related to the momentum transfer in processes such as (6.115) and (6.116), which are quite different from those with which we started. A dispersion relation of this form might therefore be expected to generate crossing connections between the scattering amplitudes in the various channels of our general S-matrix.

Unfortunately, the path here sketched out runs through extremely rugged mathematical territory on the way from principle to practice. In the first place, we must take account of the various selection rules and matrix elements implicit in the spin, parity, strangeness, isotopic spin and other symmetry properties of genuine particles; the true S-matrix for fermions must involve, for example, spinors, as in (6.107), and cannot be the simple scalar function that is usually discussed in the analyticity theory. Again, any formula for the scattering amplitude of a relativistic process must contain another energy parameter, s or t, on which the integrand and its singularities in a dispersion relation in u would have to depend. This dependence is likely to be very complicated algebraically, in order to reproduce the complex geometrical structure of the physical regions in a Mandelstam diagram such as fig. 9. We are led into the study of functions of two or more complex variables, with *double dispersion relations* as consequences of analyticity.

Finally, we need explicit information about the nature and location of all the singularities of the scattering amplitude; indeed we know from the general theory of functional analysis that these in themselves completely determine this function everywhere, and hence would solve our whole problem. How do such singularities arise?

As an elementary example, consider nucleon–nucleon scattering through the exchange of mesons—in fact, the Yukawa force studied in §1.11. The first-order perturbation formula for the S-matrix is given by the Feynman diagram.

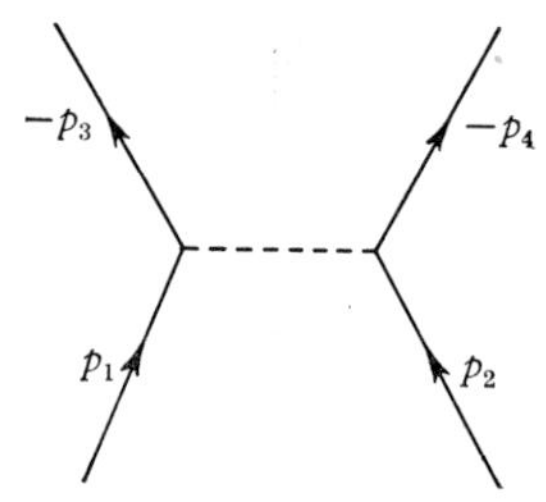

We ignore vertex factors like $\mathbf{g}^2$, and because of momentum/energy conservation at each vertex the state of the intermediate boson is fixed by the initial momenta of the interacting nucleons. In the analyticity theory, the overall conservation factor $\delta(p_1+p_2+p_3+p_4)$ is dropped from the S-matrix (cf. (4.155)), so that by the rules of §3.8 we are left simply with the propagator for the exchanged boson. In the relativistic regime, this is given by (6.48), i.e.

$$\mathcal{T}(12\to 34) \sim \frac{1}{(p_1+p_3)^2 - m^2 + \mathrm{i}\delta}$$

$$= \frac{1}{t - m^2 + \mathrm{i}\delta} \tag{6.119}$$

in the notation of (6.111).

This result is really the relativistic analogue of (1.111), being simply the Born approximation for scattering by the Yukawa potential. But we now see that it has deeper significance; considered as a function of the variable t, the scattering amplitude has a pole at $t = m^2$. There must, by symmetry, be a similar pole when $u = m^2$, i.e. *when the barycentric energy of the two colliding nucleons equals the rest mass of a meson.* The point at which the 'virtual' boson becomes 'real' dominates and determines the form of the scattering amplitude for the simple exchange process.

But this is only the first term in an infinite perturbation series. The true scattering amplitude must surely contain a contribution from *square diagrams* such as fig. 11 (*a*), in which two mesons are exchanged during the process. Given sufficient energy, both internal boson lines might become 'real', we must expect a singularity at $t = 4m^2$ from this diagram. It turns out that this is a branch point rather than a simple pole—but the study of the *Landau singularities* generated in turn, at higher and higher energies along the real axis, by graphs of higher and higher order, is an art unto itself.

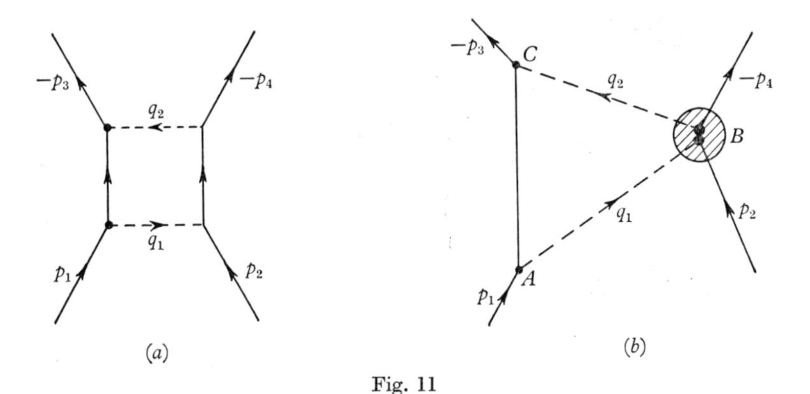

Fig. 11

This diagram also illustrates another feature of S-matrix theory. Suppose we redraw it as in fig. 11 (*b*). The meson lines q_1 and q_2 are now real, so that the event at B would be observable as, say, nucleon–pion scattering. Indeed, the processes at A and C, where this pion was produced and subsequently absorbed, could have occurred in another part of the universe; the S-matrix for the event at B will depend only on its energy-momentum parameters $(q_1, p_2, q_2, -p_4)$ however these came to be. In other words the *connectedness* of such diagrams may be related to the separation of the overall S-matrix into terms which factorize into products of simpler functions, whose singularities can be identified. This possibility is a mathematical consequence of the physical principle that the forces responsible for strong interactions are of short range.

It is impossible in this short account to do justice to the mathematical ingenuity that has gone into the study of the analytic S-matrix nor the useful physical results that have been thereby derived. Unfortunately, the final goal of this activity—a self-contained theory of strong interactions without perturbation expansions—seems continually elusive. Connections such as (6.119) between 'forces' and 'particles' are very instructive, but we do not yet know whether they really go beyond a systematization of the basic principle implicit in Yukawa's famous hypothesis. The attempt to construct a complete description of the S-matrix out of relativistic invariance, unitarity, causality, etc. without appeal to dynamical principles, seems to require certain further mathematical assumptions or axioms which have no obvious physical significance.

Typical of such assumptions is the *Mandelstam representation*. From a detailed analysis of the square diagram, fig. 11 (*a*), one can derive the following double dispersion relation for the scattering amplitude as a function of the invariant energy parameters (s, t, u):

$$F(s,t) = \frac{1}{\pi^2}\iint_{4m^2}^{\infty} \frac{F_{12}(s',t')}{(s'-s)(t'-t)}\,ds'\,dt' + \frac{1}{\pi^2}\iint_{4m^2}^{\infty} \frac{F_{23}(t',u')}{(t'-t)(u'-u)}\,dt'\,du' + \frac{1}{\pi^2}\iint_{4m^2}^{\infty} \frac{F_{31}(u',s')}{(u'-u)(s'-s)}\,du'\,ds', \quad (6.120)$$

where the integration, in each case, is only over the 'physical' regions of the variables (s', t', u') of a diagram such as fig. 9. The conjecture is that the distribution of singularities in the complete S-matrix is similarly restricted. This has in fact been proved to all orders in perturbation theory—but cannot be shown in general. Implicit in such formulae is one of the assumptions needed to derive the elementary Hilbert transform (6.118)—that $f(w)$ tends to zero as $w \to \infty$. Without information about the asymptotic behaviour of scattering amplitudes at infinite (complex) energies we are more or less back at square one.

CHAPTER 7

THE ALGEBRA OF SYMMETRY

> *Thomas Nunn*, Breeches-Maker, No. 29 Wigmore Street, Cavendish Square, has invented a System on a mathematical Principle, by which Difficulties are solved, and Errors corrected; its usefulness for Ease and Neatness in fitting is incomparable, and is the only perfect Rule for that Work ever discovered. Several hundreds (Noblemen, Gentlemen, and Others) who have had Proof of its Utility, allow it to excel all they ever made Trial of.
>
> (*Advt.* 1815)

7.1 Symmetry operations

In his search for material systems that might be capable of mathematical analysis, the theoretical physicist devotes particular attention to those which, by nature or design, exhibit *symmetry*. Thus, he studies isolated 'elementary' particles, spherical nuclei and atoms, diatomic molecules, the benzene hexagon, regular crystals, etc. whose natural symmetry allows enormous simplification of the mathematical prescription and analysis of their properties.

It is proper, therefore, that the fundamental algebraic theory of symmetry—the theory of *groups*—should play a major part in advanced quantum theory. It is argued, indeed, that this algebra is truly fundamental, providing a primitive connection between such universally acknowledged principles as the isotropy of empty space and the observable quantized parameters of the elementary particles. Group theory is not merely a computational device for exploiting to the full the finite set of symmetries of a system such as a molecule or crystal: as we have seen in the case of relativistic invariance, the demands of symmetry under a *continuous* group of transformations—a set containing infinitely many members—can almost call the tune of the dynamics.

This chapter cannot possibly do justice to the pure theory of groups—which is an autonomous branch of mathematics in itself—nor to the multifarious applications of this theory in almost all branches of theoretical physics. But in the spirit of previous chapters, we shall try to look at some of the key principles and techniques of this large topic to indicate the sort of things that can be deduced under very simple circumstances, so that the serious student may have some idea of what to expect when he begins to read one of the numerous excellent treatises on the subject.

In our quantum-mechanical applications, we are usually concerned with a Hamiltonian $\mathscr{H}$ of a physical system. This is, of course, a function of the co-ordinates and momenta of various particles—electrons, nucleons, etc.—which we may symbolize, abstractly, by the letter $\mathbf{x}$. When we say that the system has a certain symmetry, we mean that $\mathscr{H}$ is invariant under a corresponding transformation of the co-ordinates; in algebraic language, we acknowledge the existence of an operator, S say, such that

$$\mathscr{H}(S\mathbf{x}) \equiv \mathscr{H}(\mathbf{x}). \tag{7.1}$$

We might, for example, be studying the energy levels of a diatomic molecule, where $\mathbf{x}$ would represent the set of co-ordinates $\mathbf{r}_1, \mathbf{r}_2, \ldots, \mathbf{r}_N$ of the various electrons, moving in the field of the two atomic nuclei. There is an obvious symmetry of axial rotation about the line of centres of the atoms, so that S might correspond to rotation of the co-ordinates by an arbitrary angle θ about this axis; this operation is evidently a member of a *continuous* group. If the atoms are identical, there is a simple reflection operation about the median plane; the set of co-ordinates $S\mathbf{x}$ would differ from $\mathbf{x}$ by a change of sign of, say, the z-component of each vector $\mathbf{r}_1, \mathbf{r}_2, \ldots, \mathbf{r}_N$. Again, since all electrons are identical, the Hamiltonian must be invariant under a *permutation* of the various indices, $1 \ldots N$, assigned to these co-ordinates—one of a set of operations with a finite number, $N!$, of members.

Such operations in any particular case may be quite complicated geometrically and algebraically, so that their complete description, with the assignment of distinguishing labels, may occupy several pages of text, with tables and illustrative figures. It is intuitively obvious, however, that the symbols S, T, etc. that we use for the various operations must satisfy certain formal self-consistency rules if they are to be participate in algebraic formulae. For example, the 'product', ST, means the result of operating first with T then with S, and itself stands for yet another operation, R say, of the set. Again, the *identity* operation, E, which leaves every co-ordinate unchanged, must be thought of as a member of the set, for completeness. Finally, we can always 'undo' any transformation, S, by another operation, Q say, of the same set; we can call this the *inverse* S^{-1} of the first operation, and write

$$SQ = SS^{-1} = E. \tag{7.2}$$

Together with the associative convention for products, these rules provide axioms for a pure algebraic theory in which the symbols S, etc. are said to be the *elements* of an abstract *group*, $\mathscr{G}$.

The reader will easily satisfy himself that these rules are consistent with our physical intuition and analytical formulae for the Hamiltonian of, say, the diatomic molecule mentioned above. Notice that we have *not* said anything about the commutativity of products of group elements; in general $ST \neq TS$. This also may be verified, by comparing, for example, the results of successive rotations about two different axes, or two successive permutations of labels, in alternative orders. It is difficult to think of more primitive axioms on which to erect a mathematico-physical theory; the surprising thing is the amount of meaningful information that can be obtained from the mere algebra of the symmetry operations of the system.

7.2 Representations

If we are to avoid calculus, we are almost bound to set up a matrix representation of our Hamiltonian $\mathscr{H}$, in terms of some infinite orthonormal set of functions $\phi_i(\mathbf{x})$, spanning the Hilbert space of the variables $\mathbf{x}$. In other words, we construct the matrix $\mathscr{H}_{ij}$ so that

$$\mathscr{H}(\mathbf{x})\,\phi_i(\mathbf{x}) = \sum_j \mathscr{H}_{ij}\,\phi_j(\mathbf{x}). \tag{7.3}$$

The matrix elements $\mathscr{H}_{ij}$ can be evaluated in the usual way.

Now consider the effect of any operation S upon this representation. The change of variables implied by $S\mathbf{x}$ changes each function of the basis set $\phi_i(\mathbf{x})$ into another function $\phi_i(S\mathbf{x})$; but this may itself be re-expanded in terms of the original functions, i.e.

$$\phi_i(S\mathbf{x}) = \sum_k S_{ik}\,\phi_k(\mathbf{x}), \tag{7.4}$$

where again the numbers S_{ik} can readily be calculated in any particular case. For example the operation S might read: add $\pi/2$ to the variable x. Then if $\phi_1(x) = \sin x$ and $\phi_2(x) = \cos x$ (etc., etc.), we shall have

$$\phi_1(Sx) = \sin(x+\tfrac{1}{2}\pi) = \cos x = \phi_2(x)$$

so that the element S_{12} of this matrix would be 1. In general, the orthogonality of the basis functions would be used to find the coefficients.

A familiar type of algebraic manipulation (cf. (3.29)) shows that the effect of this operation on the Hamiltonian itself may be represented as a transformation of its matrix, i.e.

$$\mathscr{H}(S\mathbf{x})\,\phi_i(\mathbf{x}) = \sum_{jkl}\{S_{il}^{-1}\mathscr{H}_{lk}S_{kj}\}\,\phi_j(\mathbf{x}), \tag{7.5}$$

where we suppose that each matrix S_{ik} has a genuine inverse. If now the Hamilton remains invariant under this transformation, as in (7.1), we have

$$\mathscr{H}_{ij} = \sum_{kl} S_{il}^{-1}\mathscr{H}_{lk}S_{kj}. \tag{7.6}$$

In other words, *the Hamiltonian matrix commutes with the matrix of every element of its symmetry group*: in a more abstract symbolism,

$$S\mathscr{H} - \mathscr{H}S = 0 \tag{7.7}$$

for every S in $\mathscr{G}$.

To each element S of the group, there corresponds a matrix S_{ik} defined as in (7.4). It is very easy to prove that the matrix of the product of two elements is just the product of the corresponding matrices, and that the matrix of the unit element E is the unit matrix. All the abstract operations of the group are thus faithfully represented by corresponding algebraic formulae in terms of these matrices: we say that these constitute a *representation* of the group.

But such a representation is certainly not unique, for the basis set ϕ_i was chosen quite arbitrarily. Suppose we had started with a different basis, related to the original set by a unitary transformation U_{ij}. The standard theory of Hilbert space tells us at once that the symmetry operation S will now be represented by a *new* matrix

$$S'_{ij} = \sum_{kl} U_{ik}^{-1}S_{kl}U_{lj}. \tag{7.8}$$

However, the systematic application of this transformation to every element of the group will not disturb such relations as

$$R = ST, \tag{7.9}$$

which will be true of the matrix representations of these elements with respect to any basis set, provided it is true of the symmetry operations for which they stand. The group is defined, abstractly, by its *multiplication table*; matrix representations connected as in (7.8) are entirely *equivalent* from this point of view. In much of what follows, it will be assumed that once we have proved that two representations are equivalent in this technical sense, then we can treat them henceforth as identical, for we know that we can always find a unitary transformation (7.8) that will turn one set of matrices into the other without upsetting the physical consequences in any way.

Suppose, therefore, that we diagonalize the Hamiltonian $\mathscr{H}$. The functions $\phi_i(\mathbf{x})$ would then be the eigenfunctions ψ_i of the equation

$$\mathscr{H}(\mathbf{x})\,\psi_i(\mathbf{x}) = \mathscr{E}_i\,\psi_i(\mathbf{x}). \tag{7.10}$$

Now apply our abstract symmetry operation S to this equation; by merely relabelling the variables or changing the whole co-ordinate system in conformity with the operation S, we find

$$\mathscr{H}(S\mathbf{x})\,\psi_i(S\mathbf{x}) = \mathscr{E}_i\,\psi_i(S\mathbf{x}), \tag{7.11}$$

whence, from (7.1),

$$\mathscr{H}(\mathbf{x})\,\psi_i(S\mathbf{x}) = \mathscr{E}_i\,\psi_i(S\mathbf{x}): \tag{7.12}$$

the function $\psi_i(S\mathbf{x})$ is itself an eigenfunction of the same Hamiltonian with the same eigenvalue $\mathscr{E}_i$.

This has most important implications for the matrix representing S in this special basis. Suppose that the level of energy $\mathscr{E}_\alpha$ is n-fold degenerate—that the corresponding eigenfunction is a member of a group of just n independent eigenfunctions $\psi_r^{(\alpha)}$ $(r = 1, \ldots, n)$ with this same eigenvalue. Then (7.12) can be satisfied only if

$$\psi_q^{(\alpha)}(S\mathbf{x}) = \sum_{r=1}^{n} S_{qr}^{(\alpha)}\,\psi_r^{(\alpha)}(\mathbf{x}), \tag{7.13}$$

where $S_{qr}^{(\alpha)}$ is an $n \times n$ matrix: the 'new' eigenfunction supposedly generated by the symmetry operation can only be a linear combination of the 'old' eigenfunctions of the same energy.

Comparing (7.13) with (7.4), we see that the matrix representation of S in this basis is now severely restricted, for it does not mix any other states into those belonging to the n-dimensional manifold spanned by $\psi_r^{(\alpha)}$. In this representation, therefore, the matrix for S is of the form

$$S_{ij} = \begin{pmatrix} S^{((1))} & \cdot & \cdot & \cdot & \cdot & \cdot & \cdot \\ \cdot & (S^{(2)}) & \cdot & \cdot & \cdot & \cdot & \cdot \\ & & \ddots & & & & \\ \cdot & \cdot & \begin{pmatrix} S_{11}^{(\alpha)} & S_{12}^{(\alpha)} & \ldots & S_{1n}^{(\alpha)} \\ S_{21}^{(\alpha)} & S_{22}^{(\alpha)} & \ldots & \ldots \\ \ldots & \ldots & \ldots & \ldots \\ S_{n1}^{(\alpha)} & S_{n2}^{(\alpha)} & \ldots & S_{nn}^{(\alpha)} \end{pmatrix} & & & \cdot & \cdot \\ & & & & & \ddots & \\ \cdot & \cdot & \cdot & \cdot & \cdot & \cdot & \cdot \end{pmatrix}. \tag{7.14}$$

All the elements are zero, except for square matrices $S^{(1)}$, $S^{(2)}$, etc. along the diagonal, each of order equal to the degeneracy of the corresponding eigenvalue. These blocks are distinct; there are no components linking the rows of $S^{(1)}$, say, with the columns passing through $S^{(2)}$.

This result we have proved for any operation S of the group $\mathscr{G}$ under which the Hamiltonian is invariant; the matrices representing all elements of the group are similarly *reducible* to this general form. But of course the eigenfunctions ψ_i are related to any arbitrary set of basis functions ϕ_i by a unitary transformation, so that our original matrix representation of the group must be equivalent to the reduced form (7.14). In such a representation the algebra of the multiplication table of the matrices is greatly simplified, for we need only multiply together the matrices in the corresponding positions on the grand diagonals—i.e. (7.9) implies

$$R_{pq}^{(\alpha)} = \sum_r S_{pr}^{(\alpha)} T_{rq}^{(\alpha)} \tag{7.15}$$

for each distinct eigenvalue $\mathscr{E}_\alpha$. We can, in fact, adopt a convention for 'addition' in which (7.14) is written in the form

$$S = S^{(1)} \oplus S^{(2)} \oplus \ldots \oplus S^{(\alpha)} \oplus \ldots. \tag{7.16}$$

The implication is that each sub-matrix acts in a different sub-space, and that, by (7.15), the product of two such matrix sums is the usual *inner product* of the component sub-matrices, i.e.

$$ST = \sum_\alpha{}^{\oplus} S^{(\alpha)} T^{(\alpha)}. \tag{7.17}$$

This connection between the degeneracy of eigenvalues and the dimensionality of reduced matrix representations of symmetry operations is the key to a typical application of group theory to quantum mechanics. The argument runs as follows. Suppose that by some means we have found the *irreducible representations* of a group—matrices such as $S^{(\alpha)}$ of minimum order n_α. According to (7.7), the complete matrix representation of each element of the group commutes with the matrix of any observable, such as the Hamiltonian, which is left invariant by the symmetry operations of the group. By a fundamental algebraic theorem (*Schur's Lemma*) it follows that the eigenvalues of the observable fall into sets, of which the degeneracies are the numbers n_α.

It will be noticed—and this is typical of the method—that we cannot by this means calculate the actual magnitudes of the various eigenvalues; the symmetry of the system implies certain equalities between various measurable properties, but does not contain all the information that would be required for a complete solution of the dynamical equations by which these numbers are eventually determined. On the other hand, the irreducible representations may be constructed directly from the basic definitions of the operations of the group, without solving any of the dynamical equations, so that much may be learnt about the system at a modest price.

7.3 Regular representations of finite groups

The construction of a matrix representation of a group through its transformations in Hilbert space is unnecessarily laborious. For a finite group, we can construct the *regular representation* directly from the basic intuitive definition of the elements of the group, without any recourse to analysis.

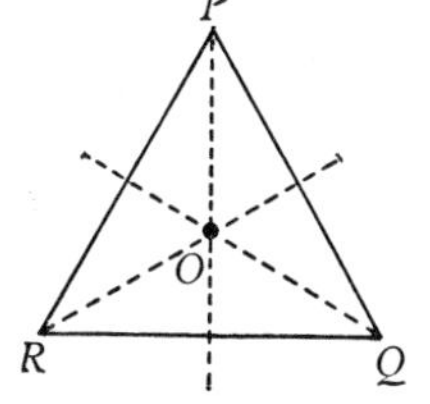

To illustrate the principle of this technique, consider the following standard example. An equilateral triangle is transformed into itself under the following operations: P, Q, R, symbolizing reflections about the lines OP, OQ, OR, respectively; C_3 symbolizing a clockwise rotation through an angle $2\pi/3$ about the centre; and $C_{\bar{3}}$, a corresponding anti-clockwise rotation. By direct experiment or geometrical intuition, we arrive at the following *multiplication table* for the elements of this group—called the D_3 group.

Table 1

	E	P	Q	R	C_3	$C_{\bar{3}}$
$E^{-1} = E$	E	P	Q	R	C_3	$C_{\bar{3}}$
$P^{-1} = P$	P	E	C_3	$C_{\bar{3}}$	Q	R
$Q^{-1} = Q$	Q	$C_{\bar{3}}$	E	C_3	R	P
$R^{-1} = R$	R	C_3	$C_{\bar{3}}$	E	P	Q
$C_3^{-1} = C_{\bar{3}}$	$C_{\bar{3}}$	Q	R	P	E	C_3
$C_{\bar{3}}^{-1} = C_3$	C_3	R	P	Q	$C_{\bar{3}}$	E

This table represents, abstractly the whole structure of the group. We do not, from now on, need to know the geometrical significance of the symbols, which might, indeed, refer to quite different events in a different realm of thought. For example, if we interpret P, Q, R as interchanges of pairs, and C_3, $C_{\bar{3}}$ as cyclic interchanges, then this is the multiplication table for the group of permutations on three objects: the same algebra could arise in the antisymmetrization of the wave functions of lithium as in the mechanical vibrations of a triangular plate. Notice that the various operations are not all commutative; the element C_3 in row P, column Q symbolizes the relation $PQ = C_3$; when the product is taken in reverse order, we get $QP = C_{\bar{3}}$.

Now to construct a matrix $\Gamma_R(P)$ representing the element P, simply put a 1 in every place where P occurs in the multiplication table, and zero elsewhere—i.e.

$$\Gamma_R(P) = \begin{pmatrix} \cdot & 1 & \cdot & \cdot & \cdot & \cdot \\ 1 & \cdot & \cdot & \cdot & \cdot & \cdot \\ \cdot & \cdot & \cdot & \cdot & \cdot & 1 \\ \cdot & \cdot & \cdot & \cdot & 1 & \cdot \\ \cdot & \cdot & \cdot & 1 & \cdot & \cdot \\ \cdot & \cdot & 1 & \cdot & \cdot & \cdot \end{pmatrix}. \tag{7.18}$$

Because the inverse of any element of a group is unique, each symbol can occur only once in each row or column of the multiplication table. Each matrix, $\Gamma_R(P), \Gamma_R(Q)$, etc. is thus a permutation matrix, with only a single unit in each row or column. The table has been arranged so that the identity operation appears along the diagonal; its representative $\Gamma_R(E)$ is thus a unit matrix.

It is easy enough to show that these matrices do indeed represent the elements of the group, i.e.

$$\Gamma_R(PQ) = \Gamma_R(P)\,\Gamma_R(Q). \tag{7.19}$$

The key to the proof is that when we multiply the first row of the table by an element P, say, we rearrange the order of the symbols to produce the Pth row. This is just the sort of rearrangement generated by the matrix $\Gamma_R(P)$; multiplication by successive elements of the group produces the same rearrangements as multiplication by the corresponding permutation matrices.

It is obvious that each matrix of Γ_R is *unitary*; this is not really a specialization since it may be achieved quite easily for any matrix representation by suitable renormalization. But by contrast with the *infinite* matrices S_{ij} of the previous section, each $\Gamma_R(S)$ has only as many rows and columns as there are elements S in the group. For the purpose of reproducing the multiplication table this is quite sufficient; indeed, the regular representation has what seems the useful property of being *faithful*; each of the g abstract operations of the group is represented by a different matrix.

Nevertheless, this does not mean that Γ_R is irreducible—that every sub-matrix in (7.14), say, is of order g. The most difficult part of the theory is to show that there exists a unitary transformation that changes the representative matrix of every element S of the group into a diagonal block form similar to (7.14), i.e.

$$\Gamma_R(S) \longrightarrow \begin{pmatrix} \Gamma^{(1)}(S) & \cdot & \cdot & \cdot \\ \cdot & \Gamma^{(2)}(S) & \cdot & \cdot \\ \cdot & \cdot & \Gamma^{(3)}(S) & \cdot \\ \cdot & \cdot & \cdot & \text{etc.} \end{pmatrix} \tag{7.20}$$

When the matrices reduced in this way are multiplied together, as in (7.19), only corresponding sub-matrices come together; by (7.15), for each value of α,

$$\Gamma^{(\alpha)}(ST) = \Gamma^{(\alpha)}(S)\,\Gamma^{(\alpha)}(T). \tag{7.21}$$

Thus, these matrices also obey the multiplication table of the group; if they cannot be reduced further, we say that $\Gamma^{(\alpha)}$ is an *irreducible representation* of $\mathscr{G}$.

Consider now the following apparently trivial substitution: let *every* element of the group be represented by the *same* one-dimensional matrix—the number 1. It is easy to see that this will satisfy the multiplication table—the fact that $C_3 = PQ$, say, is correctly expressed by the arithmetical relation $1 = 1.1$. Indeed, any abstract group whatsoever allows formally of this representation, however trivial it may seem: the *identity representation* is certainly a solution to the problem of finding an irreducible representation of the group, and must be included in any complete list. It is very important to accept at this stage that faithfulness is not the supreme virtue; we often encounter representations in which the same number, or the same matrix, may stand for several different elements of the group.

Now the regular representation is of order g, and can be shown to contain all the irreducible representations of the group, including the identity representation. Hence the order of every irreducible representation is less than g. Moreover, we may find, in the reduction to (7.20), that several of the blocks could have been made identical by a suitable unitary transformation—in other words, some of the $\Gamma^{(\alpha)}$ turn out to be equivalent representations, within the convention of (7.8), and hence may be treated as identical. We may thus use the convention (7.16) and write

$$\Gamma_R = p_1 \Gamma^{(1)} \oplus p_2 \Gamma^{(2)} \oplus \ldots \oplus p_n \Gamma^{(n)}, \tag{7.22}$$

where now each irreducible representation $\Gamma^{(\alpha)}$ occurs p_α times. The number n of different irreducible representations, and the order n_α of each such representation, is thus finite, and certainly less than g.

But pure mathematical theory shows that the irreducible representations that occur in the regular representation are the *only* non-equivalent irreducible representations of the group. Going back now to (7.14), where an *infinite* matrix representation of the group was reduced to block diagonal form by transformation to the eigenfunctions of the Hamiltonian, we can assert that each of these block matrices $S_{pr}^{(\alpha)}$ must be equivalent to one or other of the irreducible representations $\Gamma^{(\alpha)}(S)$ of the corresponding group elements. Thus, the problem of classifying the eigenvalues of the Hamiltonian is reduced to finding the relatively small number of irreducible representations of the symmetry group of the system.

7.4 The orthogonality theorem

The task of constructing the various irreducible representations of a group is greatly facilitated in practice by a few fundamental principles. For example, we have at our disposal several powerful arithmetical tools in the shape of *orthogonality theorems*, of which the following is the most general:

Let $\Gamma_{ip}^{(\alpha)}(S)$ be the ipth matrix element in a unitary irreducible representation of the element S of the group: then

$$\sum_S \Gamma_{ip}^{(\alpha)}(S)\, \Gamma_{jq}^{(\beta)}(S^{-1}) = \frac{g}{n_\alpha} \delta_{\alpha\beta} \delta_{ij} \delta_{pq}. \tag{7.23}$$

In other words, choose a 'site' in the matrix of a group element, and of its inverse, in two irreducible representations; multiply the numbers at these two sites together; then add these products for all elements of the group. If the two representations are not equivalent, the sum vanishes. If they are equivalent, this sum still vanishes unless, in effect, we are simply multiplying each matrix element by the corresponding element in its inverse. In that case, the sum is equal to the order of the group divided by the order of the particular irreducible representation.

This theorem evidently puts so many restrictions on the various matrix elements that we can sometimes almost guess the form of the irreducible representations. To give an elementary example consider the group with multiplication table 1 (p. 219). We have already noticed the 'identity' representation in which each element is just 1. This is obviously irreducible; is there any other irreducible representation of order unity? If this is to be unitary, each element of the group can only be represented by a square root of unity—i.e. by $+1$ or -1. But if such a representation is not to be equivalent to the 'identity', the theorem tells us that the sum of the products of each of these numbers with unity must vanish; there must be as many 'positive' as 'negative' elements in the group. We obviously use $+1$ for the identity operation E. On the other hand, out of a total of six elements, the three operations P, Q and R are similar (as we shall see, they belong to the same *class*), and ought to be given the same sign. These must therefore be represented by -1 whilst C_3 and $C_{\bar{3}}$, like E, are represented by $+1$.

Although the proof of (7.23) is rather too elaborate to give in full, the general argument is instructive. The left-hand side has the mathematical structure of an *outer product* or *direct product* of group elements. As we shall show in §7.6, the symbol $R \otimes S$ may be used for an abstract operation that behaves in some respects like multiplying by S and in other ways by R. In any matrix representation of the group elements, we could construct the array

$$(R \otimes S)_{ij,\,kl} \equiv R_{ij} S_{kl}, \tag{7.24}$$

much as we construct components of a cartesian tensor by taking dyadic products of vectors.

Now consider the object

$$\mathbf{A} = \sum_S S \otimes S^{-1}. \tag{7.25}$$

This commutes with every element of the group, i.e. for a fixed R,

$$\begin{aligned} R\mathbf{A} &= \sum_S RS \otimes S^{-1} \\ &= \sum_S (RS) \otimes S^{-1}(R^{-1}R) \\ &= \sum_{(RS)} (RS) \otimes (RS)^{-1} R = \mathbf{A}R, \end{aligned} \tag{7.26}$$

since a sum over (RS) simply runs through all the elements of the group in another order.

As an explicit representation of the elements of the group in (7.25) we might have used the direct sum of two irreducible representations $\Gamma^{(\alpha)}$ and $\Gamma^{(\beta)}$ as in (7.16). The left-hand side of the orthogonality theorem (7.23) would then appear as one of the components of **A**. But looking at (7.26), we find that we have constructed a matrix A (not necessarily square) such that

$$A\Gamma^{(\alpha)}(R) = \Gamma^{(\beta)}(R)\,A, \tag{7.27}$$

for every element R of the group. By comparison with (7.8), this suggests that the two representations are equivalent. Schur's lemma tells us that if $\Gamma^{(\alpha)}$ and $\Gamma^{(\beta)}$ are *not* equivalent, then every component of A vanishes, whilst if they are equivalent unitary irreducible representations A can only be a multiple of the unit matrix. These are the algebraic results embodied in (7.23).

The language of 'tensors' and 'dyadics' used in the interpretation of (7.24) suggests another way of looking at (7.23). We may think of the group as generating a space of g dimensions, with 'axes' labelled by the elements R, S, T, etc. The numbers that occur at a particular matrix site of an irreducible representation may be thought of as components of a vector in that space—e.g. the number $\Gamma^{(\alpha)}_{ip}(S)$ is 'the component along the S-axis of the vector labelled $\Gamma^{(\alpha)}_{ip}$'. The theorem then merely asserts the orthogonality of these vectors: the components of the matrices of the irreducible representations of the group provide a new basis of orthogonal axes in this abstract space.

A further important principle follows at once. Each irreducible representation $\Gamma^{(\alpha)}$ of order n_α has just n_α^2 components. But the number of orthogonal vectors spanning the space cannot exceed its dimensionality g, the order of the group. Hence, counting only non-equivalent representations, we must have

$$\sum_{\alpha} n_\alpha^2 \leqslant g. \tag{7.28}$$

It turns out, in fact, by another theorem, that equality is always achieved in this relation.

This rule greatly restricts the possible irreducible representations of a group. Thus, for example, we have already found two non-equivalent irreducible representations of our standard permutation group of order 6. Each of these being of order unity, there can remain only one further irreducible representation of order 2, making up the equality

$$1^2 + 1^2 + 2^2 = 6. \tag{7.29}$$

7.5 Character and class

Despite these powerful theorems, the task of constructing explicit irreducible representations of a finite group by direct algebraic analysis of, say, the regular representation is altogether too laborious. In practice we are dealing with some physical problem such as the solution of a Schrödinger equation, where matrix representations of the group are automatically generated in the course of the calculation—for example by operation on a set of functions as in (7.4). Most of our task can be accomplished by intelligent guesswork in the choice of such functions, provided that we have some arithmetical test for the reducibility or otherwise of any actual representation that we may have produced. This would, of course, be shown by application of the orthogonality theorem—but then we should need to know already the matrix elements of the various irreducible representations of the group!

Indeed, the fact that two apparently different sets of matrices may be equivalent representations shows that these matrix elements themselves are relatively arbitrary. The test for reducibility can only involve quantities that remain invariant under unitary transformations such as (7.8). For an arbitrary representative matrix, the simplest invariant is its *trace*; in the theory of group representations, the trace of the matrix representing a given group element is called the *character* of that element in the given representation:

$$\chi(S) \equiv \mathrm{Tr}\,(S_{ij}) \equiv \sum_i S_{ii} \tag{7.30}$$

in the notation of (7.4).

It is well known that the trace of a matrix is invariant under a unitary transformation. If two representations of a group are equivalent, then they will assign the same character to an element of the group; by a converse theorem, which I shall not prove, *if two representations assign the same set of characters to the elements of a group, then the two representations must be equivalent.* This provides us with an immediate arithmetical comparison test between some arbitrary representation of the group and a known irreducible representation. Indeed we do not need to know the matrices of the irreducible representation in full, but only the characters of the various group elements in that representation.

Now of course the character of S in a given representation depends upon the representation. But suppose the representation is reducible as in (7.16):

$$S = S^1 \oplus S^2 \oplus \ldots S^r \oplus \ldots. \tag{7.31}$$

The trace of a *direct sum* of matrices is obviously the sum of the traces of the constituents. Thus

$$\chi(S) = \chi^1(S) + \chi^2(S) + \ldots + \chi^r(S) + \ldots. \tag{7.32}$$

The character of an element in a reducible representation is just the arithmetical sum of the characters (of that element) in the representations into which the original representation may be decomposed. In the case of the regular representation, for example, (7.22) implies

$$\chi_R(S) = p_1\chi^{(1)}(S) + p_2\chi^{(2)}(S) + \ldots + p_n\chi^{(n)}(S), \tag{7.33}$$

where $\chi^{(\alpha)}(S)$ is the trace of S in the irreducible representation $\Gamma^{(\alpha)}$, which occurs p_α times in Γ_R.

This parallelism between the arithmetic of group characters and the abstract algebra of representations is the practical key to the door between the physical properties of the system and the mathematical properties of finite groups. It is easy enough to compute the character of each element in some representation Γ generated analytically in connection with the solution of the dynamical equations. If we know the characters of all elements in all the irreducible representations of the group, we have only to solve a set of equations such as (7.32) or (7.33) to discover the number of times that each irreducible representation occurs in Γ. We thus have a means of analysing the behaviour of the various functions that have been used in constructing Γ in terms of the various irreducible representations of the group. We can say things like 'This set of eigenfunctions is not accidentally degenerate, and transforms like the representation $\Gamma^{(3)}$ under a rotation of the axes'.

At first sight, however, (7.33) would seem to yield far too many equations; the group has g elements, which by (7.28) is usually much larger than the number, n, of different irreducible representations. Fortunately, many of these equations are redundant.

As we have already noticed, many operations of a group are intuitively similar—for example, the reflections P, Q, R about the bisectors of the triangle in the elementary example of § 7.3. We say that two elements of a group, S and T, belong to the same *class* if there exists another element X such that

$$S = X^{-1}TX. \tag{7.34}$$

This agrees with our intuitive notions, for it tells us that S is what T

would become if we systematically relabelled all our symmetry operations by the transformation implied by X: in our example,

$$P = C_3^{-}QC_3, \tag{7.35}$$

showing that a reflection about OQ becomes a reflection about OP if we move all labels round the triangle by a rotation through $2\pi/3$.

But by elementary matrix algebra, two matrices S and T related by a similarity transformation (7.34) have the same trace. Thus, in any representation of the group, *all elements belonging to the same class have the same character*:

$$\chi(S) = \chi(T). \tag{7.36}$$

In attempting to solve (7.32) or (7.33), we shall get just the same equation from the element T as we have already obtained from S. The number of *independent* equations of this type is no more than the number of classes to which the elements of the group may be assigned.

The reader will not now, perhaps, be surprised to learn of a theorem proving that *the number of classes in a group is exactly equal to the number of irreducible representations*. Thus, a set of equations such as (7.33) always has a unique solution for the numbers p_α.

Indeed, we can exhibit this solution explicitly by an appeal to the orthogonality theorem (7.23). The characters of the two irreducible representations $\Gamma_{ip}^{(\alpha)}(S)$ and $\Gamma_{jq}^{(\beta)}(S^{-1})$ are obtained by putting $i = p$, $j = q$ and summing over each of these indices. Because we are using a unitary representation, the character of S^{-1} must be the complex conjugate of the character of S. The sum over all elements of the group can now be reduced to a sum over classes. Paying due attention to the numbers of terms in each summation, we get

$$\sum_k \chi_k^{(\alpha)} \chi_k^{(\beta)*} N_k = g\, \delta_{\alpha\beta}, \tag{7.37}$$

where the kth class has N_k elements, each with character $\chi_k^{(\alpha)}$ in the irreducible representation $\Gamma^{(\alpha)}$. In other words, the numbers $\chi_k^{(\alpha)}$ form an orthogonal array of components, of order n.

Suppose that we have a representation Γ in which the elements of the kth class, $\mathscr{C}_k$, have character χ_k. From (7.32) and (7.37) we can show by elementary matrix algebra that the decomposition of Γ into a direct sum

$$\Gamma = a_1 \Gamma^{(1)} \oplus a_2 \Gamma^{(2)} \oplus \ldots a_n \Gamma^{(n)} \tag{7.38}$$

contains the irreducible representations $\Gamma^{(\alpha)}$ with multiplicity

$$a_\alpha = g^{-1} \sum_k N_k \chi_k^{(\alpha)*} \chi_k. \tag{7.39}$$

All the important information about the abstract structure of the group is contained in its *character table*—a square array of numbers $\chi_k^{(\alpha)}$

with rows labelled by the irreducible representations $\Gamma^{(\alpha)}$ and columns by classes of group elements $\mathscr{C}_k$.

From this result, incidentally we may verify a useful general principle: *the number of times an irreducible representation* $\Gamma^{(\alpha)}$ *appears in the regular representation is equal to the order* n_α *of the matrices of* $\Gamma^{(\alpha)}$. This stems from the definition of the regular representation, as in (7.18). Except for the identity operation, every element of the group is represented in Γ_R by a matrix with zeros along the diagonal. Thus, the characters of all elements except E are zero. But from (7.34) E always constitutes a class by itself, with only one element. Thus, in (7.39), the right-hand side reduces to the character of E in $\Gamma^{(\alpha)}$, which is, of course, just n_α. This result, in its turn, requires the equality sign in (7.28).

To illustrate the theory of characters, let us return to our group D_3. We have already shown that this has three irreducible representations, and three classes. The characters of the one-dimensional representations $\Gamma^{(1)}$ and $\Gamma^{(2)}$ follow immediately from the discussion of orthogonality in §7.4. Also, the character of E in the two-dimensional representation must be 2. Only $\chi_2^{(3)}$ and $\chi_3^{(3)}$ now remain to be determined; from the orthogonality conditions (7.37) we easily deduce the values -1 and 0 respectively. We thus obtain with scarcely any effort, the following character table:

Table 2

Class ...	$\mathscr{C}_1$	$2\mathscr{C}_2$	$3\mathscr{C}_3$
Elements ...	E	$C_3, C_{\bar{3}}$	P, Q, R
$\Gamma^{(1)}$	1	1	1
$\Gamma^{(2)}$	1	1	-1
$\Gamma^{(3)}$	2	-1	0

To show how this might be used in a physical problem, let us think of a free atom placed in a field with symmetry D_3 about the z axis; what would be the effect on the one-electron energy levels?

Let us suppose, in an unsophisticated way, that we are interested only in the s, p and d states, whose wave functions are of the form

$$\left.\begin{aligned}\psi_s &= \phi_s(r);\\ \psi_p &= x\phi_p(r),\ y\phi_p(r),\ z\phi_p(r);\\ \psi_d &= xy\phi_d(r),\ yz\phi_d(r),\ zx\phi_d(r),\\ &\qquad (x^2-z^2)\,\phi_d(r),\ (y^2-z^2)\,\phi_d(r),\end{aligned}\right\} \tag{7.40}$$

where x and y are co-ordinates in the plane of the triangle with y axis through OP.

Since $\phi_s(r)$ is a function only of distance from the origin, it is invariant under all operations of the group, and therefore generates the identity representation $\Gamma^{(1)}$. We can drop the factors $\phi_p(r)$, $\phi_d(r)$ from the other functions for the same reason, and concentrate on the factors x, y, xy, etc.

The effect of the various elements of the group on the p-state functions, x, y, and z is easily calculated. For example, C_3 acting on x is merely a rotation in space, giving rise to

$$C_3 x = -\tfrac{1}{2}x + \frac{\sqrt{3}}{2}y, \tag{7.41}$$

whilst acting on y we get

$$C_3 y = -\frac{\sqrt{3}}{2}x - \tfrac{1}{2}y. \tag{7.42}$$

In the spirit of (7.13), we have generated a two-dimensional representative matrix for this operator:

$$C_3 = \begin{pmatrix} -\frac{1}{2} & \frac{\sqrt{3}}{2} \\ -\frac{\sqrt{3}}{2} & -\frac{1}{2} \end{pmatrix}. \tag{7.43}$$

The character of the class $\mathscr{C}_2$ in this representation is therefore -1. In a similar way, we find that the character of the group element P (which merely interchanges x and $-x$ but leaves y unaltered) must be zero, so $\chi_3 = 0$. Evidently these two functions form a basis for a representation which has exactly the same characters as the irreducible representation $\Gamma^{(3)}$. But these functions were degenerate energy levels of the free atom; this degeneracy cannot be removed by our D_3-type field. On the other hand the third p-function, $z\phi_p(r)$, must belong to a one-dimensional representation, and may be split off from the other two by the perturbing field. Notice, however, that this argument tells us nothing about the magnitude of the splitting, which depends upon the strength of the perturbation, the nature of the atomic states, etc., etc.

The application to the d-states follows the same lines; for example, the states xy and yz transform in the same way as x and y and hence belong to $\Gamma^{(3)}$. But if we study the other three functions, xy, x^2-z^2,

y^2-z^2, we get a slightly more complicated situation. Thus from (7.41) and (7.42) we get for class $\mathscr{C}_2$,

$$C_3(xy) = \left(-\tfrac{1}{2}x+\frac{\sqrt{3}}{2}y\right)\left(-\frac{\sqrt{3}}{2}x-\tfrac{1}{2}y\right)$$
$$= \tfrac{3}{4}(x^2-z^2)-\tfrac{3}{4}(y^2-z^2)-\tfrac{1}{2}xy, \qquad (7.44)$$

showing that the representation is indeed three-dimensional. Applying the same operation to x^2-z^2 and y^2-z^2, we construct a matrix whose trace is zero; in this representation $\chi_2 = 0$. On the other hand, a trivial calculation of the effect of the element P on each of these functions shows that $\chi_3 = 1$. The representation, with characters $(3, 0, 1)$, must be reducible. By use of (7.39)—or indeed, by inspection of the character table—we deduce that it is the direct sum $\Gamma^{(1)} \oplus \Gamma^{(3)}$. In other words, a transformation exists such that these three d-functions may be rearranged to yield one function that behaves like the identity and two other functions that behave like the pair (x, y) under symmetry operations of the group. All in all, the perturbing field cannot do more than split the five-fold degenerate d-state of the free atom into one single level and two separate doublets.

In any further computations on this problem, we should need to know the actual functions generating the various irreducible representations. It is easy to discover, for example, that the pair (xy, x^2-y^2) transform according to $\Gamma^{(3)}$, whilst the third function, $x^2+y^2-2z^2$, remains invariant under the operations of the group. The systematic procedure for constructing such functions is to use a *projection operator*, which combines the irreducible representation matrix for each element of the group with the corresponding abstract element. The formula

$$O_{pq}^{(\alpha)}\psi(\mathbf{x}) = \frac{n_\alpha}{g}\sum_S \Gamma_{pq}^{(\alpha)}(S^{-1})\,\psi(S\mathbf{x}) \qquad (7.45)$$

produces out of any function $\psi(\mathbf{x})$ the pth of a set of n_α functions transforming like $\Gamma^{(\alpha)}$. But we must know that $\psi(\mathbf{x})$ belongs to a set that generates this representation, if the result of such projection is not to be just zero. The justification and exemplification of this technique we must leave to the reader or to any standard treatise.

7.6 Product groups and representations

The full power of group theory is seen when we go from *simple* systems —e.g. the behaviour of a single electron in a given electrostatic field— to the study of *composite* systems.

Suppose, to start with, that we have two distinguishable particles, described by two independent Hamiltonians $\mathscr{H}$, $\mathscr{H}'$ with different symmetry properties. To describe the states of the system, we use a set of product wave functions, such as

$$\Phi_{i,i'} = \phi_i^{(1)}(\mathbf{x})\,\phi_{i'}^{(2)}(\mathbf{x}'), \tag{7.46}$$

where $\phi_i^{(1)}(\mathbf{x})$ belongs to a set of orthogonal functions in the co-ordinates $\mathbf{x}$ of the first particle, etc. Following the lines of (7.4), we can discover the effect of applying to the product wave function an operation R out of the group $\mathscr{G}$ and S' out of $\mathscr{G}'$; the result is

$$\begin{aligned}\{R\otimes S'\}\,\Phi_{i,i'} &= \phi_i^{(1)}(R\mathbf{x})\,\phi_{i'}^{(2)}(S'\mathbf{x}')\\ &= \sum_{kk'} R_{ik}\phi_k^{(1)}(\mathbf{x})\,S'_{i'k'}\phi_{k'}^{(2)}(\mathbf{x}')\\ &= \sum_{k,k'} R_{ik}\,S'_{i'k'}\,\Phi_{k,k'}.\end{aligned} \tag{7.47}$$

The symbol $R\otimes S$ may be thought of as an element of a group represented by a matrix with components $R_{ik}S'_{i'k'}$. We have already used this type of *direct product* of group elements in (7.24)–(7.26).

Now follow the lines of (7.13) and use the eigenfunctions of the Hamiltonian of each particle as basis functions. If $\mathscr{H}$ and $\mathscr{H}'$ are invariant under $\mathscr{G}$ and $\mathscr{G}'$ respectively, each operation of each group may be reduced to a direct sum of irreducible representations, as in (7.14) and (7.16).

Taking a direct product of the two direct sums, we get

$$R\otimes S' = \sum_{\alpha,\alpha'}{}^{\oplus} R^{(\alpha)}\otimes S'^{(\alpha')}, \tag{7.48}$$

showing that we have by this means reduced the representation of the *product group* $\mathscr{G}\otimes\mathscr{G}'$ to a direct sum of the direct products of the various irreducible representations in $\mathscr{G}$ and $\mathscr{G}'$.

In general, such a representation cannot be reduced further: the classification of the eigenvalues of the total Hamiltonian can go no further than we reach by choosing a level (or degenerate set of levels) for the first particle and another level for the second particle and adding their energies. The irreducible representations of the product group are generated by taking direct products of the various irreducible representations of each factor group

$$\{\Gamma\otimes\Gamma'\}^{(\alpha,\beta)} = \Gamma^{(\alpha)}\otimes\Gamma'^{(\beta)}. \tag{7.49}$$

Since the arithmetic of characters is the same as the 'direct algebra' of group representations, we find that the characters of the product group are just products of the corresponding characters in Γ and Γ'; from the definitions (7.24) and (7.30) we have

$$\chi(R \otimes S') = \chi(R) . \chi'(S') \tag{7.50}$$

in any representations of the two groups. It is not difficult, therefore, to build up a complete character table of the product group.

This technique for the construction of more complicated groups by taking direct products is obviously of value in cases where the symmetry of a system is increased by the introduction of new operations. Our trigonal group D_3, for example, may be extended by the introduction of an inversion operation changing z into $-z$. This operation, together with the identity, belongs to a group of only two elements—but its product group with D_3 now has 12 elements and a 6×6 character table, which can easily be computed by the use of (7.50). We thus have a procedure for building up character tables and irreducible representations, step by step, until we arrive at such complex groups as arise in the crystallography of cubic crystals or the classification of many-particle wave functions. Indeed, one of the important abstract properties of such a group is its ability to be *factorized* into *sub-groups* in various ways, for this too imposes useful arithmetical conditions on the character table and representations.

But the system with which we began this section—two independent particles with different symmetries—is not very interesting in itself. What happens when the two Hamiltonians are invariant under the *same* group, as it might be in the case of two electrons in a hydrogen molecule or helium atom? Out of the g^2 elements of the product group $\mathscr{G} \otimes \mathscr{G}$, we can use the g elements 'along the diagonal' to represent the group $\mathscr{G}$ itself; for if $\Gamma(R)$ and $\Gamma'(R)$ are any two representations of the same element R, we easily verify that

$$\Gamma(RS) \otimes \Gamma'(RS) = \{\Gamma(R) \otimes \Gamma(R)\}\{\Gamma(S) \otimes \Gamma'(S)\}, \tag{7.51}$$

showing that the matrix $\Gamma(R) \otimes \Gamma'(R)$ behaves like the group element R in all such products. Thus, out of any two representations, Γ and Γ', we can construct a new representation of the group,

$$\Gamma'' = \Gamma \otimes \Gamma'. \tag{7.52}$$

Following the lines of (7.49) we now consider the product of two *irreducible* representations $\Gamma^{(\alpha)}$ and $\Gamma^{(\beta)}$ of $\mathscr{G}$. The product representation has dimensionality $n_\alpha n_\beta$, which may often be larger than the maximum order of any irreducible representation of $\mathscr{G}$. *In these circumstances, this product must be reducible—i.e. we can find two or more non-zero integers* q_1, q_2, *etc. such that*

$$\Gamma'' = \Gamma^{(\alpha)} \otimes \Gamma^{(\beta)} = q_1 \Gamma^{(1)} \oplus q_2 \Gamma^{(2)} \oplus \ldots \oplus q_n \Gamma^{(n)}. \tag{7.53}$$

This is one of the most profound and valuable principles in the whole of group theory. Notice that it is not incompatible with (7.49), which states that the direct products of the irreducible representations of two groups are irreducible representations of their product group. The matrices chosen in (7.51) and (7.52) are representatives of only *some* of the g^2 elements of the full product group; elements represented by $\Gamma^{(\alpha)}(R) \otimes \Gamma^{(\beta)}(S)$ with $R \neq S$ would *not* all be reducible in the manner of (7.53).

Since the arithmetic of characters (7.50) is still valid, we can write

$$\chi_k'' = \chi_k^{(\alpha)} \chi_k^{(\beta)}, \tag{7.54}$$

and use (7.39) to calculate explicitly the factors q_1, etc. in the decomposition of the product representation (7.53). The resulting formula

$$q_\gamma = g^{-1} \sum_k N_k \chi_k^{(\gamma)*} \chi_k^{(\alpha)} \chi_k^{(\beta)} \tag{7.55}$$

is particularly useful in all applications of the theory of group representations.

As a simple example of this technique, we go back again to the group D_3, whose character table is set out on p. 228. By (7.54), the product representation $\Gamma^{(3)} \otimes \Gamma^{(3)}$ has characters $(4, 1, 0)$ for the three classes of the group. The obvious decomposition (we may generate this by (7.55) if we wish) is

$$\Gamma^{(3)} \otimes \Gamma^{(3)} = \Gamma^{(1)} \oplus \Gamma^{(2)} \oplus \Gamma^{(3)}. \tag{7.56}$$

We have already noted that the pair of functions (x, y) transform according to $\Gamma^{(3)}$. To give an air of reality to the algebra, let us introduce two electrons into the p-states of the atom in the trigonal field. We then have four functions xx', xy', yx', yy' (multiplied, of course, by $\phi_p(r)\,\phi_p(r')$) to form a basis for the product representation. These may be recombined to exemplify the decomposition (7.56):

$$\left.\begin{array}{rll} (xx'+yy') & \text{transforms according to} & \Gamma^{(1)}, \\ (xy'-yx') & \text{transforms according to} & \Gamma^{(2)}, \\ (xx'-yy', xy'+yx') & \text{transform according to} & \Gamma^{(3)}. \end{array}\right\} \tag{7.57}$$

Of course the total antisymmetry principle for fermion wave functions will have to be satisfied in each case by the assignment of spin functions of appropriate symmetry.

What has happened here is that by imposing the symmetry of the Hamiltonian on the two-electron wave function, and relaxing any requirement on the wave functions of the separate electrons, we allow four levels that were originally degenerate to split into a doublet and two singlets. Such a splitting of one-particle eigenstates in complex systems is a typical consequence of allowing for the indistinguishability of the particles, which do not, therefore, have to obey individual symmetry rules. The use of product representations gives immediate information about the new classification scheme for the many-particle eigenstates.

This theorem is also extremely useful in the evaluation of the matrix elements of any perturbing Hamiltonian $\mathscr{H}'$ between the eigenstates of a Hamiltonian $\mathscr{H}$. As we can easily see from (7.13), two eigenfunctions $|\psi_i\rangle$ and $|\psi_j\rangle$ that do not belong to the same irreducible representation of the symmetry group of $\mathscr{H}$ must be orthogonal, for they cannot, except accidentally, be degenerate. In general, therefore,

$$\langle\psi_i|\psi_j\rangle = 0 \quad \text{if} \quad \Gamma^{(i)} \neq \Gamma^{(j)}. \tag{7.58}$$

The perturbation $\mathscr{H}'$ applied to $|\psi'\rangle$ produces a sum of eigenfunctions

$$\mathscr{H}'|\psi_j\rangle = \sum_k \langle\psi_k|\mathscr{H}'|\psi_j\rangle . |\psi_k\rangle. \tag{7.59}$$

But $\mathscr{H}'|\psi_j\rangle$ could itself be used as a basis function for a representation Γ of the symmetry group of $\mathscr{H}$. In this representation the irreducible representation $\Gamma^{(k)}$ would appear only if the coefficient $\langle\psi_k|\mathscr{H}'|\psi_j\rangle$ were not zero.

But $\mathscr{H}'$, being a function of, or operator on, the co-ordinates of the system, may be used as a basis for another representation Γ' of the same group. For example, in our model of an atom in a trigonal crystal field, we might be thinking of optical transitions induced by an electric dipole perturbation polarized along the x axis; i.e.

$$\mathscr{H}' = ex, \tag{7.60}$$

which generates the (irreducible) representation $\Gamma^{(3)}$ of the D_3 group. On the other hand, the original eigenfunction $|\psi_j\rangle$ generates the irreducible representation $\Gamma^{(j)}$, so that the left-hand side of (7.59), considered simply as a product of basis functions, would generate $\Gamma' \otimes \Gamma^{(j)}$. Let us reduce this representation to the form

$$\Gamma' \otimes \Gamma^{(j)} = r_1 \Gamma^{(1)} \oplus r_2 \Gamma^{(2)} \oplus \ldots \oplus r_k \Gamma^{(k)} \oplus \ldots \tag{7.61}$$

using (7.55) say, to calculate the coefficients r_k.

The representation (7.61) must, however, be equivalent to the representation Γ generated by the sum (7.59). If, therefore, the coefficient r_k in (7.61) is zero, the matrix element $\langle\psi_k|\,\mathscr{H}'\,|\psi_j\rangle$ must certainly be zero. *In other words a perturbation transforming like* Γ' *cannot produce transitions between states belonging to the irreducible representations* $\Gamma^{(k)}$ *and* $\Gamma^{(j)}$ *unless the product representation* $\Gamma'\otimes\Gamma^{(j)}$ *contains* $\Gamma^{(k)}$. This is the formal basis of most *selection rules* that depend on the symmetry properties of the wave functions and perturbations. It must be emphasized that this principle does not assert that transitions allowed by it necessarily occur.

To illustrate this principle, let us apply the perturbation (7.60) to a state of our trigonal system transforming like $\Gamma^{(3)}$. From (7.56) we observe that this generates a representation containing all three irreducible representations of the group, so that it may cause transitions from this state to any other state of the system.

On the other hand, one can show from (7.54), that the same perturbation applied to the state $x^2+y^2-2z^2$, which transforms according to the identity representation $\Gamma^{(1)}$, merely reproduces the representation

$$\Gamma^{(3)}\otimes\Gamma^{(1)} = \Gamma^{(3)}, \tag{7.62}$$

and could not, therefore, cause transitions to any states belonging to $\Gamma^{(1)}$ or $\Gamma^{(2)}$. These particular results could, of course, have been proved very readily by inspection of the sign variations of the functions in the integrand of the matrix element, but the method gives direct answers when applied systematically to much more complicated cases.

7.7 Translation groups

In solid state physics we make a special study of systems having *lattice translational symmetry*. This implies a finite group whose structure is particularly simple, but which may be generalized in an obvious way to exemplify the notion of an *infinite*, or *continuous* group.

For simplicity let us consider only the one-dimensional case—for example, the linear chain of § 1.3. The Hamiltonian (1.25) is invariant under the operation T_m which corresponds to the operation 'add m times the lattice spacing, to the co-ordinate-label of every variable'. It is trivial that these operations form a group, with the simple multiplication law

$$T_m T_n = T_{m+n}. \tag{7.63}$$

For exact symmetry under this group, we must assume that our chain is closed into a loop, having circumference of length N lattice spacings.

The translation group of this lattice has N elements; any element with index greater than N, such as might have been arrived at by application of (7.63), is automatically identified with an existing element:

$$T_{N+l} \equiv T_l. \tag{7.64}$$

From this it follows, by an elementary application of the laws of arithmetic, that the Nth power of any element is just the unit element:

$$\{T_l\}^N = E. \tag{7.65}$$

Indeed, this group has an obvious and trivial representation in which the integer m represents T_m, and multiplication of group operations is represented by addition *modulo*(N). Since such addition is commutative, the group itself must be *Abelian*; every element commutes with every other.

Let us try to set up matrix representations of this group in the standard way, with a character table, etc. Looking for elements that might satisfy the class condition (7.34), we know that

$$T_l = X^{-1} T_l X \tag{7.66}$$

because T_l commutes with every element X of the group. Thus, there are N classes, one for each separate operation of the group. But the number of irreducible representations is equal to the number of classes; by (7.28) this is only possible if each representation is one-dimensional. For an Abelian group, therefore, the representative of any element in an irreducible representation is just its character in that representation.

Here we may use (7.65). The character of the identity element must be unity. The whole problem is solved if we assign to each element the appropriate power of one of the Nth roots of unity. In other words, the character table of this group has N rows and columns with

$$\chi_l^{(n)} = \{e^{2\pi i n/N}\}^l = e^{2\pi i n l/N} \tag{7.67}$$

as the representative of the element T_l in the irreducible representation $\Gamma^{(n)}$.

This is the group theoretical derivation of *Bloch's theorem*. In the conventional notation of (1.28) (but with the lattice spacing $a = 1$ for simplicity) we define a wave number

$$k = 2\pi n/N \tag{7.68}$$

as the label for an eigenstate $\psi_k(x)$ transforming according to the irreducible representation $\Gamma^{(n)}$. Then by (7.67) we have

$$\psi_k(x+l) \equiv T_l \psi_k(x) = e^{ikl} \psi_k(x). \tag{7.69}$$

The generalization to three dimensions, as in §1.4, is trivial. Because translations along different lattice directions commute with one another, the group of all such translation operations is still Abelian, and hence by (7.66), etc. all irreducible representations are one-dimensional. In general, therefore, the eigenfunctions of a Hamiltonian with the translational symmetry of the lattice form a basis for a representation which is always reducible to a simple diagonal matrix—i.e. the eigenfunctions may be transformed to a set in which for each state $\psi_{\mathbf{k}}(\mathbf{r})$ there exists a wave-vector $\mathbf{k}$ such that

$$\psi_{\mathbf{k}}(\mathbf{r}+\boldsymbol{l}) = \mathrm{e}^{\mathrm{i}\mathbf{k}.\boldsymbol{l}}\,\psi_{\mathbf{k}}(\mathbf{r}) \tag{7.70}$$

for any translation $\boldsymbol{l}$ of the lattice.

From this point, in a treatise on the application of group theory to solid state physics, we should enter the field of *Brillouin zones* and *electronic band structure*. This subject, which is, of course, of great technical importance, depends upon the conjunction of the translational symmetry of the lattice with the *point group* symmetry of most simple crystals, to form a *space group*. But although the details of such applications are of considerable complexity, the general principles remain essentially those set out and exemplified above.

7.8 Continuous groups

Suppose now that we follow §1.5, and let the spacing of the linear chain of the previous section tend to zero. The operation T_l of the finite translation group now becomes a translation, $T(x)$ say, through a distance x, which is now a continuous variable. But such an operation still satisfies the group axioms; for example, instead of (7.63) we have a product rule

$$T(x)\,.\,T(x') = T(x+x'). \tag{7.71}$$

A *continuous group* such as this obviously has a non-denumerable infinity of elements; yet most of the results obtained by representation theory are still valid. For example, this group being Abelian, all the irreducible representations are one-dimensional. The analogy of (7.67) leads us to assign the character

$$\chi^{(k)}(x) = \mathrm{e}^{\mathrm{i}kx} \tag{7.72}$$

to the element $T(x)$, in an irreducible representation that we might label $\Gamma^{(k)}$. This character is, of course, also the *representative* of $T(x)$ in $\Gamma^{(k)}$—the operation of translation through the distance x has the effect simply of multiplying a basis function in this representation by a one-dimensional matrix, which is just this phase factor.

In general, k could be any real number. But to avoid difficulties of integrating over an infinite domain, and yet to preserve the translational symmetry through the end points, we usually retain the cyclic boundary conditions of the linear chain. If the range of x is a loop of circumference L, then (7.64) becomes

$$T(L+x) \equiv T(x). \tag{7.73}$$

This condition imposed on (7.72) restricts the 'allowed values' of k to the set

$$k = 2\pi n/L, \tag{7.74}$$

where n is an integer. This again is perfectly familiar in the continuum limit of field theory discussed in § 1.5. Notice that this restricts the irreducible representations of the group to a *denumerably* infinite set.

The orthogonality of the characters of this group can be verified by replacing the sum over group-elements in (7.37) by an integral over the continuous variable x. The range of integration, L, now takes the place of g, the number of elements in the group, i.e.

$$\int_0^L \chi^{(k)}(x)\,\chi^{(k')*}(x)\,\mathrm{d}x = L\,\delta_{kk'}. \tag{7.75}$$

For characters defined by (7.72) and (7.74) this is obviously true. Here is Fourier's theorem in yet another guise.

Now, one would think at first that the range of definable continuous groups, like that of continuous functions, would include such a diversity of possible behaviour that very little could be said about them in general, or that very lengthy specifications would have to be given in each case. But continuity itself is an important *topological* property, which imposes quite considerable obligations on the elements of any such group. If a parameter such as x in (7.71) is continuous, there are other operations of the group 'as close as one likes' to any given operation; in 'epsilontic' language, we must be able to write

$$T(x+\epsilon) \to T(x) \quad \text{as} \quad \epsilon \to 0. \tag{7.76}$$

For generality, consider an abstract group, with operations $R(\alpha_1, \alpha_2, \ldots, \alpha_n)$ depending on n continuous parameters $\alpha_1 \ldots \alpha_n$. We can easily establish a convention such that when all these parameters are zero we have defined the identity operation

$$E = R(0, 0, \ldots, 0). \tag{7.77}$$

Now let us look at operations 'in the neighbourhood' of the identity. For example, we vary the parameter α_1, by a small amount ϵ_1. The continuity of the group, as in (7.76) implies that we can write

$$R(\epsilon_1, 0, \ldots, 0) = R(0, 0, \ldots, 0) + \mathrm{i}\epsilon_1 I_1(0, 0, \ldots, 0), \tag{7.78}$$

with a convention of 'addition', akin to ordinary matrix addition. In other words, the effect of this element of the group is to produce a small change, proportional to the infinitesimal quantity ϵ_1, but otherwise characterized by an operator I_1 which is independent of the exact values of the parameters.

Following this line of argument, we define the *infinitesimal generators* of the group:

$$I_r = \lim_{\epsilon_r \to 0} \frac{1}{\mathrm{i}\epsilon_r}\{R(0, 0, \ldots, \epsilon_r, \ldots, 0) - R(0, 0, \ldots, 0, \ldots, 0)\}. \quad (7.79)$$

Each of these operations, of which there are only as many as there are parameters defining the group, is a 'constant' operation, requiring only to be defined at the special point (7.77), and hence may be specified or represented with relative simplicity. Almost all the properties of a *Lie group*—a continuous group with certain analytical restrictions which we need not specify here—can be derived from the defining properties of its generators.

Consider, for example, the translation operator $T(\alpha)$ applied to some arbitrary function of the basic variable x—i.e., by definition,

$$T(\alpha)f(x) = f(x+\alpha). \quad (7.80)$$

Hence, from (7.79),

$$\begin{aligned} If(x) &= \lim_{\epsilon \to 0}\left[\frac{1}{\mathrm{i}\epsilon}\{T(\epsilon) - T(0)\}f(x)\right] \\ &= \lim_{\epsilon \to 0}\frac{1}{\mathrm{i}\epsilon}\{f(x+\epsilon) - f(x)\} \\ &= -\mathrm{i}\frac{\mathrm{d}f}{\mathrm{d}x}. \end{aligned} \quad (7.81)$$

The generator of this group, in this representation, is just the momentum operator, p—the space derivative

$$I = -\mathrm{i}\frac{\mathrm{d}}{\mathrm{d}x}. \quad (7.82)$$

Given the generators of a Lie group, we can find all the elements by successive applications of the product rule. In a one-parameter group, for example, suppose we wish to arrive at $R(\alpha)$. Then we can choose a large number N, and define an infinitesimal $\epsilon = \alpha/N$. We get to $R(\alpha)$ by applying the operation

$$R(\alpha) \approx E + \mathrm{i}\epsilon I \quad (7.83)$$

N times; i.e.

$$R(\alpha) \approx \left\{E + \mathrm{i}\frac{\alpha}{N} I\right\}^N. \tag{7.84}$$

In the limit, as N tends to infinity, this becomes the exact result

$$R(\alpha) = \exp\{\mathrm{i}\alpha I\}, \tag{7.85}$$

where, of course, the exponential of an operator is a shorthand for the usual series expansion in powers of the operator, just as in (3.33). Where the group has n parameters, (7.85) is easily generalized to

$$R(\alpha_1, \alpha_2, \ldots, \alpha_n) = \exp\left\{\mathrm{i} \sum_{r=1}^{n} \alpha_r I_r\right\}, \tag{7.86}$$

using (7.79).

This result may seem surprising; how can the group element $R(\alpha_1 \ldots \alpha_n)$, which is a 'function' of these parameters, be determined everywhere by its 'first partial derivatives' at the point $(0, 0, \ldots, 0)$? The reason is that the basic group property itself—that the product of two operations of the group must itself be an operation of the group (e.g. (7.71))—imposes far-reaching conditions on this 'function' everywhere that it is defined.

This abstract formula is true, of course, in any representation of the group. Thus, by constructing a representation of the symbols $I_1, I_2 \ldots I_n$ we generate a representation of every element of the group. Our study of the 'physical' consequences of a particular symmetry is thus reduced to an analysis of the algebraic properties of these generators.

As an elementary example of (7.85) for a one-parameter group, let us suppose that I is some constant, k. We immediately arrive at a one-dimensional irreducible representation of the translation group

$$T(\alpha) = \mathrm{e}^{\mathrm{i}k\alpha}, \tag{7.87}$$

which we have already discovered in (7.72).

On the other hand, in (7.82) we generated another representation of the operation I for this group, suitable for application to arbitrary functions of the variable x. By (7.85), we have

$$\begin{aligned} T(\alpha) f(x) &= \left[\exp\left\{\mathrm{i}\alpha\left(-\mathrm{i}\frac{\mathrm{d}}{\mathrm{d}x}\right)\right\}\right] f(x) \\ &= \left[\exp\left\{\alpha \frac{\mathrm{d}}{\mathrm{d}x}\right\}\right] f(x) \\ &= f(x+\alpha), \end{aligned} \tag{7.88}$$

by expanding out the exponential and arriving at the usual Taylor expansion for $f(x+\alpha)$. This is of course the defining property (7.80) of the operation $T(\alpha)$.

Finally, to show the physical significance of this bit of algebra, let us suppose that we have a system whose Hamiltonian $\mathscr{H}$ commutes with an observable represented by a Hermitian operator I. By (3.36), this operator does not vary with time in the Heisenberg representation, and hence its expectation value is a *constant of the motion*. But the prescription (7.85) allows us to generate from I a continuous group, whose elements $R(\alpha)$ are unitary operators. But since $\mathscr{H}$ commutes with I, it commutes with every element of this group; hence, by (7.7) the Hamiltonian is invariant under all transformations induced by this group. The converse can also be proved—if $\mathscr{H}$ is invariant under a continuous group of operations $R(\alpha)$ then this group has a generator which is a constant of the motion.

We thus establish a connection, at the very deepest level, between an *invariance* of the Hamiltonian and a *conservation* law for a physical quantity. The obvious example is the familiar relationship between *invariance under spatial translation*, as represented by group operations such as (7.80), and the *conservation of linear momentum*, defined by the corresponding generator (7.82). Again, the *gauge transformation* of §1.12 can be shown to be generated by the *total charge operator* (1.129) which must therefore be conserved.

A major aim of the theory of elementary particles is to make the fullest use of all such natural symmetries and invariances, with group theory as a guide to rigour and completeness at all steps in the argument.

7.9 The rotation group

The conservation of *angular* momentum is quite as important in classical mechanics as the conservation of *linear* momentum. In quantum theory we learn to associate these two physical laws with the 'Galilean' invariance of the Hamiltonian under rotations and translations of the co-ordinate axes. Before introducing the more general laws of relativistic invariance we assert that there are no privileged positions or directions in ideal space.

A rotation about a fixed point is a physical operation which obviously has all the properties of an element of a continuous group. The algebraic theory of the *rotation group* is of fundamental importance throughout quantum physics.

First let us study the two-dimensional rotations, which form a simple sub-group of the complete group of rotations in three dimensions. The operation of rotating the system about a fixed axis through an angle ϕ obviously belongs to an Abelian group, which is, indeed, isomorphous with the 'linear loop' translation group of (7.71). The variable x now becomes the angle ϕ, with range $L = 2\pi$. The irreducible representations are all one-dimensional: by (7.72) and (7.74), the integer m labels the representation

$$R^{(m)}(\phi) = e^{im\phi}. \tag{7.89}$$

Let us generate another representation by allowing $R(\phi)$ to rotate the x and y axes of a function $f(x, y)$ in the plane. As in (7.80), we may write

$$R(\phi)f(x,y) = f(x',y'), \tag{7.90}$$

where

$$\begin{pmatrix} x' \\ y' \end{pmatrix} = \begin{pmatrix} \cos\phi & -\sin\phi \\ \sin\phi & \cos\phi \end{pmatrix} \begin{pmatrix} x \\ y \end{pmatrix}. \tag{7.91}$$

The infinitesimal generator (7.79) in this representation is given by

$$\begin{aligned} I_z f(x,y) &= \lim_{\phi\to 0} \frac{1}{i\phi}[f(x',y') - f(x,y)] \\ &= \frac{1}{i}\left[\frac{d}{d\phi} f(x\cos\phi - y\sin\phi,\, x\sin\phi + y\cos\phi)\right]_{\phi=0} \\ &= i\left(y\frac{\partial}{\partial x} - x\frac{\partial}{\partial y}\right) f(x,y) \\ &= L_z f(x,y): \end{aligned} \tag{7.92}$$

this group is generated by the familiar operator for the component of angular momentum (in units of $\hbar$) normal to the plane.

Yet another representation is provided by the transformation matrix

$$R(\phi) = \begin{pmatrix} \cos\phi & -\sin\phi \\ \sin\phi & \cos\phi \end{pmatrix} \tag{7.93}$$

used in (7.91). By differentiation with respect to ϕ, just as in (7.92), we obtain the infinitesimal generator in another form

$$I_z = \begin{pmatrix} 0 & i \\ -i & 0 \end{pmatrix}: \tag{7.94}$$

we recognize one of the Pauli spin matrices (6.55). Notice how the representation (7.93) of the elements of the group is regenerated by application of the fundamental formula (7.85) to this case.

We now have various clues to the structure of the rotation group in three dimensions. From (7.92), for example, we deduce that the group has three infinitesimal generators, given by the three components of the angular momentum in elementary quantum mechanics. By (7.86), a general element of the group takes the form

$$R(\alpha_x, \alpha_y, \alpha_z) = \exp\{\mathrm{i}(\alpha_x L_x + \alpha_y L_y + \alpha_z L_z)\}, \tag{7.95}$$

where L_x and L_y are obtained from L_z by a cyclic permutation of the variables. Since these symbols transform like the components of a cartesian vector, the set of parameters $(\alpha_x, \alpha_y, \alpha_z)$ must define a vector whose direction and magnitude prescribe the direction of an axis and the angle of rotation about that axis.

This formula is, of course, practically impossible to evaluate numerically because the various operators L_x, L_y and L_z do not commute with one another. It is well known in elementary quantum mechanics that they satisfy the conditions

$$[L_x, L_y] = \mathrm{i}L_z, \quad [L_y, L_z] = \mathrm{i}L_x, \quad [L_z, L_x] = \mathrm{i}L_y. \tag{7.96}$$

These commutation relations must hold, in fact, if we are to perform the multiple integration leading from the various infinitesimal operations of the group (7.78) to the general exponential formula (7.86). Without such *compatibility conditions* we could not be sure of arriving at the same element of the group if we chose to allow the parameters $\alpha_x, \alpha_y, \alpha_z$ to grow from zero in a different order.

Indeed, the basic principle of the theory of Lie groups is that the commutator of any pair of infinitesimal generators can always be expressed as a linear combination of generators, with constant coefficients:

$$[I_p, I_q] = \sum_{r=1}^{n} C^r_{pq} I_r, \tag{7.97}$$

where the *structure constants* C^r_{pq} do not depend on the representation. Conversely, the structure of the group itself is determined by these constants, which may be used to define a *Lie algebra* for the generators. These relationships are the analogue, for a continuous group, of the multiplication table (§7.3) of a finite group, being valid, in the abstract, for any representation of the various symbolic elements. This is the abstract justification for the statement in §6.7 that the commutation relations (6.79) define the physics of the Dirac matrices, which are a representation of the generators of the Lorentz group.

A much more mundane representation of the rotation group would be obtained by introducing the familiar *Euler angles* (ψ, θ, ϕ). A vector with cartesian components (x, y, z) is then transformed by multiplication by a matrix such as

$$R(\psi,\theta,\phi) = \begin{bmatrix} \cos\psi\cos\phi\cos\theta - \sin\psi\sin\phi & -\cos\psi\sin\phi\cos\theta - \sin\psi\cos\phi & \cos\psi\sin\theta \\ \sin\psi\cos\phi\cos\theta + \cos\psi\sin\phi & -\sin\psi\sin\phi\cos\theta + \cos\psi\cos\phi & \sin\psi\sin\theta \\ -\cos\phi\sin\theta & \sin\phi\sin\theta & \cos\theta \end{bmatrix}. \tag{7.98}$$

This rather formidable expression is, of course, a generalization of (7.93). Differentiating for the infinitesimal generators, we get

$$I_\psi = \begin{bmatrix} \cdot & \mathrm{i} & \cdot \\ -\mathrm{i} & \cdot & \cdot \\ \cdot & \cdot & \cdot \end{bmatrix}; \quad I_\theta = \begin{bmatrix} \cdot & \cdot & -\mathrm{i} \\ \cdot & \cdot & \cdot \\ \mathrm{i} & \cdot & \cdot \end{bmatrix}; \quad I_\phi = \begin{bmatrix} \cdot & \mathrm{i} & \cdot \\ -\mathrm{i} & \cdot & \cdot \\ \cdot & \cdot & \cdot \end{bmatrix}. \tag{7.99}$$

Each of these might have been obtained from (7.94) by simply adding another row and column of zeros to stand for the co-ordinate axis about which the rotation is taking place in each case. Thus, I_θ is a representation of L_y, and generates the rotation $R_y(\theta)$ through an angle θ about the y axis, whilst I_ψ and I_ϕ generate rotations $R_z(\psi)$ and $R_z(\phi)$ about the z axis. Notice that the representation (7.98) does not require all three generators L_x, L_y and L_z; on the other hand it cannot be put directly into the form (7.86) because the definition of the Euler angles requires the rotations to be performed in the definite order $R_z(\psi)$, $R_y(\theta)$, $R_z(\phi)$ if we are to arrive at (7.98). Here is a case where the commutation relations would play an important role as compatibility conditions for alternative orderings of the operations.

7.10 Irreducible representations of the rotation group

The appearance of the angular momentum operators in (7.92) and (7.95) immediately recalls to mind the existence of the *spherical harmonics*, $Y_j^m(\theta, \phi)$, which are simultaneous eigenfunctions of the total angular momentum operator J, defined by

$$J^2 = L_x^2 + L_y^2 + L_z^2, \tag{7.100}$$

and of the component L_z, measured along the axis of spherical polar co-ordinates. For a given positive integral value of j, we say that the total angular momentum has magnitude $j\hbar$, and that its z-component is quantized into $2j+1$ steps, as the integer m takes the values $j, j-1, \ldots, -j$.

The functions Y_j^m therefore form a basis, of $(2j+1)$ dimensions, in which L_z is a diagonal matrix:

$$L_z Y_j^m = m Y_j^m. \tag{7.101}$$

By manipulation of the commutation relations, one can show that the operators

$$L_{\pm} = L_x \pm \mathrm{i} L_y \tag{7.102}$$

have the effect of transforming the function Y_j^m into one or other of $Y_j^{m\pm1}$, but do not introduce functions out of any other set with a different value of j. The other components of angular momentum, L_x and L_y, may thus be represented as matrices of $(2j+1)$ dimensions in this basis. Using the formal expression (7.95), we thus generate a representation $D^{(j)}$, of order $(2j+1)$, for every rotation of the whole group.

To test the irreducibility of this representation, we construct the character table. For the two-dimensional rotations we noted in (7.66) that every element belonged to a separate class, which is, of course, implied by the one-dimensional representations (7.89), i.e.

$$\chi^{(m)}(\phi) = \mathrm{e}^{\mathrm{i}m\phi}. \tag{7.103}$$

From this result, or directly from (7.95) and (7.101), we infer that $R_z(\phi)$ is represented in $D^{(j)}$ by a matrix with these numbers along the diagonal, i.e.

$$\begin{aligned} \chi^{(j)}(\phi) &= \sum_{m=-j}^{j} \mathrm{e}^{\mathrm{i}m\phi} \\ &= \frac{\sin(j+\frac{1}{2})\phi}{\sin\frac{1}{2}\phi}. \end{aligned} \tag{7.104}$$

Any three-dimensional rotation, $R(\boldsymbol{\alpha})$, may, however, be regarded geometrically as a rotation through a particular angle ϕ, say, about a specific axis, whose direction may be brought into coincidence with the z axis by another group operation $R(\boldsymbol{\beta})$. This is expressed algebraically by a relation of the form

$$R(\boldsymbol{\alpha}) = R^{-1}(\boldsymbol{\beta})\, R_z(\phi)\, R(\boldsymbol{\beta}) \tag{7.105}$$

which is just the condition (7.34) that $R(\boldsymbol{\alpha})$ should belong to the same class as $R_z(\phi)$, and hence have the same character (7.104). In the notation of (7.95), the character of the element depends only on the magnitude of the vector $(\alpha_x, \alpha_y, \alpha_z)$ and not on its direction.

The orthogonality of these characters, as in (7.75), over the range $0 < \phi < 2\pi$ can be verified, and various other tests for irreducibility of the representation triumphantly confirmed. But the real utility of (7.104) is in the decomposition of product representations of the rotation group.

Let us follow (7.53), and look for a decomposition in the form

$$D^{(j)} \otimes D^{(j')} = \sum_{j''}^{\oplus} q_{j''} D^{(j'')}. \tag{7.106}$$

The characters (7.104) admit of the following elementary algebraic identity:

$$\begin{aligned} \chi^{(j)}(\phi)\,\chi^{(j')}(\phi) &= \sum_{m=-j}^{j} \sum_{m'=-j'}^{j'} \mathrm{e}^{\mathrm{i}(m+m')\phi} \\ &\equiv \sum_{j''=|j-j'|}^{j+j'} \left\{ \sum_{m''=-j''}^{j''} \mathrm{e}^{\mathrm{i}m''\phi} \right\} \\ &= \sum_{j''=|j-j'|}^{j+j'} \chi^{(j'')}(\phi). \end{aligned} \tag{7.107}$$

Comparing this with (7.54), we arrive at the fundamental theorem for the *addition of angular momenta*: the *product* of two irreducible representations of the rotation group may be reduced into a *sum* of such representations, of the form

$$D^{(j)} \otimes D^{(j')} = D^{(j+j')} \oplus D^{(j+j'-1)} \oplus \ldots \oplus D^{(|j-j'|)}. \tag{7.108}$$

We say, physically, that two objects with angular momenta $j\hbar$ and $j'\hbar$ may be combined to form a system with total angular momentum $j''\hbar$, which can vary in magnitude from the sum to the difference of the magnitudes of the separate contributions.

In analytical terms, this theorem tells us that when we express the product of two spherical harmonics as a linear combination of spherical harmonics, we get a sum with a finite number of terms. This follows from the definitions of the product representation implied by (7.46) and (7.52). The basis functions for $D^{(j)} \otimes D^{(j')}$ must be the various products of the basis functions for $D^{(j)}$ and $D^{(j')}$ respectively. But each of these new basis functions must itself be capable of decomposition into a sum of functions each transforming according to one of the irreducible representations on the right-hand side of (7.108). We must be able to write

$$Y_j^m(\theta,\phi)\, Y_{j'}^{m'}(\theta,\phi) = \sum_{j''=-|j-j'|}^{j+j'} \left\{ \sum_{m''=-j''}^{j''} C_{jj'j''}^{mm'm''}\, Y_{j''}^{m''}(\theta,\phi) \right\}. \tag{7.109}$$

The coefficients $C_{jj'j''}^{mm'm''}$ are variously called *Wigner coefficients* or *Clebsch–Gordan coefficients*, according to the precise definition of the basis functions, etc. and are tabulated for many different values of the labels. Their importance in the theory of atomic and nuclear spectra must be familiar to most readers of this book.

In the crystal field problem exemplified under (7.40) and (7.57), we expressed the s, p, d wave functions of a free atom in terms of basis functions transforming as irreducible representations of the trigonal group D_3. We could equally well have expressed the crystal field or the perturbation (7.60) as a sum of spherical harmonics, and calculated the splitting of levels, selection rules, etc. on the basis of irreducible representations of the rotation group. For such purposes, a systematic theory of representations of the rotation group in the form of successively higher-order vector and tensor functions is of great utility. This is the context of the *Wigner–Eckhart theorem* which may be invoked to calculate the ratios of numerous transition-matrix elements without having to evaluate a corresponding number of definite integrals. The argument relies on the theory of selection rules (e.g. (7.61)), on the decomposition of the crystal field according to representations of the rotation group, and on the existence of complete tables of the coefficients in (7.109).

It is also worth remarking that a typical matrix element, such as (7.59), in a problem involving the perturbation of atomic or nuclear levels, involves an integral of a function of the form $\psi_k \mathscr{H}' \psi_l$. If the two wave-functions are eigenstates of given angular momentum, and if we have decomposed $\mathscr{H}'$ also into functions transforming according to various $D^{(j)}$, then this integrand is typically of the form of a product of *three* spherical harmonics. By a second application of (7.109), this can be expressed as a sum of spherical harmonics of various orders, with various coefficients similar to those of (7.109). Physically speaking, these *Racah coefficients* define the rules for the combination of three angular momentum vectors to make a fourth; arithmetically, they can be generated, and have been tabulated, by systematic applications of representation theory, using projection operators as in (7.45), orthogonality conditions as in (7.37), etc.

7.11 Spinor representations

In § 6.5 we discussed the properties of *spinors*—two-component functions with a law of transformation associated with a Lorentz transformation of co-ordinate space. We saw, for example, that we might associate with a simple rotation θ about the x axis the transformation (6.60) with matrix

$$Q_x(\theta) = \begin{pmatrix} \cos\frac{1}{2}\theta & \mathrm{i}\sin\frac{1}{2}\theta \\ \mathrm{i}\sin\frac{1}{2}\theta & \cos\frac{1}{2}\theta \end{pmatrix}. \tag{7.110}$$

Since a similar matrix exists for any arbitrary spatial rotation this must be an element of a representation of the rotation group in the 'spinor space'.

The explicit form of such a representation can easily be found by constructing the infinitesimal generators of the group. From (7.79), we discover

$$L_x = \frac{1}{\mathrm{i}}\left[\frac{\partial Q_x(\theta)}{\partial\theta}\right]_{\theta=0} = \frac{1}{2}\begin{pmatrix} \cdot & 1 \\ 1 & \cdot \end{pmatrix} = \tfrac{1}{2}\sigma_x, \tag{7.111}$$

where we have produced just half the Pauli spin matrix σ_x defined in (6.55). It takes little wit to infer from the commutation relations (7.96), which are independent of the representation, that the other generators are represented by the other 2×2 Pauli matrices:

$$L_y = \tfrac{1}{2}\sigma_y, \quad L_z = \tfrac{1}{2}\sigma_z. \tag{7.112}$$

A general rotation in space is then represented by (7.95), which could be written

$$R(\boldsymbol{\alpha}) = \exp\left(\tfrac{1}{2}\mathrm{i}\boldsymbol{\alpha}\cdot\boldsymbol{\sigma}\right) \tag{7.113}$$

in an obvious notation. If we wanted to extend this argument to representations of the Lorentz group, including time-like components, the formalism of (6.62) would be an obvious starting point.

We have shown in §7.10 that the rotation group has irreducible representations $D^{(j)}$ of order $2j+1$, where j is an integer. The representation (7.113) is also, in fact, irreducible, but is of only *two* dimensions. It is reasonable to suppose—and it can be proved—that it has all the properties of $D^{(j)}$ with now the *half integral* quantum number $j = \frac{1}{2}$.

Moreover, irreducible representations of the rotation group are associated with eigenstates of the angular momentum operator; states transforming according to $D^{(\frac{1}{2})}$ must have angular momentum $\frac{1}{2}\hbar$. This is the group-theoretical statement of the physical observation that 'an electron or nucleon has spin $\frac{1}{2}$'.

The basis of the representation $D^{(j)}$ for integral values of j was provided by $2j+1$ spherical harmonics Y_j^m. The rotation operator (7.110) was defined as a matrix acting on the spinor components $\psi_\uparrow$ and $\psi_\downarrow$ whose 'angular variation' would be labelled $Y_{\frac{1}{2}}^{\frac{1}{2}}$ and $Y_{\frac{1}{2}}^{-\frac{1}{2}}$ in the notation of (7.101). These symbols cannot be represented explicitly as functions of polar co-ordinates θ, ϕ but otherwise they must have the various properties implied by the subsequent equations of § 7.10. Thus, the decomposition of product representations of the rotation group, as in (7.108), follows the same laws, whether the quantum numbers j and j' are integral or half integral, and Wigner and Racah coefficients may be defined, just as in (7.109), for all such cases.

It is important to notice, however, that in the complete enumeration of irreducible representations of the rotation group the spinor representation with $j = \frac{1}{2}$ can occur in two different forms according as we choose to insert in (7.110), say, the angle θ or the angle $\theta + 2\pi$. The representation is *double-valued*. This point is discussed in § 6.5, and although it causes no great difficulty a convention has to be established and care must be taken in the formal group-theoretical manipulations.

7.12 *SU*(2)

Spinor wave functions, with two components, were introduced into physics to explain how a free electron could have just two states of angular momentum. Spin, as a physical observable, was then found to be associated with relativistic invariance under the Lorentz group, as shown in § 6.5. It is well known, however, that elementary particles are capable of existing in several different states of charge which in other respects are almost identical. The neutron and proton, for example, differ only in their electromagnetic properties, whilst there are many families of mesons and hyperons with positive, zero and negatively charged members. It is natural enough to think of these as different eigenstates of a single field, with the charge as a quantum number or eigenvalue of a well-defined operator.

In § 1.12 we introduced a simple formalism which could describe both positively and negatively charged bosons. A complex scalar field, ϕ was introduced with an appropriate Lagrangian density. In effect, this corresponds to two field operators—the real and imaginary parts of ϕ—and could thus allow the existence of the two types of excitation. But these two parts are not independent, for the formalism is invariant under a transformation of the type

$$\phi \rightarrow e^{i\alpha}\phi; \quad \phi^* \rightarrow e^{-i\alpha}\phi^*. \tag{7.114}$$

It can be shown that such a transformation of the wave functions is required by any change of gauge of the electromagnetic potentials, of the kind discussed in §6.3; the physics of the electromagnetic interactions of the particles is thereby left unaltered.

From a formal group-theoretical point of view, we might argue as follows: The quantity to be conserved is the total charge, measured by the operator Q of (1.129). As shown in §7.8, this operator must be the generator of a continuous group, whose typical element is given by (7.85), i.e.

$$U(\alpha) = \exp\{i\alpha Q\}. \tag{7.115}$$

In the occupation-number representation, Q is diagonal, so that the effect of this transformation on the wave function for a single boson is just what we have written in (7.114). We may say that the *gauge transformation* is isomorphic with the group $U(1)$—the group of multiplications by a complex number of modulus unity, i.e. the group of one-dimensional *unitary transformations*. This group is also, of course, the same as the group of rotations about a fixed axis, (7.93).

This formalism is a little too restrictive. The analogy of spin suggests a more subtle possibility. Consider the case of the proton and neutron. Let us suppose that these are two states of a single 'nucleon field', defined simply as being eigenstates of an *isospin* operator I_3. By convention, we say that the pure 'proton' state is defined so as to satisfy

$$I_3 |p\rangle = +\tfrac{1}{2} |p\rangle, \tag{7.116}$$

whilst the pure 'neutron' state $|n\rangle$ has eigenvalue $-\frac{1}{2}$ for this operator; the electric charge in each case would be measured by $(1+I_3)$.

But $|p\rangle$ and $|n\rangle$ now span two dimensions of Hilbert space, where I_3 would have the same representation as the Pauli spin operator $\frac{1}{2}\sigma_z$ of (6.55). The hypothesis is that this is merely one component of a *total isospin operator* $\mathbf{I}$, whose magnitude is defined by

$$I^2 = I_1^2 + I_2^2 + I_3^2, \tag{7.117}$$

and that physically observable systems must be simultaneous eigenstates of I and of I_3. The analogy with spin is the assumption that generators $\tau_i = 2I_i$ exist, with commutation relations exactly the same as those of the σ_i of (7.111) and (7.112).

At first sight, this looks extremely odd. What has the charge of a particle to do with angular momentum? The connection is, in fact, pure analogy based upon a mathematical coincidence.

Consider an arbitrary element of the group of transformations generated by the isospin components I_i. Following the example of

(7.113), this may be written as a 2×2 unitary matrix in the representation (7.116):

$$U(\boldsymbol{\alpha}) = \exp(\mathrm{i}\boldsymbol{\alpha}\cdot\mathbf{I}), \tag{7.118}$$

where $\boldsymbol{\alpha}$ is an arbitrary vector with three components α_1, α_2, α_3. Alternatively, we might have represented the transformation as in (7.98) and (7.110), by a general rotation defined by three Euler angles (ψ, θ, ϕ). In either case the matrix would be a function of just three independent real variables.

But as one may readily verify algebraically, the most *general* unitary matrix of rank 2 has only three independent real parameters. The transformation (7.118) may therefore be defined alternatively as an element of $SU(2)$—a *special* unitary group because only 2×2 matrices with determinant $+1$ are included.

The accident is that this group, which is worthy of investigation in its own right, is formally identical with the three-dimensional rotation group. The generators I_i happen to have the same algebraic properties (7.96) as a set of angular momentum operators L_x; there the connection between charge and spatial rotations ends. The isospin operator generates an internal symmetry property of the particles, which has nothing to do with ordinary kinematics.

Nevertheless, we must be thankful for the algebraic apparatus of spin, etc. when we want to explore the consequences of generalizing from the $U(1)$ symmetry of (7.115) to the $SU(2)$ symmetry defined by (7.118). Consider, for example, the problem of classifying states containing just two nucleons. A single nucleon state function transforms according to the irreducible representation $D^{(\frac{1}{2})}$ in the notation of §7.11. The product of two such wave functions therefore transforms according to the product representation, which may be reduced, by (7.108), to

$$D^{(\frac{1}{2})} \otimes D^{(\frac{1}{2})} = D^{(1)} \oplus D^{(0)}. \tag{7.119}$$

This tells us that two nucleons of different charge are not independent of one another in their behaviour, as they would be if they were ultimately distinguishable. The decomposition (7.119) implies the existence of two classes of states, analogous to the 'singlet' and 'triplet' spin states of a pair of electrons. Thus the antisymmetric combination of a proton and a neutron wave function may be said to have total isospin zero, and has different dynamical properties from the symmetric combination with $I = 1$. The latter combination, with $I_3 = 0$, belongs to the triplet state, and therefore has the same properties—except for electromagnetic interactions—as a pair of protons

($I_3 = +1$) or a pair of neutrons ($I_3 = -1$). The hypothesis that *strong interactions are charge independent* implies that the complete formula analogous to (6.107) would contain only two coupling constants, corresponding to scattering processes in the singlet and triplet isospin states respectively, without regard to whether we were dealing with *pp*, *nn*, or *pn* interactions in particular.

However, the dependence of the algebraic theory of $SU(2)$ on the isomorphism with the rotation group is too specialized. Let us look at this group in its own right, from a slightly different point of view. The basic element of this group is simply a unitary matrix.

$$U_{\alpha\beta} = \begin{pmatrix} a & b \\ -b^* & a^* \end{pmatrix}, \tag{7.120}$$

where a and b are any two complex numbers such that

$$aa^* + bb^* = 1. \tag{7.121}$$

This may be supposed to operate on column vector u_α, with just two components u_1 and u_2; we show that u_α transforms according to the 2-dimensional irreducible representation '**2**', by writing

$$u_\alpha \to U_{\alpha\beta} u_\beta, \tag{7.122}$$

with the summation convention for the indices.

We are familiar, however, with the existence of vectors that transform *contravariantly* in these circumstances. Let us give these the label v^γ with an upper index. Then

$$v^\gamma \to [U^{-1}]_{\gamma\delta} v^\delta = U^*_{\gamma\delta} v^\delta, \tag{7.123}$$

because the matrix is unitary. We might say that v^γ transforms according to the *conjugate representation* **2***; although in the case of $SU(2)$ this is equivalent to **2**, as one may check by finding matrices that transform $U_{\alpha\beta}$ into $U^*_{\alpha\beta}$ by interchanging the order of rows, etc.

It is natural now to consider the transformation properties of *tensors*, transforming as products of vectors. Thus, by (7.122) and (7.123), we could write

$$T^\gamma_\alpha = u_\alpha v^\gamma \to U_{\alpha\beta} U^*_{\gamma\delta} u_\beta v^\delta = [U_{\alpha\beta} U^*_{\gamma\delta}] T^\delta_\beta. \tag{7.124}$$

Thus, the set of numbers $[U_{\alpha\beta} U^*_{\gamma\delta}]$ would be arranged as a 4×4 matrix which would be an element in the product representation $\mathbf{2} \otimes \mathbf{2}^*$ of the group.

But because the matrix (7.120) is unitary, we can construct various invariants out of the components of the tensor T^{γ}_{α}. We know, for example, that the trace of this matrix is unchanged by the transformation, and hence may be said to belong to the identity representation, **1**, of our group. Another such invariant is the antisymmetric combination of off-diagonal elements, which we prefer to write

$$T^1_2 - T^2_1 \equiv \epsilon_{\alpha\gamma} T^{\alpha}_{\gamma}, \tag{7.125}$$

thus introducing the *totally antisymmetric symbol* defined by

$$\epsilon_{11} = \epsilon_{22} = 0; \quad \epsilon_{12} = -\epsilon_{21} = 1. \tag{7.126}$$

The proof that (7.125) is invariant under the transformation (7.124) depends simply on writing out the various terms and discovering that they all vanish except those reproducing the same combination of components multiplied by unity in the form of (7.121).

Let us now decompose the set of four numbers T^{γ}_{α} into symmetric and antisymmetric combinations:

$$T^{\gamma}_{\alpha} = \tfrac{1}{2}\{T^{\gamma}_{\alpha} + T^{\alpha}_{\gamma}\} + \tfrac{1}{2}\{T^{\gamma}_{\alpha} - T^{\alpha}_{\gamma}\}. \tag{7.127}$$

The first bracket constitutes a symmetric matrix, with only three independent components. The second bracket is an antisymmetric matrix, with only the single independent parameter, the quantity $\frac{1}{2}\epsilon_{\alpha\gamma} T^{\alpha}_{\gamma}$. But this, as we have seen, transforms according to the irreducible representation **1**. The components of the symmetric part must therefore transform according to a representation of only three dimensions, which we label **3**. This is in fact irreducible; we have derived, by another method, the formula (7.119), which we now write

$$\mathbf{2} \otimes \mathbf{2} = \mathbf{3} \oplus \mathbf{1}. \tag{7.128}$$

The method is quite general. A tensor $T^{\alpha\beta\ldots}_{\gamma\delta\ldots}$, with p upper indices and q lower indices, generates a representation of the group, of order 2^{p+q}. But this can be reduced by the operation of antisymmetrizing with respect to any two indices—thus

$$T^{\beta\ldots}_{\delta\ldots} = \epsilon_{\alpha\gamma} T^{\alpha\beta\ldots}_{\gamma\delta\ldots} \tag{7.129}$$

generates a representation of order 2^{p+q-2} contained within the original representation. This applies not merely to the combination of an upper with a lower index. Because the conjugate representations of $SU(2)$ are not distinct, we can always shift an index from the upper to the lower position by a simple equivalence transformation. Thus, the representation generated by our tensor will be reducible unless it is symmetric in *all* pairs of indices. This, however, greatly reduces the number of independent components of the tensor. Indeed, a totally symmetric array with $p+q$ indices, each taking 2 values, can only have $(p+q+1)$ independent components, and so generates the irreducible representation $(\mathbf{p}+\mathbf{q}+\mathbf{1})$.

This theory, of course, merely reproduces the results already obtained for this group through its isomorphism with the rotation group. Thus, the general decomposition of the product of irreducible representations, given by (7.108) now becomes

$$(\mathbf{2j}+\mathbf{1})\otimes(\mathbf{2j}'+\mathbf{1}) = (\mathbf{2j}+\mathbf{2j}'+\mathbf{1})\oplus(\mathbf{2j}+\mathbf{2j}'-\mathbf{1})\oplus\ldots\oplus(\mathbf{2}\,|\mathbf{j}-\mathbf{j}'|+\mathbf{1}), \tag{7.130}$$

which may be proved by consideration of the various terms that arise when the product of two totally symmetric tensors is expressed as a sum of totally symmetric expressions by devices such as (7.127) and (7.129). Indeed, if we actually found explicit representations of the group elements in this language we could reconstruct the Wigner coefficients of (7.109). This is another context in which the Wigner–Eckhart theorem plays an important role.

7.13 $SU(3)$

The classification of states transforming under various representations of $SU(2)$ follows the familiar pattern of atomic and nuclear states of spin and angular momentum. This pattern does not, however, fit the observations for elementary particles. If we go beyond the simple isospin/charge conservation rules and look for more complex selection rules involving groups of particles with only approximate equality of mass, we find that we can assign further quantum numbers—baryon number, strangeness, hypercharge, etc.—which are not independent of one another. The solution to this puzzle was, of course, the suggestion that the symmetry underlying the pattern was of the type $SU(3)$.

By definition, this is the group of all unitary matrices of rank 3 and positive determinant. The above analysis of the irreducible representations of $SU(2)$ can in fact be generalized quite easily to this more complex case.

We must start with column vectors with three components, transforming according to a 3×3 unitary matrix analogous to (7.120). The vector u_α, transforming according to (7.122), provides a basis for the basic irreducible representation $\mathbf{3}$, whilst a 'contravariant' vector v^γ

transforms according to (7.123), giving rise to the irreducible representation **3***. The main difference between $SU(2)$ and $SU(3)$ is that **3** and **3*** are not equivalent; there does not exist an equivalence transformation (cf. (7.8)) which can turn the unitary matrix U into its complex conjugate U^*.

For this reason, the distinction between upper and lower indices in tensors such as $T^{\alpha\beta\cdots}_{\gamma\delta\cdots}$ must be preserved. But any pair of indices on the same line may be reduced one step by a process analogous to (7.129):

$$T^{\nu\cdots}_{\gamma\delta\cdots} = \epsilon_{\alpha\beta\gamma} T^{\alpha\beta\nu\cdots}_{\delta\cdots}, \tag{7.131}$$

where $\epsilon_{\alpha\beta\gamma}$ is the totally antisymmetric symbol in three dimensions, analogous to (7.126), i.e. $\epsilon_{123} = \epsilon_{231} = \epsilon_{312} = -\epsilon_{213}$, etc. The relation (7.131) is invariant under the operations of the group; this is because the symbols $\epsilon_{\alpha\beta\gamma}$ themselves transform as components of a tensor of the third rank, but remain unchanged thereby.

This process allows us to replace a pair of upper indices by one lower index, or vice versa. Another reduction process is to take a *trace*—contract an upper with a lower index by the summation implied by

$$T^{\beta\cdots}_{\delta\cdots} = T^{\alpha\beta\cdots}_{\alpha\delta\cdots}. \tag{7.132}$$

This relation, also, is not disturbed by the transformations of the group.

The upshot is that any tensor with p upper indices and q lower indices may form a basis for a representation of the group; but this representation will be reducible unless all contractions, etc. of the form (7.131) and (7.132) yield zero. The representation will be irreducible if the tensor is separately symmetric in all its upper and all its lower indices, i.e.

$$T^{\alpha\beta\cdots}_{\gamma\delta\cdots} = T^{\beta\alpha\cdots}_{\gamma\delta\cdots} = T^{\beta\alpha\cdots}_{\delta\gamma\cdots}, \quad \text{etc.} \tag{7.133}$$

and if all traces such as (7.132) vanish.

This irreducible representation of $SU(3)$ might be labelled (p, q); but one can calculate that the number of independent tensor components is

$$\mu = \tfrac{1}{2}(p+1)(q+1)(p+q+2). \tag{7.134}$$

The convention is, therefore, to label the representation by this dimensionality, just as in the case of $SU(2)$. Thus, the representation $(1, 1)$, whose basis would be a traceless 3×3 tensor T^{β}_{α} has just eight independent components, and is therefore labelled **8**. Similarly, the case $p = 0$, $q = 2$ corresponds to a symmetric tensor of the type $T_{\alpha\beta}$, which belongs, of course, to the representation **6**. Notice that this is not equivalent to the case $p = 2, q = 0$, which would be the conjugate representation **6*** with basis $T^{\alpha\beta}$.

We could now begin to write down decompositions of product representations—for example, equations like

$$\mathbf{3} \otimes \mathbf{3} = \mathbf{6} \oplus \mathbf{3}^* \tag{7.135}$$

and

$$\mathbf{3} \otimes \mathbf{3}^* = \mathbf{8} \oplus \mathbf{1} \tag{7.136}$$

—which would be essential for any useful applications. Suppose, for example, that we had come to the conclusion that all strongly interacting particles are made up of quarks—more elementary entities that transform according to the representation **3** of $SU(3)$. Putting three quarks together, we get wave functions that transform according to the representation

$$\mathbf{3} \otimes \mathbf{3} \otimes \mathbf{3} = \mathbf{10} \oplus \mathbf{8} \oplus \mathbf{8} \oplus \mathbf{1}, \tag{7.137}$$

using rules such as (7.135) and (7.136). We expect therefore, to find two groups of 8, and one group of 10, similar particles—for example the 8 spin-$\frac{1}{2}$ baryons $(n, p, \Sigma^-, \Sigma^0, \Sigma^+, \Xi^-, \Xi^0, \Lambda^0)$. The completion of a spin-$\frac{3}{2}$ baryon decuplet **10** by the discovery of the Ω^- particle was, of course, the signal triumph of the $SU(3)$ theory.

Within the framework of $SU(3)$ symmetry, all particles transforming according to a particular irreducible representation are essentially identical, in that they belong to eigenstates that are degenerate with one another for strong interactions. But in fact these states will be distinguished from one another by other properties, such as electrical charge. This we saw in the case of $SU(2)$, where the proton and neutron states, although both belonging to the representation **2**, are distinguished by the values of the component I_3 of isospin, and hence behave differently under the influence of an electromagnetic field. We may say that the $SU(2)$ symmetry is *broken* by this further interaction. We could have treated I_3 as a perturbation, generating the Abelian group $U(1)$ [think of the properties of L_z, a single component of angular momentum, in relation to the rotation group about the z axis, as in (7.89)], for which, of course, any representation of $SU(2)$ is reducible to a sum of one-dimensional representations as in (7.115).

The analogous theory for $SU(3)$ is somewhat more complicated. A 3×3 unitary matrix is a function of eight independent variables, so that the group has in fact eight generators according to the prescription (7.79). But these in fact have various commutation relations, of the form of (7.97), and may be represented by a set of simple 3×3 matrices analogous to the Pauli spin matrices—for example

$$\left.\begin{aligned}
&\lambda_1 = \begin{pmatrix} \cdot & 1 & \cdot \\ 1 & \cdot & \cdot \\ \cdot & \cdot & \cdot \end{pmatrix}; \quad \lambda_2 = \begin{pmatrix} \cdot & -i & \cdot \\ i & \cdot & \cdot \\ \cdot & \cdot & \cdot \end{pmatrix}; \quad \lambda_3 = \begin{pmatrix} 1 & \cdot & \cdot \\ \cdot & -1 & \cdot \\ \cdot & \cdot & \cdot \end{pmatrix}; \\
&\lambda_4 = \begin{pmatrix} \cdot & \cdot & 1 \\ \cdot & \cdot & \cdot \\ 1 & \cdot & \cdot \end{pmatrix}; \quad \lambda_5 = \begin{pmatrix} \cdot & \cdot & -i \\ \cdot & \cdot & \cdot \\ i & \cdot & \cdot \end{pmatrix}; \\
&\lambda_6 = \begin{pmatrix} \cdot & \cdot & \cdot \\ \cdot & \cdot & 1 \\ \cdot & 1 & \cdot \end{pmatrix}; \quad \lambda_7 = \begin{pmatrix} \cdot & \cdot & \cdot \\ \cdot & \cdot & -i \\ \cdot & i & \cdot \end{pmatrix}; \quad \lambda_8 = \frac{1}{\sqrt{3}}\begin{pmatrix} 1 & \cdot & \cdot \\ \cdot & 1 & \cdot \\ \cdot & \cdot & -2 \end{pmatrix}.
\end{aligned}\right\} \tag{7.138}$$

Of these λ_1, λ_2, λ_4, λ_5, λ_6, λ_7 are non-diagonal, and are obtained by adding a new row and column of zeros to σ_x and σ_y. But λ_3 and λ_8 are both diagonal, and hence commute with one another. One can, therefore find simultaneous eigenstates of both λ_3 and λ_8, and classify the states belonging to an irreducible representation of $SU(3)$ by quantum numbers which are essentially eigenvalues of these operators. The existence and properties of such operators can be derived from the Lie algebra of the group, and they form the basis for the famous diagram of eight particles arranged in a hexagonal pattern—*The Eight-Fold Way.*

INDEX